SEDIMENTATION VELOCITY ANALYTICAL ULTRACENTRIFUGATION

Interacting Systems

SEDIMENTATION VELOCITY ANALYTICAL ULTRACENTRIFUGATION

Interacting Systems

Peter Schuck

National Institute of Biomedical Imaging and Bioengineering,
National Institutes of Health, Bethesda, Maryland

Huaying Zhao

National Institute of Biomedical Imaging and Bioengineering,
National Institutes of Health, Bethesda, Maryland

CRC Press
Taylor & Francis Group
Boca Raton London New York

CRC Press is an imprint of the
Taylor & Francis Group, an **informa** business

CRC Press
Taylor & Francis Group
6000 Broken Sound Parkway NW, Suite 300
Boca Raton, FL 33487-2742

First issued in paperback 2019

ISBN-13: 978-1-138-03528-7 (hbk)
ISBN-13: 978-0-367-88669-1 (pbk)

Library of Congress Cataloging-in-Publication Data

Names: Schuck, Peter (Peter W.), author. | Zhao, Huaying (Biophysicist), author.
Title: Sedimentation velocity analytical ultracentrifugation : interacting systems / Peter Schuck and Huaying Zhao.
Description: Boca Raton, FL : CRC Press, Taylor & Francis Group, [2017] | Includes bibliographical references and index.
Identifiers: LCCN 2017016200| ISBN 9781138035287 (hardback ; alk. paper) | ISBN 1138035289 (hardback ; alk. paper) | ISBN 9781315268705 (e-book) | ISBN 1315268701 (e-book) | ISBN 9781351976831 (e-book) | ISBN 1351976834 (e-book) | ISBN 9781351976848 (e-book) | ISBN 1351976842 (e-book) | ISBN 9781351976824 (e-book) | ISBN 1351976826 (e-book)
Subjects: LCSH: Ultracentrifugation. | Macromolecules--Analysis. | Nanoparticles--Analysis. | Sedimentation analysis.
Classification: LCC QH324.9.C4 S385 2017 | DDC 572--dc23
LC record available at https://lccn.loc.gov/2017016200

Visit the Taylor & Francis Web site at
http://www.taylorandfrancis.com

and the CRC Press Web site at
http://www.crcpress.com

With Love, to Leo — Master of Life, Energy, and Happiness.

Peter Schuck

To Dr. Dorothy Beckett, whose impressively thorough research style always inspires me

Huaying Zhao

Contents

Foreword

It is a pleasure and an honor to write the Foreword to the third part of this comprehensive trilogy by Peter Schuck and co-authors. With the completion of this volume they will have assembled, brought up-to-date and greatly extended the classic text on this technique that was published by Howard Schachman in 1959 (1), and will have put together a guide to the important science (and art) of analytical ultracentrifugation that will serve the field and its practitioners for many years to come.

Schachman's opus might be considered to represent the culmination and documentation of the first 'golden age' of the ultracentrifuge, in which the applications of analytical ultracentrifugation became widespread within the physical biochemistry community. That period began in the late 1940s and early 1950s and was made possible by the commercial development of the first stable and reliable instrument and accessories for analytical ultracentrifugation, in the form of the Spinco Model E Analytical Ultracentrifuge and its associated rotors, controls and optical systems. The leading practitioners of this first cycle of commercially supported methodology included Schachman himself, of course, but also David Waugh, David Yphantis, Gerson Kegeles, Kensal Van Holde, Robert Baldwin, Victor Bloomfield and William Harrington, among others. All these workers participated in the development of new methods to make analytical ultracentrifugation more flexible and adaptable to a wider range of macromolecular problems and systems, and all also applied their methods to the first wave of quantitative characterization of a wide variety of the macromolecular systems that underlie modern molecular biology.[1]

In the late 1970s and 1980s analytical ultracentrifugation began to fall out of style, based in part on the development of other biophysical methods for determining the structures and interactions of the proteins and nucleic acids that comprise the molecular underpinnings of biological systems, but also because Spinco stopped making and servicing the Model E and its components. During this 'dark' period the field was kept alive by a cadre of dedicated ultracentrifuge practioners, led by David Yphantis, who organized NSF-sponsored workshops on analytical ultracentrifugation at the University of Connecticut to introduce new people to the field, and kept the instruments going by setting up and operating an informal system of parts and methodological knowledge exchange to support the continued

---

[1]These ultracentrifuge enthusiasts were a happy, interactive and non-competitive lot. In a brief essay written for the Festschrift volume(s) of *Biophysical Chemistry* honoring David Yphantis, I sketched a snapshot of David Waugh's laboratory as it existed at MIT in the 1950s (2). As described by Schachman in another article in the Yphantis Festschrift, the other 'ultracentrifuge laboratories' of that era had similar environments (3).

operation of the world's aging stable of Model Es. These efforts were also facilitated by a number of former Spinco service representatives, who set up their own small service companies and flew from lab to lab to help keep most of the Spinco machines functional.

Despite the difficulties of this dark period, a loyal cadre of younger scientists, trained by Yphantis and his generation and led by — among others — Peter Schuck and the other authors of the present methodological trilogy, continued to appreciate the unique power of the ultracentrifuge as a biophysical tool, and in the 1990s Beckman Instruments bought Spinco and began to produce new, simpler and more user-friendly analytical ultracentrifuges, the Beckman Models XL, which are widely used today and made the continuation of the field possible.

What is it about the analytical ultracentrifugation technique that makes it such a unique and flexible tool for biophysics and molecular biology? Clearly it cannot compete with x-ray crystallography and nuclear magnetic resonance as a method to solve the detailed atomic level structures of biological macromolecules. But what the ultracentrifuge could (and can) do was to take advantage of its centrifugal design to amplify the force of gravity, apply this force to move and separate macromolecules and their complexes in solution under physiological conditions of temperature, macromolecular concentrations and solvent environment, and provide ingenious optical techniques to observe these movements in real time.

The utility of analytical ultracentrifugation for macromolecular studies is due, in part, to the happy coincidence that the molecular masses of biological macromolecules and their complexes (ranging from 10^4 to 10^7 Daltons) and their biological concentrations within cells (ranging from $\sim 10^{-8}$ to $\sim 10^{-5}$ moles/liter) are well matched to the centrifugal forces that can be generated in rotors of about one foot in diameter spinning at controlled rates ranging up to 60,000 rpm. Advances in rotor design also made it possible to achieve these forces without fatiguing the metal of the rotors to the point where they exploded. Thus solution boundaries between macromolecular components and complexes could be established in ultracentrifuge cells with quartz windows, and the migration of, and concentration changes across, these boundaries could be monitored in real time using clever absorbance and interferometric (and now fluorescence) optical systems. This made it possible to determine the sedimentation coefficients (s) of defined molecular species, as well as their diffusion constants (D) as the boundaries spread, and thus — by applying hydrodynamic interpretations — determine their molecular weights and frictional properties (as well as their levels of homogeneity) under conditions that could be maintained close to those of 'real' biological environments.

Of course measuring s and D doesn't tell one everything one wants to know about these particles and their complexes, but it is a step in the right direction. Building on the clever application of other methods in parallel, ultracentrifugation reveals a great deal about these systems that we cannot easily learn in other ways. The methods by which these physical parameters can be determined in the analytical ultracentrifuge are described in detail by Schuck et al. in Part I of this series ("Basic Principles of Analytical Ultracentrifugation"), and the (relatively) straightforward application of these methods to systems of 'non-interacting' macromolecules is described by Schuck in Part II under the subtitle: "Discrete Species and Size-Distributions of Macromolecules and Particles."

By definition a non-interacting macromolecule, as the term is used here, means one that does not bind specifically or non-specifically to the other macromolecular components present (at least under the conditions of the experiment). However, this usage doesn't mean that these components can be considered to be 'unaware' of the presence of the other macromolecules in solution, nor that the measured parameters are not perturbed by the measurement technique, gentle though it appears. Thermodynamic issues of excluded volume, macromolecular crowding and (for charged systems) Donnan effects must be considered. In addition instrument-specific problems — including concentration dependencies during sedimentation that are induced by the sector shape of the ultracentrifuge cell, the effects of pressure differences [from one atmosphere (atm) at the top of the cell to up to ~ 1000 atms at the bottom] on macromolecules and complexes, and the pile-up of macromolecules at the bottom of the cell during sedimentation, all perturb sedimentation velocity measurements and analyses of the resulting data and must be taken into account as described in Part II.

Of course the situation gets even more complicated when we turn to the 'interacting systems of macromolecules' that are discussed in the present volume (Part III) of the series. Now, in addition to all the problems of non-interacting systems described in Part II, the macromolecules can and do bind specifically and non-specifically to one another with varying affinities by means of non-covalent interactions, and the presence or absence of these interactions will depend on the concentrations of the components in the solution. These concentrations will change across the ultracentrifuge cell as a function of sedimentation time and position in the cell and also, of course, as a function of the addition of small molecule binding ligands and perturbants of various other sorts.

If, as described in this volume, the investigator has developed a robust molecular model of the specific and non-specific binding interactions between the components of the solution, and has knowledge of the binding parameters as a function of component concentrations as the different macromolecular components sediment in the ultracentrifuge cell, then the methods that Schuck and Zhao present here can be applied to predict (at least approximately) the time-dependent distribution of macromolecules in the ultracentrifuge cell. These methods are based in part on numerical solutions of the Lamm equations for the system as defined by the interaction model, but also — for rapidly equilibrating systems — involve the use of 'effective particle theory' to develop reasonable estimates of at least the more robust hydrodynamic parameters on the basis of defined simplifying assumptions. In addition a complete analysis further requires — through additional Lamm equations — the inclusion of the 'non-interacting' effects (excluded volume, concentration changes due to sedimentation in the sector-shaped cell, etc., as described by Schuck in Part II) on the time-dependence of the sedimentation process.

In principle such analyses will work, and it is very valuable to have a complete theoretical description of the interplay of all these interactions for well-defined systems undergoing sedimentation velocity in the ultracentrifuge cell. In practice, however, one rarely has sufficient background information about real biological systems, and to apply these approaches one needs to gather significant additional information about the system in advance. This is where the molecular intuition of the investigator comes into play, based on his/her prior knowledge of the system, because just taking an uncharacterized cellular extract and loading it into an ultracentrifuge cell and attempting to apply these modeling approaches *ab initio* will obviously not yield useful results. Nevertheless, ultracentrifugal analyses of such complex systems on the basis of the methodologies laid out in these monographs by Schuck et al. *can* be successfully performed, and *will* provide us with information that would be hard to obtain any other way, *if* we know something about the system to start with.

An example of a system of interacting macromolecules about which one can obtain significant and unique molecular information by sedimentation velocity approaches — again if one knows something about the system before starting — is the DNA replication complex of bacteriophage T4. (I should emphasize that while this happens to be a system that I am very familiar with because our research group has studied it for many years, the same points could be equally well-illustrated with the work of many other investigators on many other macromolecular systems.)

As originally shown by the laboratories of Bruce Alberts and Nancy Nossal, the T4 DNA replication system comprises eight discrete types of T4-coded protein subunits that can be assembled *in vitro* onto a model DNA replication fork to form a functional replisome complex that can perform leading and lagging-strand DNA synthesis at rates and fidelities characteristic of the *in vivo* system (4,5). The complete *in vitro* system contains a nucleic acid framework onto which assemble two oppositely-directed polymerases, a polymerase clamp-clamp loader sub-assembly that controls the stability of the polymerases during replication, and a helicase-primase complex that opens the DNA ahead of the leading-strand polymerase to expose the replication templates and form RNA primers to initiate the synthesis of 'Okasaki' DNA fragments on the lagging strand template. The complex also includes a single-stranded DNA binding protein that binds cooperatively to the single strand DNA templates that are transiently exposed by the helicase complex and regulates the interactions of the other components of the system.

Details can be found elsewhere (6,7,8), but it is worth noting that this system provides a wonderful biophysical 'playground' on which sedimentation velocity analyses have been crucial for the establishment of many of the molecular details of the functioning sub-assemblies of the overall complex. For example, such analyses were ideal for determining the subunit stoichiometry of the various sub-assemblies of the T4 replication system, including showing that the T4 replication helicase is a hexamer of identical subunits held together in a ring-shaped configuration by six ATP molecules that lie between and hold together adjacent subunits; that this helicase is functionally stabilized by a single primase subunit that binds to the hexameric helicase ring at the advancing replication fork; and that the functional helicase is loaded onto the replication fork with the cooperation of a single subunit of the helicase loader protein. Similar approaches also showed that the T4 clamp loader complex contains five subunits of two different types, and that the replication clamp itself consists of three subunits.

Furthermore, because sedimentation rates are slow, complexes that associate and dissociate with time constants of less than 10 minutes or so can be considered to remain at equilibrium with co-sedimenting constant concentrations (after taking radial dilution into account) of the constituent proteins of the complex during a sedimentation velocity run, while complexes that dissociate more slowly can be considered to be stable species on the ultracentrifuge time scale. It turned out that in the T4 system this 'time-boundary' between rapidly and slowly dissociating systems falls at a convenient point to permit us to use these approaches to establish appropriate subunit assembly mixing protocols, because the slowly dissociating metastable aggregates that resulted from the use of incorrect order-of-addition pathways could be easily discriminated from "correct" assembly reaction sequences within which complex equilibria could be maintained (8). Again, this T4 DNA replication story is only one example of how the use of sedimentation velocity approaches has helped scientists all over the world untangle and more fully describe the molecular mechanisms of the central processes of biology.

In summary, the analytical ultracentrifuge continues to be a crucial tool for the study of interacting macromolecular systems, and the materials contained in these three volumes will long provide an invaluable guide for biophysically oriented investigators who wish to apply this important tool, both correctly and imaginatively.

Peter H. von Hippel, University of Oregon

References

1. H.K. Schachman, "Ultracentrifugation in Biochemistry", Academic Press, New York, 1959.

2. P.H. von Hippel, Graduate student days at MIT, Biophys. Chem., vol. 108(1), pp. 17-22, 2004.

3. H.K. Schachman, Those wonderful early years with the Model E ultracentrifuge and David Yphantis, Biophys. Chem., vol. 108(1), pp. 9-16, 2004.

4. N. Nossal and B.M. Peterlin, DNA replication by bacteriophage T4 proteins. The T4 43, 32, 44-62, and 45 proteins are required for strand displacement synthesis at nicks in duplex DNA, J. Biol. Chem., vol. 254(13), pp. 6032-6036, 1979.

5. N.K. Sinha, C.F. Morris and B.M. Alberts, Efficient in vitro replication of double-stranded DNA templates by a purified T4 bacteriophage replication system. J. Biol. Chem., vol. 255(9), pp. 4290-4303, 1980.

6. F. Dong, E.P. Gogol and P.H. von Hippel, The phage T4-coded DNA replication helicase (gp41) forms a hexamer upon activation by nucleoside triphosphate, J. Biol. Chem., vol. 270(13), pp. 7462-7473, 1995.

7. T.C. Jarvis, L.S. Paul and P.H. von Hippel, Structural and enzymatic studies of the T4 DNA replication system. I. Physical characterization of the polymerase accessory protein complex, J. Biol. Chem., vol. 264(21), pp. 12709-12716, 1989.

8. D. Jose, S.W. Weitzel and P.H. von Hippel, Assembly and subunit stoichiometry of the functional helicase-primase (primosome) complex of bacteriophage T4 DNA, Proc. Natl. Acad. Scis. USA, vol. 109(34), pp. 13596-13601, 2012.

Preface

Reversible, non-covalent interactions between macromolecules or particles are important in a wide range of fields, including, for example, material sciences, biotechnology, colloidal chemistry, immunology, structural biology and cell biology. In particular, the study of reversibly interacting systems of macromolecules is a crucial challenge in the goal to obtain physical understanding of biological processes in a cell. Interactions can span many orders of magnitude of strength and orientation. Volume exclusion and hydrodynamic interactions are obligatory repulsive interactions in the crowded cellular environment [3]. In the eye lens, for example, these are balanced by weakly attractive interaction potentials that help to prevent aggregation of crystallins [4, 5]. Weak transient binding events are also important, for example, in the context of multi-valent pattern recognition of the immune system [6–8]. Slightly stronger, short-lived and cooperative interactions are at play in many cases in the formation of adaptor protein complexes in signal transduction and other dynamic multi-protein complexes [9]. Specific high-affinity interactions often facilitate the assembly of cellular structures, and enable specific recognition of sequences in transcription [10, 11]. Finally, very high-affinity and usually long-lived complexes are formed, for example, in antibody-antigen interactions [12]. These are just a few examples from a myriad of such interactions in the cell. Ubiquitous motifs in the biological context are multi-site and multi-component interactions, cooperativity, and and frequently the formation of structurally polymorph multi-protein assemblies [9, 13]. The study of such interactions poses a formidable challenge.

By observing the mass-driven separation of molecules in a centrifugal field, sedimentation velocity (SV) analytical ultracentrifugation (AUC) is uniquely suited for the study of reversible interactions of purified macromolecules in solution, across the entire spectrum, from strongly attractive to repulsive interactions, with equilibrium dissociation constants ranging from the order of 10 pM [14] to 10 mM [7], and with the potential to characterize macromolecules or particles from below kDa [15, 16] into the GDa range [17] in molar mass. An important virtue is the strong hydrodynamic resolution which often allows the size-distribution of macromolecules and their complexes to be detected, and/or their dynamic co-sedimentation process to be observed, such that the number of different complexes, their stoichiometry, and affinity may be determined. It is applicable to self-association and multi-component hetero-association processes alike, and can generally be carried out without the introduction of extrinsic labels. This makes SV-AUC one of the most powerful techniques to study reversible macromolecular assembly processes in solution.

This has been recognized early on in the development of AUC, and after nearly a century of methodological evolution, a wealth of knowledge has been created through the collective work of several generations of scientists. Excellent monographs on AUC and the sedimentation of interacting systems have been published by Svedberg and Pedersen [18], Schachman [19], Fujita [20, 21], Nichol [22], Cann [23], and Williams [24], prior to the hiatus of AUC in the 1970s and 1980s. The renaissance of the technique in the 1990s has brought different instrumentation, significant advances in statistical fluid dynamics of sedimentation, new theoretical models for dynamically linked co-sedimentation, vastly more computational capabilities both in hardware and in numerical methods leading to very different data analysis strategies, new methods for the preparation of high-quality protein samples, and a host of new fields of applications. Therefore, our goal was to provide a modern framework for AUC that is solidly rooted in the foundations built by the scientific giants earlier in the 20th century, but from a current viewpoint embedding the new capabilities and interests.

To this end, the first volume of this series *Basic Principles of Analytical Ultracentrifugation* (referred to as Part I) [1] comprehensively describes the basic physics and thermodynamic principles of sedimentation, instrumentation and optical detectors, their sensitivity, calibration and noise characteristics, centrifugal run parameters, experimental design principles, and offers useful data tables. This provides the conceptual and experimental basis for carrying out meaningful AUC experiments. The second volume *Sedimentation Velocity Analytical Ultracentrifugation: Discrete Species and Size-Distributions of Macromolecules and Particles* (referred to as Part II) [2] provides the theoretical background and principles for mathematical data analysis. It revolves around the central problem of unraveling sedimentation from diffusion and polydispersity given noisy sedimentation velocity data. The transport method is introduced, and one- and multi-dimensional or multi-component sedimentation coefficient distributions are defined, with critical view of their information content. A final chapter discusses their practical application.

Using these essential tools, in the present work, we can tackle our original goal — studying the rich phenomenology of sedimentation of interacting systems in theory and practice. In our view the most fascinating aspects are the many seemingly counter-intuitive features of sedimentation boundaries of rapidly reversible systems, which we attempted to elucidate in a physical molecular perspective, presented alongside the mathematical theory. A second aspect kept in mind throughout is the question, to what extent theoretical results can translate into reliable data analysis, given noisy experimental data.

The outline of the book is the following: In Chapter 1 we establish the master equations with prototypical examples of self-associating and hetero-associating systems. This leads to the recognition that blindly solving partial-differential equations for the sedimentation process without further insight into the characteristic phenomenology is very limited. In Chapter 2, we take a more empirical look at the sedimentation patterns and their major determinants. We show that the sedimentation behavior can be classified according to the chemical equilibration times relative to the sedimentation time, and examine the physical principles underlying dynamically coupled sedimentation. This is used quantitatively in Chapter 3 to develop insightful approximate models of the sedimentation process that focus on different aspects of the sedimentation boundaries, including their overall transport, the boundary pattern, boundary polydispersity, and diffusional spread. This allows us to establish the important link to the transport method and, in Chapter 4, the connection to the diffusion-deconvoluted sedimentation coefficient distributions of non-interacting species of Part II. If properly interpreted, these turn out to be a major tool also for the study of interacting systems. Chapter 5 is devoted to hydrodynamic nonideality in the statistical fluid mechanics picture of sedimentation, and its practical impact on the sedimentation process and data analysis of interacting systems. Finally, Chapter 6 combines the previous considerations in the systematic discussion of strategies for experimental design and practical data analysis.

As in the previous volumes, to facilitate the practical applications of the concepts discussed, the book is sparingly expanded with specially marked textboxes cross-referencing functions in the public domain software SEDFIT and SEDPHAT. These programs can be downloaded from the website of our laboratory at the National Institute of Biomedical Imaging and Bioengineering.

In fact, the original motivation for a systematic presentation arose from our annual workshops on data analysis for AUC and related biophysical techniques at the National Institutes of Health. Going beyond the course material, we intended to make the book as self-consistent and comprehensive as possible, with many citations to the original literature for further reading, such that it allows a growing number of new researchers interested in AUC to become acquainted with all aspects of this powerful methodology. Additionally, we hope it will also prove generally useful as reference for experienced colleagues reading up on specific aspects of sedimentation velocity analytical ultracentrifugation, and as a basis for continued expansion of this technique.

Bethesda, March 2017

This work was supported by the Intramural Research Program of the National Institute of Biomedical Imaging and Bioengineering at the National Institutes of Health.

SYMBOL DESCRIPTION

A_2 second virial coefficient

$a(r,t)$ radial- and time-dependent signal

$a_\lambda(r,t)$ radial- and time-dependent signal at wavelength λ

a_{fast} signal amplitude of the reaction boundary determined from integration of $c(s)$

a_{slow} signal amplitude of the undisturbed boundary determined from integration of $c(s)$

$a_{\text{pop},i}(\{c_{k,\text{tot}}\})$ effective loading signal of species i as a function of total loading concentrations of all components

$\beta(t)$ time-dependent baseline signal offset that is radially constant ('RI noise')

b bottom radius (distance from center of rotation to the distal end of the solution column)

$b(r)$ radial-dependent baseline signal offset that is temporally constant ('TI noise')

c concentration

c_0 loading concentration at time t=0

c_p plateau concentration

$\tilde{c}_{\text{A}}$ in effective particle model, the concentration of the secondary component A co-sedimenting in the reaction boundary

c_u the concentration of the secondary component sedimenting in the undisturbed boundary

$\dddot{c}$ component concentration at the phase transition in the effective particle model

$c(s)$ sedimentation coefficient distribution

$c^{(p)}(s)$ sedimentation coefficient distribution with Bayesian prior

$\{c_{k,\text{tot}}\}$ set of total loading concentrations

$\chi_k(r,t)$ spatio-temporal evolution of the concentration distribution of species k, in molar units

$\bar{\chi}_k(r,t)$ total constituent concentration of component k, in molar protomer units

$\hat{\chi}(v,w)$ concentration in velocity-inverse time coordinates of Gilbert–Jenkins theory

$\chi_{1,ni}$ Lamm equation solution of an ideal non-interacting species at 1 signal unit loading concentration

χ_1 monomer concentration in molar units

χ_w total weight concentration

$\chi_{\text{nd}}(r,t)$ evolution of the sedimentation profile of an initially uniform solution of non-diffusing, inert particles

$\delta(x)$ Dirac delta-function

Δt time interval

$\delta_{i,j}$ Kronecker symbol, $\delta_{i,j} = 1$ if $i = j$, else $\delta_{i,j} = 0$

d optical pathlength

$d\hat{c}/dv$ differential velocity distribution in rectangular geometry

D translational diffusion coefficient

D^0 ideal translational diffusion coefficient in the limit of infinite dilution

$\bar{D}_k$ constituent (average) translational diffusion coefficient of component k

$\bar{D}_{\text{B}}^*$ apparent diffusion coefficient of macromolecular component B in the constant bath theory

di subscript labeling dimer parameters in self-association

$\varepsilon_{i,\lambda}$	molar signal increment of species i at wavelength λ	$\bar{k}_s$	average non-ideality coefficient of sedimentation
$\Delta\varepsilon_{AB}$	molar extinction coefficient change in complex **AB** due to hypo- or hyperchromicity, or changes in fluorescence quantum yield	$k_{s,ji}$	mutual non-ideality coefficient of sedimentation for species j and i
f	translational friction coefficient	k_D	non-ideality coefficient of diffusion
f^0	translational friction coefficient in the limit of infinite dilution	λ	wavelength or generalized signal
f_c	translational friction coefficient at finite concentration	l.h.s.	left-hand side (of an equation)
f_0	translational friction coefficient of the equivalent compact, smooth sphere with the same mass and density as the particle	ls-$g^*(s)$	apparent sedimentation coefficient distribution of non-diffusing particles $g^*(s)$ determined by least-squares fit of data by step functions
f/f_0, f_r	frictional ratio (in long and short notation)	m	meniscus radius (distance from the center of rotation to the proximal end of the solution column)
j_{sed}	sedimentation flux		
j_{diff}	diffusion flux	M	molar mass
$H(x)$	Heaviside stepfunction, equals 0 for $x < 0$ and 1 for $x > 0$	M^*	apparent molar mass
		M_b	buoyant molar mass
h	height of the sector	N_i	total number of species
j	flux	N_k	total number of components
k	enumeration of macromolecular species or component	N_κ	total number of complexes
κ	enumeration of macromolecular complexes	mo	subscript labeling monomer parameters in self-association
k_B	Boltzmann constant	ω	rotor angular velocity
K	association equilibrium constant	ϕ'	effective partial specific volume
		Φ	volume fraction of macromolecules or particles relative to the total solution volume
$K_{1,i}$	association equilibrium constant for monomer-oligomer reaction	$p(s)$	prior distribution in the Bayesian regularization of sedimentation coefficient distributions
K_D	dissociation equilibrium constant		
K_D^*	effective 'dissociation equilibrium constant' $K^{-1/(n-1)}$ or monomer concentration in monomer-n-mer self-association	q_i	chemical reaction flux of species i
		$\hat{q}$	reaction flux in velocity/inverse time coordinates of Gilbert–Jenkins theory
k_{off}	chemical off-rate constant		
k_{on}	chemical on-rate constant	ρ	solvent density
k_s	non-ideality coefficient of sedimentation	ρ_0	standard density (of water at $20°\text{C}$ in atmospheric pressure)

r radius (distance from the center of rotation)

$\bar{r}$ second moment position of the boundary

R gas constant

R_S Stokes radius

$r.h.s.$ right-hand side (of an equation)

rms root mean square

$rmsd$ root mean square deviation

s sedimentation coefficient

s^0 ideal sedimentation coefficient in the limit of infinite dilution

$\bar{s}_k$ constituent (average) sedimentation coefficient of component k

$\bar{s}_B^*$ apparent sedimentation coefficient of macromolecular component B in the constant bath theory

$s_{A\cdots B}$ reaction boundary velocity in effective particle theory

s_{asy} average sedimentation coefficient of the asymptotic boundary in Gilbert–Jenkins theory

s_{fast} sedimentation coefficient of the reaction boundary determined from integration of $c(s)$

$s_{fast}(\{c_{k,tot}\})$ isotherm of the reaction boundary s-value as a function of component loading concentrations

S_i^k stoichiometry of component i in complex (or species) k

$S_{A\cdots(B)}$ stoichiometry of effective particle phase A$\cdots$(B)

s_w^t instantaneous weighted-average sedimentation coefficient

$s_{w,\lambda}(\{c_{k,tot}\})$ signal weighted-average sedimentation coefficient as a function of total loading concentrations of all components

$s_w^{(EPT)}$ s_w for rapidly reversibly systems based on effective particle theory

SSR   sum of squared residuals

σ_λ error of data acquisition at signal λ

s_w signal weighted average sedimentation coefficient

$s_{20,w}$ sedimentation coefficient corrected to standard conditions of water at 20°C

s_{xp} experimental sedimentation coefficient uncorrected for buffer density and viscosity

t time

τ lifetime or fractional time of a state

T absolute temperature

v linear velocity in the approximation of rectangular geometry with constant force

$\bar{v}$ partial-specific volume

$\bar{v}_{xp}$ partial-specific volume under experimental conditions

$\bar{v}_0$ partial-specific volume under standard conditions

WSSR   weighted sum of squared residuals

w weight concentration

x spatial coordinate in the approximation of rectangular geometry with constant force

CHAPTER 1

Exact Description of Ideally Sedimenting Associating Systems

I N this section, we focus on the description of sedimentation processes where the particles of interest exhibit interactions that can reversibly change their assembly or conformational state. This includes solutions with a single macromolecular component exhibiting self-association reactions, where homo-oligomers are formed that can dynamically dissociate back into subunits. Similarly, this includes systems with chemical binding reactions between different macromolecular components that lead to the formation of hetero-oligomers, which, in turn, can spontaneously break apart into their components. At the same time, solutions are assumed dilute enough for the limit of 'ideal sedimentation' to hold, such that repulsive interactions from volume exclusion and hydrodynamic interactions can still be neglected. The reduction of interactions to solely reflect either assembled or dissociated states neglects intermediate relative spatial and energetic configurations, which will be considered later in conjunction with hydrodynamic interactions and macromolecular distance distributions (Chapter 5). The utility of the 'ideal sedimentation' approximation is dictated by the time-scale of sedimentation, as well as the concentration range of the available detection systems.

Practical examples of systems exhibiting reversible chemical reactions in the 'ideal sedimentation' regime include many protein-protein and protein-nucleic acid interactions, assemblies of multi-protein complexes and molecular machines, protein-small molecule interactions, interactions involving carbohydrates, reactions between compounds of supra-molecular chemistry, and most other reactions with dissociation equilibrium constants in the micromolar range and below.

We are condensing particle characteristics to the same limited, macroscopic set of parameters as in Part II (Section 1.2) [2]. Thus, the nature of the macromolecules or particles will not play any role in the present work other than their (effective) mass, density, translational friction coefficient, and binding properties, which together determine their sedimentation behavior. For more background on how the sedimentation parameters derive, for example, the question of what can be understood as the effective sedimenting particle, considering the contributions of macromolecule-solvent and co-solute interactions, see Chapter 2 of Part I [1]. All points discussed there apply equally to interacting systems.

Sedimentation velocity offers a unique opportunity to study reversible binding reactions free in solution due to its geometric configuration where complexes can be hydrodynamically resolved, with high size-resolution generated by the universal mass-based driving force, but where the complexes are not separated during their migration from a bath of slower-sedimenting unbound species of the constituent components.[1] As a consequence, any reversibly formed complexes can be kept populated and remain in an association/dissociation reaction reflecting kinetic and equilibrium properties of the interaction. Such dynamic association/dissociation reactions can drastically change the sedimentation behavior of the macromolecular components. This allows us to use SV to elucidate the assembly scheme, measure the energetics of complex formation, learn about the life-time of complexes, and gain information on the hydrodynamic shape of transiently formed complexes.

Throughout this discussion, we will assume the reader is familiar with the experimental aspects of analytical ultracentrifugation outlined in Part I [1], the basic theory of sedimentation, as well as the toolbox for the analysis of non-interacting systems comprehensively described in Part II [2]. These will provide the foundation for the study of interacting systems. In particular, the Lamm equation — master equation for sedimentation and diffusion — will serve as the starting point for the description of the reaction/diffusion/sedimentation process.

1.1 LAMM EQUATIONS OF INTERACTING SYSTEMS

We begin by recapitulating the sedimentation and diffusion of a suspension of non-interacting particles, discussed in Part II, Section 2.2.1 [2]. The balance of sedimentation and diffusion fluxes in the radial geometry of centrifugation leads to the Lamm equation

$$\frac{\partial \chi_i}{\partial t} = -\frac{1}{r}\frac{\partial}{\partial r}\left(\chi_i s_i \omega^2 r^2 - D_i \frac{\partial \chi_i}{\partial r} r\right) \tag{1.1}$$

---

[1]This is in contrast, for example, to conventional chromatography, where samples are injected as a migrating lamella, such that complexes and unbound species with different velocities may separate, leading to complex dissociation.

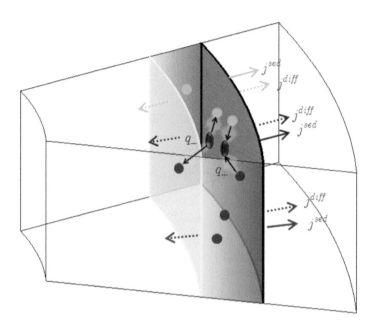

Figure 1.1 Schematics of the transport fluxes in a volume element of the solution column for a bimolecular reaction of two species (red and blue) forming a complex (purple). The unidirectional sedimentation fluxes (solid arrows) and the bidirectional diffusion fluxes (dotted arrows) have magnitudes for each species as indicated previously in Part I, Fig. 2.4 [1]. In addition, for reacting systems we have chemical interconversion in each volume element (highlighted with the black, green shadowed arrows) causing formation (q_+) and loss (q_-) of species. These create reaction fluxes that are part of the mass balance for each species in the indicated volume element.

which holds separately for each species $i = 1 \ldots N_i$ if the species do not interact.[2] In the presence of chemical reactions, the local concentration change will have an additional flux term that accounts for the time- and position-dependent rate of conversion between chemical species. This is illustrated in Fig. 1.1, which shows sedimentation and diffusion fluxes in and out of a volume element, in a picture previously invoked in the sedimentation of non-interacting species (Part II, Fig. 2.4 [2]), but now depicting additionally the creation of species within the same volume element due to a chemical reaction. Just like production of assemblies, dissociation events will take place and need to be accounted for. All reaction events

[2]In the present context of sedimentation, we regard a macromolecular 'species' as a chemical species with distinct assembly state, i.e., an ensemble of particles with well-defined uniform physical attributes such as mass, shape, conformation, and chemical properties. This term is used to differentiate a species from the macromolecular 'components,' the latter being understood in the present context as an ensemble of macromolecular 'subunits' that may be capable of assuming different assembly states or macroscopically distinct conformational states, e.g., a particular polypeptide chain that may undergo binding reactions. Thus, for example, a solution of a single macromolecular component can have different macromolecular species, linked by conformational changes or self-association reactions. On the other hand, if there are no interactions or conformational changes, components sediment as single species.

affecting concentration of species i can be summed up into the net chemical reaction flux $q_i(r, t)$, and the Lamm equation takes the form[3]

$$\frac{\partial \chi_i}{\partial t} + \frac{1}{r}\frac{\partial}{\partial r}\left(\chi_i s_i \omega^2 r^2 - D_i \frac{\partial \chi_i}{\partial r} r\right) = q_i . \tag{1.2}$$

The local reaction fluxes are determined by the local chemical kinetics, and will depend on the local composition and interactions of the particular system as a function of radius and time. Concrete examples follow below, but we can sketch already their general form as a sum over contributions from chemical rate equations among all species in the system

$$q_i = \left(\frac{\partial \chi_i}{\partial t}\right)_{\text{react}} = \sum_{\text{all reactions } \{p\} \to \kappa, \, i \in \{p\}} S_i^\kappa \left[k_{\text{off},\kappa}\chi_\kappa - k_{\text{on},\kappa}\prod_{j \in \{p\}} \chi_j^{S_j^\kappa}\right]$$
$$- k_{\text{off},i}\chi_i + k_{\text{on},i}\prod_{j \in \{l\}} \chi_j^{S_j^i}, \tag{1.3}$$

where the sum on the first line is over all reactions among sets of species $\{p\}$ that produce a complex κ to which i contributes with a stoichiometry S_i^κ, and the second line considers the (unique) reaction among the constituent species $\{l\}$ that lead to the production of species i itself. k_{on} and k_{off} denote the chemical rate constants, linked to the equilibrium association constant $K = k_{\text{on}}/k_{\text{off}}$ and equilibrium dissociation constant $K_D = k_{\text{off}}/k_{\text{on}}$. Eq. (1.2) is a coupled system of Lamm equations that describes the concerted sedimentation, diffusion, and chemical interconversion of the entire system. It can be solved by the numerical methods sketched in Appendix A [25–28].

Unless explicitly mentioned otherwise, we want to assume in the following that all components are in chemical equilibrium at the start of the experiment, and uniformly loaded across the solution column.[4] This initial condition makes it possible to predict, for any given set of chemical rate and equilibrium constants, the local concentration of all species including reaction products at the beginning of the experiment. We also limit the discussion to reversible reactions, due to the potentially ill-defined state at the start of centrifugation of systems with irreversible reactions. Alternatively, one can ensure that there are no reaction products at the start of the experiment by physically separating the reactants initially, using a band-forming centerpiece (Part I, Section 5.2.1.3 [1]) as in analytical zone centrifugation and active enzyme centrifugation [29–36].

---

[3]For a more detailed derivation, see [21].

[4]Experimentally, this may not always be trivial to fulfill when studying interactions with long-lived complexes that are pre-formed in a stock solution [14]. It may require careful planning of the sample preparation, modifications of experimental timing, and kinetic control experiments [14].

The Lamm equation for reversibly reacting systems Eq. (1.2) can be solved in SEDPHAT for various interaction schemes selected in the Model menu using parameters specified in the Global Parameters. The evolution of all species concentration profile $\chi_i(r, t)$ can be calculated for a given template of spatio-temporal grid points, or be generated from scratch for standard sedimentation conditions using the menu function Generate ▷ Sedimentation Velocity Boundary xp.

1.2 SLOW AND FAST LIMITS

It is interesting to look at limiting cases of slow and fast kinetics of the chemical reaction, compared to the typical time-scale of the SV experiment. As we will see later, in practice these two cases can describe most of the phenomenology observed, with only a narrow transition region of intermediate kinetics.

Given that centrifugation starts from chemical equilibrium conditions, it is straightforward to see that slow reactions, where all $k_{\mathrm{off},i} \to 0$, lead to vanishing reaction fluxes $q_i \to 0$, such that Eq. (1.2) becomes identical with the Lamm equation Eq. (1.1) for non-interacting species. As a consequence, the sedimentation profiles can be studied with the tools developed for non-interacting mixtures. The total concentrations of each species remains constant and reflects the species loading concentration.

If there is no chemical conversion during sedimentation, then the interpretation of the sedimentation process can proceed as that of non-interacting species. This is true even without the assumption that chemical equilibrium is established prior to centrifugation. Species may still be hydrodynamically resolved and identified [37]. However, if the assumption of initial chemical equilibrium holds, then the relative concentration of the different species can be interpreted in a secondary analysis step in the context of mass action law. For example, species populations as a function of concentration may be assembled into an isotherm (Section 6.5.2.2). If loading concentrations are chosen such that all species are populated, we can expect the isotherm analysis to reveal the equilibrium binding constants. Similarly, the signal-weighted average s_w-value as a function of loading composition will follow an isotherm that can be analyzed by models based on mass action law (Section 6.5.1). Finally, for slow systems the measurement of complex stoichiometries by multi-component SV analysis [38, 39] is as straightforward as that for non-interacting systems (Part II, Chapter 7 [2]).

The opposite case of fast reactions, compared to the time-scale of sedimentation, is more interesting. Let us study the limiting case of infinitely fast reactions such that all species are in chemical equilibrium at all times. It is convenient to subdivide all species into those representing the free state of components k, and those representing the complexes κ that are assembled from the free species in stoichiometries S_k^κ (where $S_k^\kappa = 0$ for complexes κ to which k does not contribute).

The concentrations of the complexes obey the mass action law

$$\chi_\kappa = K_\kappa \prod_{k=1}^{N_k} \chi_k^{S_k^\kappa} \tag{1.4}$$

at all times and all positions throughout the solution column, since the reaction is rapidly establishing local chemical equilibrium. Accordingly the change of complexes with time is linked to that of the free species

$$\frac{\partial \chi_\kappa}{\partial t} = \sum_k S_k^\kappa \chi_k^{-1} \frac{\partial \chi_k}{\partial t} K_\kappa \prod_k \chi_k^{S_k^\kappa} = \chi_\kappa \left(\sum_k S_k^\kappa \chi_k^{-1} \frac{\partial \chi_k}{\partial t} \right), \tag{1.5}$$

which can be inserted in Eq. (1.2) to solve for the reaction flux of the complexes

$$q_\kappa = \chi_\kappa \left(\sum_k S_k^\kappa \chi_k^{-1} \frac{\partial \chi_k}{\partial t} \right) + \frac{1}{r} \frac{\partial}{\partial r} \left(\chi_\kappa s_\kappa \omega^2 r^2 - D_\kappa \frac{\partial \chi_\kappa}{\partial r} r \right). \tag{1.6}$$

Due to mass conservation requirements, the reaction fluxes of the free species are determined entirely by the reaction fluxes of the complexes they form

$$q_k = -\sum_{\kappa=1}^{N_\kappa} S_k^\kappa q_\kappa, \tag{1.7}$$

such that Eq. (1.6) can be inserted in (1.7) and in the Lamm equation (1.2) of the free species of component k

$$\frac{\partial \chi_k}{\partial t} + \sum_\kappa S_k^\kappa \frac{\partial \chi_\kappa}{\partial t}$$
$$= -\frac{1}{r} \frac{\partial}{\partial r} \left[\left(\chi_k s_k + \sum_\kappa S_k^\kappa \chi_\kappa s_\kappa \right) \omega^2 r^2 - \left(D_k \frac{\partial \chi_k}{\partial r} + \sum_\kappa S_k^\kappa D_\kappa \frac{\partial \chi_\kappa}{\partial r} \right) r \right]. \tag{1.8}$$

This form makes it easy to change concentration units from species concentrations to total constituent component concentrations, defined as

$$\bar{\chi}_k = \chi_k + \sum_\kappa S_k^\kappa \chi_\kappa, \tag{1.9}$$

such that for each component the Lamm equation

$$\frac{\partial \bar{\chi}_k}{\partial t} = -\frac{1}{r} \frac{\partial}{\partial r} \left(\bar{s}_k(\{\bar{\chi}_j\}) \bar{\chi}_k \omega^2 r^2 - \bar{D}_k(\{\bar{\chi}_j\}) \frac{\partial \bar{\chi}_k}{\partial r} r \right) \tag{1.10}$$

is valid with a local instantaneous population-average sedimentation coefficient

$$\bar{s}_k(\{\bar{\chi}_j\}) = \frac{\chi_k s_k + \sum_\kappa S_k^\kappa \chi_\kappa s_\kappa}{\chi_k + \sum_\kappa S_k^\kappa \chi_\kappa}, \tag{1.11}$$

(also referred to as constituent sedimentation coefficient), and a local instantaneous gradient-average constituent diffusion coefficient

$$\bar{D}_k(\{\bar{\chi}_j\}) = \frac{D_k \dfrac{\partial \chi_k}{\partial r} + \sum_\kappa S_k^\kappa D_\kappa \dfrac{\partial \chi_\kappa}{\partial r}}{\dfrac{\partial \chi_k}{\partial r} + \sum_\kappa S_k^\kappa \dfrac{\partial \chi_\kappa}{\partial r}}. \tag{1.12}$$

Thus, from these equations the view arises that for a system in instantaneous chemical equilibrium each chemical component sediments locally like an imaginary particle k^* that represents the coupled system of the free species in equilibrium with all its complexes, as illustrated in Fig. 1.2.[5] Formally, this imaginary particle locally follows the same mass balance of transport fluxes as in the Lamm equation for ideal non-interacting particles, and no reaction fluxes need to be considered, but the components migrate with locally composition-dependent average sedimentation coefficients $\bar{s}_k$ and locally gradient-averaged diffusion coefficients $\bar{D}_k$. These coefficients depend on all species' concentrations and couple the Lamm equations of the different components. They can be evaluated, in principle, along with all complex concentrations, on the basis of the local total concentrations of all components using mass action law (Eq. 1.4).

It is important to recognize that the $\bar{s}_k$- and $\bar{D}_k$-values that govern local migration do not jointly define a meaningful molar mass, as in the Svedberg equation for non-interacting species [40]

$$\frac{M_b}{RT} = \frac{s}{D}, \tag{1.13}$$

where M_b denotes the buoyant molar mass of a sedimenting and diffusing particle (for an introduction see Part I, Section 1.1.2 [1]). Due to the dependence on local concentrations and concentration gradients, the resulting boundary shapes generally do not follow those of non-interacting particles. Conversely, 'apparent' s-, D-, or M_b-values extracted from experimental sedimentation boundaries of rapidly interacting systems cannot be naïvely interpreted in the same framework as non-interacting particles, with the exception of limiting cases. However, carefully selected tools from the sedimentation coefficient distribution analysis of non-interacting systems can still form the basis of a rigorous analysis, as sedimentation will still be an exponential migration with time, and diffusion will still take place on

---

[5]This 'imaginary particle' is different from the 'effective particle' model developed in Sections 2.4.1 and 3.2, which subdivides the sedimentation into a reaction boundary containing an interacting mixture, and an undisturbed boundary containing only single, truly non-interacting species. The effective particle model will utilize certain approximations (including absence of diffusion), knowledge of some phenomenology (the occurrence of two boundaries for two-component systems), and only describe the average boundary velocities. In turn, it will lead to more productive and simple predictions about the sedimentation behavior of the interacting systems. In contrast, the imaginary particle arising formally from Eq. (1.10) is exact, but requires numerical solutions of the Lamm equation, and is therefore less useful to predict and understand relationships between experimental observables.

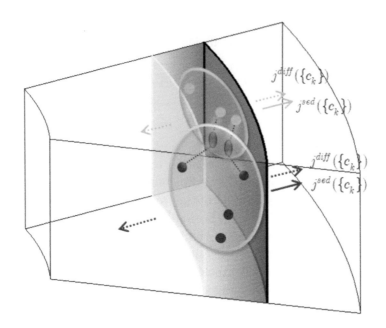

Figure 1.2 Schematics of sedimentation and diffusion of a reacting system in instantaneous chemical equilibrium. The particles shown are the same as in Fig. 1.1, but due to the instantaneous reactions the reaction fluxes do not need to be explicitly considered. Instead, the sedimentation of the system follows that of two ideally sedimenting 'imaginary particles' that are comprised of each free component and its complexes, as indicated by the blue and red circles, respectively. Their migration is coupled, as indicated by the overlap of the circles, and the sedimentation and diffusion coefficient changes as a function of local composition of all solutes (Eqs. (1.11) and (1.12)).

a $\sqrt{t}$-scale in time. In addition, as we will see later, rapidly interacting systems exhibit characteristic multi-modal diffusing boundary structures [41, 42], and exhibit polydispersity in sedimentation coefficients [43], amenable to the framework of modern sedimentation coefficient distributions (Part II, Chapter 5 [2]). The transport method (Part II, Sections 2.3 and 3.4 [2]) can also be applied to great benefit.

Self-association processes provide interesting special cases of interacting systems, since here we have only a single component with multiple oligomeric species n-fold the monomer size. Each follows the mass action law

$$\chi_n = K_n \chi_1^n \,, \tag{1.14}$$

with χ_1 denoting the molar monomer concentration, and $K_n = 0$ for oligomers that do not occur. Following Eq. (1.9), the constituent concentration for the Lamm equation (1.10) is the total macromolecular weight concentration

$$\chi_w = \bar{\chi} = \chi_1 + \sum_{n>1} n\chi_n = \chi_1 + \sum_{n>1} nK_n\chi_1^n \,, \tag{1.15}$$

which may be measured equivalently in units of molar protomer concentration, in mg/ml units, or in optical signal units (assuming the signal coefficient not to depend

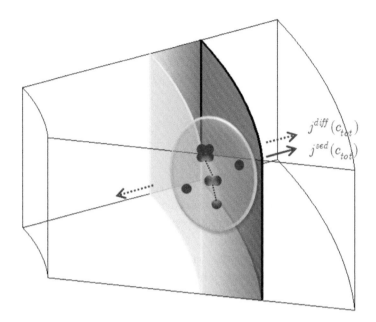

Figure 1.3 The special case of a rapid self-association, such as an instantaneous monomer-dimer-tetramer equilibrium sketched here. Similar to Fig. 1.2, the system sediments like a single 'imaginary particle,' but with sedimentation and diffusion coefficients being dependent on the total concentration.

on the oligomeric state). This leads to a simpler Lamm equation

$$\frac{\partial \chi_w}{\partial t} = -\frac{1}{r}\frac{\partial}{\partial r}\left(s_w(\chi_w)\chi_w\omega^2 r^2 - D_g(\chi_w)\frac{\partial \chi_w}{\partial r}r \right) \qquad (1.16)$$

for the weight concentration with weight-average sedimentation coefficient (expressed in monomer concentration)

$$s_w = \frac{s_1 + \displaystyle\sum_{n>1} s_n n K_n \chi_1^{n-1}}{1 + \displaystyle\sum_{n>1} n\chi_1^{n-1}} \qquad (1.17)$$

and gradient-average diffusion coefficient

$$D_g = \frac{D_1 + \displaystyle\sum_{n>1} D_n n^2 K_n \chi_1^{n-1}}{1 + \displaystyle\sum_{n>1} n^2 K_n \chi_1^{n-1}}. \qquad (1.18)$$

Once the total component concentration is predicted at any radius and time, given mass action law (Eq. 1.14), the monomer and oligomer concentrations can be determined, and locally instantaneous weight-average sedimentation and gradient-average diffusion coefficients can be calculated [44–47].[6],[7] This is illustrated in Fig. 1.3. The idea that sedimentation and diffusion proceeds like that of ideal non-interacting species with locally concentration-dependent coefficients has also been generalized to repulsive concentration-dependent interactions, as will be reviewed in Section 5.1.3.

1.3 EXAMPLES

1.3.1 Conformational Change

The precise form that the Lamm equations with reaction terms will take depends on the system under study. The most elementary reaction is a macromolecule A that sediments with s_A but can undergo a conformational change to a state A* that has different shape with an s-value s_A^*. Let us denote as k_f the forward rate constant for the transition A→A*, k_b the backward rate constant for A*→A, and $K_{\text{conf}} = k_f/k_b$ the equilibrium constant. The concrete form of the Lamm equation (1.2) for this molecule is

$$
\frac{\partial \chi_A}{\partial t} + \frac{1}{r}\frac{\partial}{\partial r}\left(\chi_A s_A \omega^2 r^2 - D_A \frac{\partial \chi_A}{\partial r} r\right) = -K_{\text{conf}} k_b \chi_A + k_b \chi_A^*
$$
$$
\frac{\partial \chi_A^*}{\partial t} + \frac{1}{r}\frac{\partial}{\partial r}\left(\chi_A^* s_A^* \omega^2 r^2 - D_A \frac{s_A^*}{s_A}\frac{\partial \chi_A^*}{\partial r} r\right) = +K_{\text{conf}} k_b \chi_A - k_b \chi_A^*
$$

$$(1.19)$$

when expressed with the parameters s_A, s_A^*, D_A, K_{conf}, and k_b, and making use of the fact that the mass does not change, hence $s_A/D_A = s_A^*/D_A^*$.

For slow reactions relative to the time-scale of sedimentation, which typically means $k_b < 10^{-4}/\text{sec}$, we will observe sedimentation patterns very close to those of two non-interacting species. As discussed in great detail in Part II [2], the heterogeneity in the sedimentation coefficients will cause additional broadening of the sedimentation boundaries, which, if misinterpreted in a single-species model, will lead to underestimates of the molar mass. In contrast, a distribution model such as $c(s)$ [49] (Part II, Section 5.4) would be more appropriate and provide a better estimate for the molar mass,[8] even though the s-values of the two species are

---

[6]Since the total weight concentration can be measured, in principle, at any time and radius, the Lamm equation for rapidly self-associating systems Eq. (1.16) lends itself to non-standard initial conditions, including initialization by experimentally measured profiles [48]. This is not possible for heterogeneous interacting systems, unless total component concentrations can be measured separately for each component.

[7]Cox [44] has reported this result first, and used it in finite difference simulations of sedimentation. This approach was also used later by Claverie [45] in finite element simulations, extended to a moving grid in [46]. In the context of modern direct boundary analysis approaches, it was reviewed in [47].

[8]This is true even if the scale relationship assumes a constant frictional ratio, rather than the more appropriate scaling law of a constant M. This is because the s-values are not very far apart, and differences in D-values (or M-values) from different scaling laws are very small. For a detailed discussion see Section 5.4.9 of Part II [2].

likely not resolved from noisy experimental data without additional Bayesian prior knowledge.

If, on the other hand, the interconversion is in a rapid equilibrium at all times, i.e., typically $k_b > 10^{-2}/\text{sec}$, we can write $\chi_A^* = K_{\text{conf}}\chi_A$, and in analogy to the development Eqs. (1.6)–(1.8), we have

$$q_A^* = \chi_A^* + \frac{1}{r}\frac{\partial}{\partial r}\left(\chi_A^* s_A^* \omega^2 r^2 - D_A^*\frac{\partial \chi_A^*}{\partial r}r\right) \tag{1.20}$$

leading to

$$\left(\frac{\partial \chi_A}{\partial t} + \frac{\partial \chi_A^*}{\partial t}\right) + \frac{1}{r}\frac{\partial}{\partial r}\left[\left(\chi_A s_A + \chi_A^* s_A^*\right)\omega^2 r^2 - \left(D_A\frac{\partial \chi_A}{\partial r} + D_A^*\frac{\partial \chi_A^*}{\partial r}\right)r\right] = 0. \tag{1.21}$$

With the imaginary particle combining both conformational states as $\chi_{\text{all}} = \chi_A + \chi_A^*$, we obtain the Lamm equation

$$\frac{\partial \chi_{\text{all}}}{\partial t} = -\frac{1}{r}\frac{\partial}{\partial r}\left(\chi_{\text{all}} s_{\text{all}}\omega^2 r^2 - D_{\text{all}}\frac{\partial \chi_{\text{all}}}{\partial r}r\right), \tag{1.22}$$

where the characteristic sedimentation and diffusion constants

$$s_{\text{all}} = \frac{s_A + K_{\text{conf}} s_A^*}{1 + K_{\text{conf}}} \quad \text{and} \quad D_{\text{all}} = \frac{D_A + K_{\text{conf}} D_A^*}{1 + K_{\text{conf}}} \tag{1.23}$$

are the ensemble averages and describe the same buoyant molar mass.

Thus, in contrast to the rapidly reacting self-associating and hetero-associating systems, we find here the important and comforting result that for rapid isomerization reactions we will observe in SV strictly only an easily measured single, ideally sedimenting species that exhibits the ensemble-average properties.[9]

---

[9]An excellent illustration of conformational changes in SV is presented by the famous example of the allosterically controlled conformation of the enzyme aspartate transcarbamoylase (ATCase) of *E. coli*. It is a $\sim$310 kDa protein that in the wildtype form exists predominantly in a $\sim$11.6 S relaxed state, but is allosterically switched by binding of a small substrate ligand and stabilized in a taut state showing a $3.1\% - 3.6\%$ lower sedimentation coefficient (in magnitude dependent on the specific ligand). These differences in s-values can be readily measured in SV in the absence of ligand and presence of an excess of ligand [50, 51].

Hypothetically, if the equilibrium of the wildtype protein conformational states were to evenly populate both forms and be in a rapid exchange, only a single species with the correct molar mass and intermediate s-value could be observed in SV. If, on the other hand, equally populated states were in slow exchange, then the SV data would show excess boundary broadening from the differential sedimentation. When this situation is simulated and re-analyzed in the context of an (impostor) single-species model, an apparent molar mass $\sim$5% lower than the true mass should be observed for the case of slow exchange, while a $c(s)$ model would represent a slightly wider peak associated with the correct molar mass.

A study by Werner, Cann, and Schachman [52, 53] demonstrated that this difference between rapid and slow interconversion of ATCase can be experimentally measured, by examining the apparent boundary spread when sub-saturating amounts of low-affinity or high-affinity ligands, respectively, are present, which suppress or permit ATCase interconversion. However, a complete theoretical description is more complex, and would more appropriately fall into the framework of heterogeneous interactions, due to the kinetics of the ligand binding coupled to the ATCase sedimentation [52, 53].

1.3.2 Monomer-Dimer Self-Association

A slightly more complicated case is the reversible homo-dimerization $A \leftrightarrow A_2$. If we denote the molar monomer concentration as χ_{mo}, the molar dimer concentration as χ_{di}, the rate constant of dimer formation as k_{12} and that of dimer dissociation to monomers k_{21}, the Lamm equation for such a system is

$$
\begin{aligned}
\frac{\partial \chi_{mo}}{\partial t} + \frac{1}{r}\frac{\partial}{\partial r}\left(\chi_{mo} s_{mo} \omega^2 r^2 - D_{mo}\frac{\partial \chi_{mo}}{\partial r} r \right) &= -2k_{12}\chi_{mo}^2 + 2k_{21}\chi_{di} \\
\frac{\partial \chi_{di}}{\partial t} + \frac{1}{r}\frac{\partial}{\partial r}\left(\chi_{di} s_{di} \omega^2 r^2 - \frac{1}{2}D_{mo}\frac{s_{di}}{s_{mo}}\frac{\partial \chi_{di}}{\partial r} r \right) &= k_{12}\chi_{mo}^2 - k_{21}\chi_{di}
\end{aligned}
\tag{1.24}
$$

(making use of the molar mass of the dimer being twice that of the monomer to express D_{di}).

Fig. 1.4 shows an example for SV data of a $\sim$100 kDa protein exhibiting a monomer-dimer self-association that is slow but not negligible on the time-scale of sedimentation [37]. It is noticeable that due to the large size of the protein in combination with the slow kinetics, we can discern two clearly separate boundary components for the monomer and dimer (in addition to a small signal from slowly sedimenting contaminating species). Data at different concentrations will exhibit variations in the relative boundary heights [37]. However, the boundaries are broader and do not show a shallow region separating the main peaks as much as would occur in non-interacting species. This is a result of a small fraction of dimers dissociating during the sedimentation experiment, which shows in the $c(s)$ analysis in non-vanishing values at intermediate s-values $\sim$6.5–7 S. The data are modeled well with Eq. (1.24), extended to account for a single contaminating species, which leads to an estimate for the dissociation rate constant of $k_{21} = 7.5 \times 10^{-5}/\text{sec}$.

For a monomer-dimer system in rapid equilibrium we arrive at a single-species Lamm equation for the measured constituent (or total weight) concentration $\bar{\chi} = \chi_{mo} + 2\chi_{di}$

$$
\frac{\partial \bar{\chi}}{\partial t} = -\frac{1}{r}\frac{\partial}{\partial r}\left(\bar{\chi} s_w(\bar{\chi})\omega^2 r^2 - D_g(\bar{\chi})\frac{\partial \bar{\chi}}{\partial r} r \right),
\tag{1.25}
$$

where the sedimentation and diffusion constants are concentration-dependent,

$$
s_w(\bar{\chi}) = \frac{s_1\chi_1 + 2s_2\chi_2}{\chi_1 + 2\chi_2} \qquad \text{and} \qquad D_g(\bar{\chi}) = \frac{D_1\chi_1 + 4D_2\chi_2}{\chi_1 + 4\chi_2}.
\tag{1.26}
$$

Of course, the well-known relationship between total and species concentrations in monomer-dimer systems

$$
\chi_1(\bar{\chi}) = \frac{1}{4K}\left(-1 + \sqrt{1 + 8K\bar{\chi}} \right), \qquad \chi_2(\bar{\chi}) = \frac{1}{2}(\bar{\chi} - \chi_1(\bar{\chi}))
\tag{1.27}
$$

holds at all radii and all times, with equilibrium association constant $K = k_{12}/k_{21}$, which allows calculating the weight-average s- and gradient-average D-values.

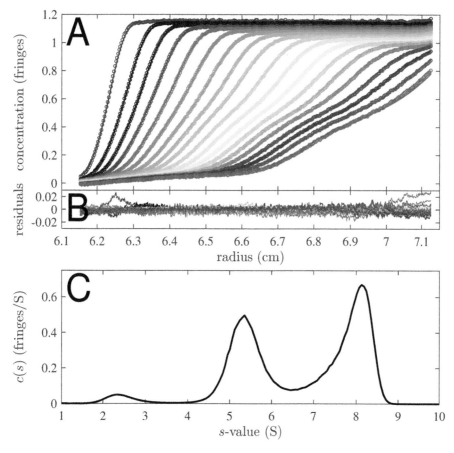

Figure 1.4 Example for the sedimentation of a slow monomer-dimer self-association of a protein IRP1 with sequence molar mass $\sim$100 kDa at 4.5 μM sedimenting at 50,000 rpm [37]. *Panel A*: Sedimentation profiles (circles, every 2nd scan shown with every 3rd data point) allow two main boundaries at $\sim$ 5.5 S and $\sim$8.3 S to be discerned, reminiscent of theoretical boundaries of stable monomer and stable dimer, but slightly broadened and not quite as separated in the transition. This is due to a slow interconversion on the time-scale of sedimentation. A close examination reveals that there is no plateau separating the two boundaries due to a subpopulation of dimers that dissociate during the sedimentation process — a result of their finite average life-time. A fit with a kinetic monomer-dimer model, combined with an extra small non-interacting species at $\sim$2 S to account for an impurity, results in a best-fit dissociation rate constant of 7.5 [6.3–9.4]$\times 10^{-5}$/sec (solid lines). This corresponds to an average lifetime $\tau = \ln 2/k_{21} \sim 2.5$ h, which may be compared to the time of the last scan of $\sim$2 h. *Panel B*: Residuals of the Lamm equation fit. *Panel C*: For comparison, the $c(s)$ distribution for the analysis of the same data.

An example of a more extended, rapidly equilibrating self-association is the monomer-dimer-tetramer system of the molecular chaperone gp57A of the bacteriophage T4 shown in Fig. 1.5. In contrast to the slow reaction of Fig. 1.4, here we see only single broad boundaries, partially due to the smaller size of the molecules but also due to the rapid kinetic exchange between the different oligomeric species. Hallmark of the interaction is that the sedimentation boundary shows a concentration-dependent velocity.

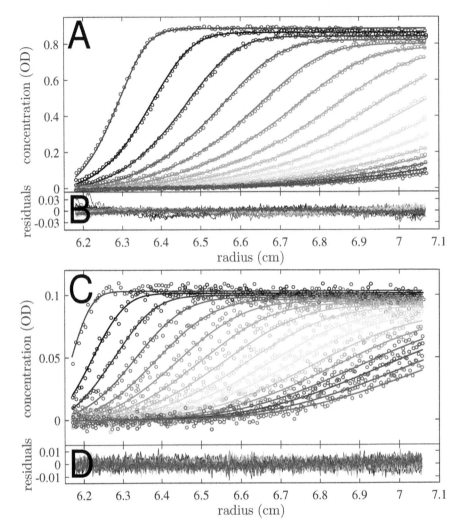

Figure 1.5 Example for the sedimentation of a rapid monomer-dimer-tetramer self-association of a ~18 kDa protein at concentrations of 42 μM (A) and 4.7μM (C) sedimenting at 50,000 rpm [54]. Every 2nd scan is shown with every 2nd data point, across approximately the same time range for both concentrations. The transition kinetics between the oligomeric states is ill-defined by the SV data, but fast compared to sedimentation. Solid lines show the best-fit global (incorporating data sets at more concentrations than those shown) with an instantaneous self-association model. *Panel A*: At a high concentration that populates initially mostly dimeric and tetrameric state, a broad boundary can be discerned. It exhibits an extended trailing part that is caused by oligomers dissociating into monomers at the lower concentrations of the diffusion-broadened boundary. *Panel B*: Residuals of the Lamm equation fit in A. *Panel C*: At a ~10-fold lower loading concentration, the majority of molecules are in the monomeric state, sedimenting at a velocity comparable with the trailing part of the high-concentration data. *Panel D*: Residuals of the fit in C.

1.3.3 Isodesmic Self-Association

In some systems there is no endpoint of the self-association, because the formation of an oligomer of size n always allows further addition of monomer to form a species of size $n + 1$. If the equilibrium between monomer and oligomer is governed by the same binding constants K for all steps n

$$\chi_n = \chi_1 \chi_{n-1} K = \chi_1^n K^{n-1}, \tag{1.28}$$

this is termed infinite isodesmic self-association (Fig. 1.6). Many multi-valent self-associating proteins exhibit this type of behavior, including many examples where self-assembly into higher-order structures such as fibrils occurs [55–57].

Even though an infinite number of species are present, the sedimentation and diffusion of a rapidly equilibrating system can be easily described as a special case of Eqs. (1.17) and (1.18) for the weight-average sedimentation coefficient and gradient-average diffusion coefficient, respectively,

$$s_w(\bar{\chi}) = \frac{\sum_{i=1}^{\infty} s_i i K^{i-1} \chi_1^i}{\bar{\chi}} \quad \text{and} \quad D_g(\bar{\chi}) = \frac{\sum_{i=1}^{\infty} D_i i^2 K^{i-1} \chi_1^i}{\sum_{i=1}^{\infty} i^2 K^{i-1} \chi_1^i}, \tag{1.29}$$

with a closed-form relationship between the monomer concentration χ_1 and the total protomer concentration $\bar{\chi}$ [58][10]

$$\bar{\chi}(\chi_1) = \frac{\chi_1}{(1 - K\chi_1)^2} \tag{1.30}$$

at all positions and all times. For the evaluation of s_w and D_g, a difficulty is that s-values for all species are required. Since the individual species cannot be measured separately, a hydrodynamic model for s_i is necessary, such as $s_i = s_0(iM)^{2/3}$ [57]. Computationally, the local transport parameters can then be approximated as a truncated sum solely covering species that are significantly populated. Useful in this regard is the fact that in isodesmic self-associations the populations of species monotonically decreases with increasing oligomer size.

---

[10]Since the total protomer concentration is dependent on the molar species concentration

$$\bar{\chi} = \sum_{i=1}^{\infty} i\chi_i = \sum_{i=1}^{\infty} iK^{i-1}\chi_1^i,$$

we can consider the product

$$\bar{\chi}(1 - K\chi_1)^2 = \bar{\chi}(1 - 2K\chi_1 + K^2\chi_1^2) = \sum_{i=1}^{\infty} iK^{i-1}\chi_1^i - 2\sum_{i=1}^{\infty} iK^i\chi_1^{i+1} + \sum_{i=1}^{\infty} iK^{i+1}\chi_1^{i+2}$$

$$= \chi_1 + 2K\chi_1^2 + \sum_{i=3}^{\infty} iK^{i-1}\chi_1^i - 2K\chi_1^2 - 2\sum_{j=3}^{\infty} (j-1)K^{j-1}\chi_1^j + \sum_{k=3}^{\infty} (k-2)K^{k-1}\chi_1^k$$

$$= \chi_1 + \sum_{i=3}^{\infty} (i - 2(i-1) + i - 2) K^{i-1}\chi_1^i = \chi_1,$$

from which follows Eq. (1.30).

Figure 1.6 Cartoon of an infinite isodesmic reaction, where oligomers of increasing size are formed by consecutive addition of free monomers, with the same binding constant governing all equilibria. Modifications of this scheme include size-dependent binding constants.

Variations of the isodesmic self-association may include the combination with cyclization [59], the case where an initial step differs from subsequent steps involving small oligomers forming the self-associating unit [60], decaying isodesmic self-association [61], isoenthalpic self-association [62], or indefinite self-association with any particular favored step [57].

1.3.4 Bimolecular Hetero-Association

For the sedimentation/diffusion/reaction process of the molecular species A and B forming a reversible complex AB the Lamm equation (1.2) will be

$$
\frac{\partial \chi_A}{\partial t} + \frac{1}{r}\frac{\partial}{\partial r}\left(\chi_A s_A \omega^2 r^2 - D_A \frac{\partial \chi_A}{\partial r} r\right) = -\frac{k_{\text{off}}}{K_D}\chi_A \chi_B + k_{\text{off}}\chi_{AB}
$$

$$
\frac{\partial \chi_B}{\partial t} + \frac{1}{r}\frac{\partial}{\partial r}\left(\chi_B s_B \omega^2 r^2 - D_B \frac{\partial \chi_B}{\partial r} r\right) = -\frac{k_{\text{off}}}{K_D}\chi_A \chi_B + k_{\text{off}}\chi_{AB} \qquad (1.31)
$$

$$
\frac{\partial \chi_{AB}}{\partial t} + \frac{1}{r}\frac{\partial}{\partial r}\left(\chi_{AB} s_{AB} \omega^2 r^2 - D_{AB} \frac{\partial \chi_{AB}}{\partial r} r\right) = +\frac{k_{\text{off}}}{K_D}\chi_A \chi_B - k_{\text{off}}\chi_{AB}\,.
$$

Although this does appear somewhat more complicated than a Lamm equation for two-step homogeneous interactions, such as the monomer-dimer self-association, heterogeneous interactions do have the advantage that the properties of the building blocks A and B can be measured in separate experiments, and that their concentrations can be varied independently in the mixtures. Further, in many cases the components studied have different spectral properties, such that the study of the mixture can be enhanced with multi-signal and/or multi-wavelength detection.

This is illustrated in Fig. 1.7, which shows an SV experiment of a mixture of SLP-76 and PLC-γ peptides conducted with interference optical and absorbance detection [28]. In this interacting system, the absorbance detection provides a signal that is specific to one of the components. The interference optical signal shows both, in addition to an offset for unmatched buffer signals, which is modeled as an extra discrete species superimposed to that of the reacting system (see Section 6.5.3). An example for global Lamm equation modeling of multiple sample mixtures is presented in Fig. 1.8, which shows a rapid heterogeneous interaction between a receptor protein Ly49C that has two symmetrical sites for an MHC-I ligand [28]. This system highlights the common problem that the precise kinetic rate constants

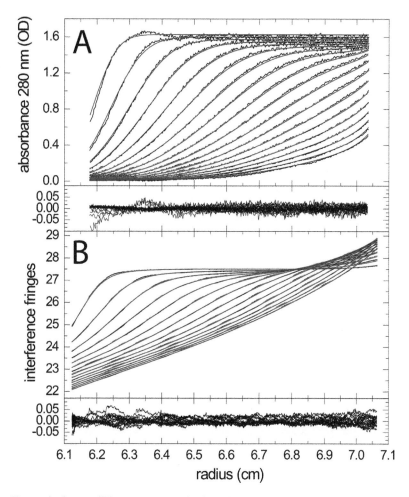

Figure 1.7 Example for an SV experiment of a bimolecular interaction, fitted by Lamm equation modeling. It shows the analysis of the interaction of peptides derived from the adaptor protein SLP-76 (11.7 kDa, 0.60 S) and PLC-γ (7.4 kDa, 0.75 S) which form complexes with 1:1 stoichiometry. SLP-76 contains only one tyrosine and no tryptophan residues, allowing its spectral discrimination from PLC-γ. Shown are absorbance (*Panel A*) and interference (*Panel B*) profiles of a mixture (65 μM PLC-γ with 25 μM SLP-76) in time intervals of 2500 sec, at a rotor speed of 59,000 rpm. The interference optical data are superimposed by the sedimentation of a small buffer component modeled as discrete species at 0.055 S. Molar mass values were kept fixed at the values predicted from amino acid sequence, and extinction and sedimentation coefficients of free peptides were determined in prior experiments, but loading concentrations were fitting parameters. The global analysis of the two data sets shown leads to best-fit estimates of $s_{AB} = 1.05$ S, $K_D = 10$ μM and $k_{off} = 10^{-5}$/sec. Reproduced from [28]. A multi-signal $c_k(s)$ analysis of the same data demonstrating the presence of 1:1 complexes and the absence of 2:1 complexes can be found in Fig. 4.6 on p. 140.

may not be well determined by the experimental data. While a good fit with rmsd of 0.0117 fringes can be achieved with the model for instantaneous reactions, a finite value of $k_{off} = 0.0023$/sec leads to a virtually indistinguishable fit with the rmsd of 0.00118 fringes. Only very slow reactions, with k_{off}-values such as 10^{-5}/sec can be clearly excluded with rmsd of 0.018 fringes [28].

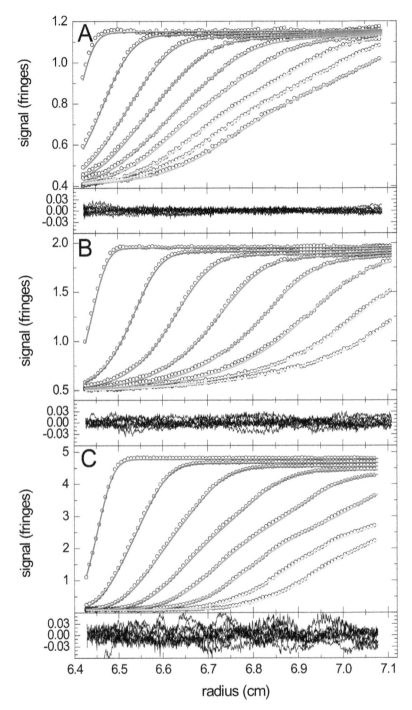

Figure 1.8 Example for a rapidly reversible interacting two-site system in SV. Experimental profiles are shown for a constant concentration of Ly49C (5 μM) with MHC class I molecules H-2Kb at 1.2 μM (A), 6.0 μM (B), and 28.7 μM (C) at 50,000 rpm. To highlight the concentration-dependent time-course of sedimentation, panels display every 10th data point of every 7th (A) or 6th (B and C) scan (circles), corresponding to similar time intervals of 1020 and 1130 s, respectively. The data were fitted globally (including data from later scans and additional mixtures that are not shown) with a two-step association model for noncooperative equivalent sites (solid lines), converging to estimates of $K_D = 1.7$ μM, $k_{\text{off}} = 0.1/\text{sec}$, $s_{\text{AB}} = 4.96$ S, and $s_{\text{ABB}} = 6.11$ S. Reproduced from [28].

In the limit of instantaneous local chemical equilibria, it is $\chi_{AB} = K\chi_A\chi_B$, and Eq. (1.31) can be written again as a function of total local constituent concentration of A, as the sum $\bar{\chi}_A(r,t) = \chi_A(r,t) + \chi_{AB}(r,t)$, and B, as the sum $\bar{\chi}_B(r,t) = \chi_B(r,t) + \chi_{AB}(r,t)$, such that for each radial position and time-point

$$\chi_A = \frac{1}{2K}\left(-1 + K(\bar{\chi}_A - \bar{\chi}_B)\right.$$
$$\left. + \sqrt{1 + 2K(\bar{\chi}_A + \bar{\chi}_B) + K^2(\bar{\chi}_A + \bar{\chi}_B)^2}\right) \quad \text{and} \quad (1.32)$$

$$\chi_B = \frac{\bar{\chi}_B}{1 + K\chi_A}.$$

This leads to the Lamm equations

$$\frac{\partial\bar{\chi}_A}{\partial t} + \frac{1}{r}\frac{\partial}{\partial r}\left(\bar{s}_A(\chi_A, \chi_B)\bar{\chi}_A\omega^2 r^2 - \bar{D}_A(\chi_A, \chi_B)\frac{\partial\bar{\chi}_A}{\partial r}r\right) = 0$$
$$\frac{\partial\bar{\chi}_B}{\partial t} + \frac{1}{r}\frac{\partial}{\partial r}\left(\bar{s}_B(\chi_A, \chi_B)\bar{\chi}_B\omega^2 r^2 - \bar{D}_B(\chi_A, \chi_B)\frac{\partial\bar{\chi}_B}{\partial r}r\right) = 0 \quad (1.33)$$

with the local composition dependent sedimentation and diffusion coefficients

$$\bar{s}_A(\chi_A, \chi_B) = \frac{s_A + \chi_B K s_{AB}}{1 + \chi_B K}$$

$$\bar{D}_A(\chi_A, \chi_B) = \frac{D_A + \chi_B K\left[1 + \left(\frac{\partial\chi_B}{\partial r}/\chi_B\right)\left(\frac{\partial\chi_A}{\partial r}/\chi_A\right)^{-1}\right]D_{AB}}{1 + \chi_B K\left[1 + \left(\frac{\partial\chi_B}{\partial r}/\chi_B\right)\left(\frac{\partial\chi_A}{\partial r}/\chi_A\right)^{-1}\right]}, \quad (1.34)$$

and symmetrical expressions for $\bar{s}_B(\chi_A, \chi_B)$ and $\bar{D}_B(\chi_A, \chi_B)$.[11,12] Here $\bar{D}_A$ is written in a way that highlights the impact of relative gradients: For example, if B is a small

---

[11]It is interesting to establish the relationship between the Lamm equations for the case of an instantaneously equilibrating monomer-dimer system Eqs. (1.25)–(1.26) and that of hetero-associations Eqs. (1.33)–(1.34) if all sedimentation and diffusion parameters are equal, i.e., $s_A = s_B = s_1$, $s_{AB} = s_2$, $D_A = D_B = D_1$, and $D_{AB} = D_2$. If we consider A and B indistinguishable, for equimolar loading concentrations we have $\chi_A = \chi_B = \frac{1}{2}\chi_1$ and $\chi_{AB} = \chi_2$, the equilibrium constant for hetero-association $K_{AB} = \chi_{AB}/(\chi_A\chi_B)$ relates to that of homo-dimerization $K_{12} = \chi_2/\chi_1^2$ as $K_{AB} = 4K_{12}$. Adding both equations of (1.33) leads to (1.25), and (1.34) is identical to (1.26) considering that $\chi_B = \frac{1}{2}\chi_1$ and $K_{AB} = 4K_{12}$, and that with $\chi_A = \chi_B$ the brackets in the expression for the diffusion coefficient turns into a factor of 2. This symmetry between homo- and hetero-dimerization breaks down for non-equimolar mixtures.

This correspondence can be useful in practice for interactions of molecules with very similar sedimentation parameters where equimolar concentrations can be guaranteed, for example, through purification of the assembled complex. An example is the study of tubulin hetero-dimer assembly of α- and β-subunits [63].

[12]For numerical solutions, this form of Eqs. (1.32) to (1.34) — and similar equations for other reaction schemes — can be more efficiently implemented as compared to the more general expression for sedimenting instantaneous equilibria in Eq. (1.10).

molecule in large molar excess over a macromolecule A, then its relative gradient will be negligible compared to that of the macromolecule [52], and the term in brackets approaches unity. Furthermore, if B is a small molecule that exhibits only a shallow gradient, then the fractional saturation of A is nearly constant, $\chi_{AB}/\chi_A = K\chi_B$. In this case, Eq. (1.34) leads to nearly constant $\bar{s}_A$- and $\bar{D}_A$-values that are in between those of A and AB. We may further ask whether under these particular circumstances the local ratios of $\bar{s}_A$ and $\bar{D}_A$ would correspond to a meaningful molar mass, if the sedimentation data report only on the macromolecular component A. Under these conditions, the Svedberg equation (1.13) leads to

$$\frac{M_b}{RT} \approx \frac{\bar{s}_A}{\bar{D}_A} \approx \frac{s_A + \chi_B K s_{AB}}{D_A + \chi_B K D_{AB}}$$
$$\approx M_A \frac{1}{1 + \chi_B K D_{AB}/D_A} + M_{AB} \frac{\chi_B K D_{AB}/D_A}{1 + \chi_B K D_{AB}/D_A}.$$

(1.35)

We can see from this that the molar mass would approximately correspond to a value between that of free A and the complex AB, dependent on the fractional saturation achieved in the experiment. In summary, this limiting case satisfactorily connects the equations of interacting systems to that of ideal non-interacting species for very small ligands. This is important as proteins always at least bind solvent and ions, while still exhibiting sedimentation behavior of 'non-interacting' particles. This aspect is further discussed in the context of the constant bath theory [64–66] described in Section 3.4 on p. 114.

1.3.5 Three-Component Systems — A Molecule with Two Independent Sites for Different Ligands

In Fig. 1.8 we have already seen data from a two-site system, where a molecule A has two independent, indistinguishable sites for B. We now consider another system with two independent sites, but for two different ligands B and C, which do not interact with each other. This is a quite important case to study, as biologically meaningful interactions usually involve more than just two components in order to enable information transfer in signaling pathways. Also, the model of two independent sites applies to bispecific antibodies, a pharmaceutically increasingly important class of regulatory modulators [67, 68].

This seemingly small change causes substantially more complicated sedimentation behavior, as we move from a two-component to a three-component system. With the reaction scheme 2A+B+C↔AC+AB↔A+ABC there are six species in solution, including three free species, two binary complexes, and one ternary complex. Fig. 1.9 shows an example for the sedimentation behavior of such a system, assuming weight-based signal increments for all species. Two different loading mixtures are shown in *Panel A* and *Panel B*. The conditions shown in *Panel A* exhibits three boundaries, two slower ones with the velocities of free species and a faster one migrating at a velocity not corresponding to the *s*-value of any species. In *Panel B*

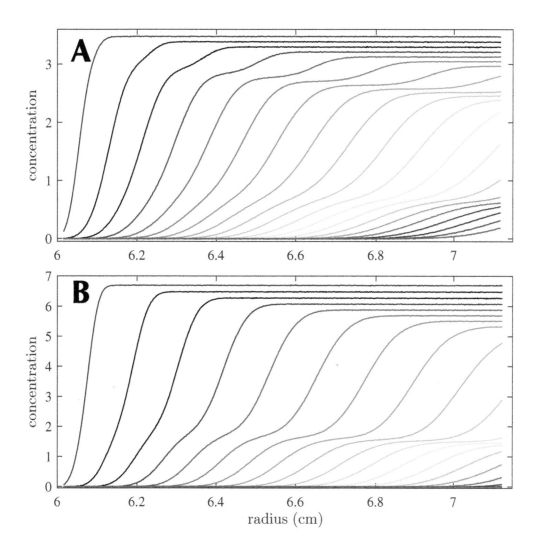

Figure 1.9 Sedimentation profiles for a three-component system describing a molecule A with independent sites for B and C, both equilibrating rapidly on the time-scale of sedimentation. Lamm equation solutions were calculated based on Eq. (1.2), assuming A and C having molar masses of 150 kDa and B being 100 kDa, with equilibrium dissociation constants $K_{AB} = K_{AC} = 1$ μM, and species s-values of $s_{\text{Afree}} = 7$ S, $s_{\text{Bfree}} = 5.5$ S, $s_{\text{Cfree}} = 8$ S, $s_{AB} = 10$ S, $s_{AC} = 11.5$ S, and $s_{ABC} = 15$ S. Sedimentation was simulated in a 12 mm solution column at 50,000 rpm, and total signals assuming mass-based signal increments of all species are plotted in 600 sec intervals. *Panel A*: Loading concentrations are $c_{A,\text{tot}} = 0.36$ μM, $c_{B,\text{tot}} = 2.8$ μM, and $c_{C,\text{tot}} = 5$ μM. *Panel B*: Loading concentrations are $c_{A,\text{tot}} = 8.7$ μM, $c_{B,\text{tot}} = 0.36$ μM, and $c_{C,\text{tot}} = 5$ μM.

only two boundaries can be recognized, the faster one again in between the s-value of any complex species but slower and broader than the fast one in *Panel A*.

This example highlights the limit of what can be understood on the basis of the coupled Lamm equation: Despite the clearly visible boundaries, we do not learn how to further interpret their structure, even for comparatively simple reacting systems. With 7 species in solution, why is it that we only see two or three boundaries? What is the mechanism that determines their pattern? We will come back to this example in Section 3.2.4 after laying the groundwork for better understanding the sedimentation process of rapidly reacting systems.

1.4 HISTORY, POTENTIAL, AND LIMITATIONS OF FITTING LAMM EQUATIONS OF INTERACTING SYSTEMS

So far, we have outlined the theoretical basis for modeling reaction-diffusion-sedimentation processes. Above examples and references show clearly that it is possible to describe experimental SV data with the theoretical models of interacting systems within the noise of the data acquisition. Dependent on data acquisition system, the sedimentation model may be extended with suitable models for the data characteristics, such as baselines, as covered in detail in Chapter 1 of Part II [2]. In principle, the application is straightforward, and practical strategies for how to arrive at such fits can be found, for example, in [28, 69, 70] and will be discussed in Section 6.5.3.

> Lamm equation modeling of interacting systems can be carried out in SEDPHAT as a global fit to data sets acquired at different concentrations and/or optical detection systems. Details are discussed in Chapter 6.

At first glance, the direct fitting of SV data from interacting systems with Lamm equation solutions appears theoretically very satisfactory. Additional appeal comes from the fact that it rigorously addresses seemingly counter-intuitive features of sedimentation patterns that were discovered in the 1950s [43, 71], which prohibited most simple alternative data analysis approaches. Historically, the development of the Lamm equation modeling approach for interacting systems has been closely linked to the numerical methods and computer programs for solving the coupled partial differential equations efficiently with sufficient precision [25, 44, 72–76], although initially the focus of the latter was mainly to examine general principles of SV rather than the evaluation of particular sedimentation parameters from given data.

Fitting of Lamm equation solutions appeared in the 1960s and 1970s. The earliest reports include the procedure by Trautman [77, 78] using a magnetic card programmable early desktop computer. This was followed by more powerful approaches described by Cox [76, 79] and Gilbert [80] — the latter consisting of iterative adjustment of parameters until the simulated Lamm equation solutions match the experimental Schlieren patterns — and an implementation of Claverie's finite element algorithm [45] for self-associating systems by Todd and Haschemeyer [81].[13] Baseline assignment was found to be the greatest remaining problem at the time [80]. While comparisons of simulations with experimental data were expected to become increasingly useful, it was recognized early that unambiguous interpretations of the sedimentation data may not always be possible [76,80,83]. In some systems, sample polydispersity was identified as a major factor determining the shape of the sedimentation boundaries [83–85].

Unfortunately, none of these methods found much practical application at the time. A general decline in the use of AUC interrupted this field in the 1980s, until it reawakened in the late 1990s, with the renaissance of AUC and mainstream interest in protein interactions [86]. By that time, computational resources were sufficiently abundant to make fitting of Lamm equation solutions to experimental data practically feasible. Current SV analysis packages for interacting systems including SEDPHAT [28,46], AKKUPROG [87], and in differential mode SEDANAL [27], and others.[14] A precondition for the application to interacting mixtures is the study of non-interacting components. However, with the exception of small particles [88,89], the success of fitting discrete non-interacting species models to samples of putatively non-interacting macromolecules is quite mixed: As discussed at length in Part II [2], the exquisite sensitivity of SV to heterogeneity can magnify the impact of minor impurities and microheterogeneity such that they can become dominant factors for the boundary spread. Neglect of polydispersity and imposter interpretation of boundaries as if caused by single discrete species can cause qualitatively wrong results in the parameters M and D.[15]

Thus, the central problem of SV analysis of distributions of non-interacting systems reappears for interacting systems: *Since the direct modeling with solutions of coupled Lamm equation systems implies only discrete states, it will be similarly susceptible to misinterpretation from unaccounted polydispersity as non-interacting systems.* Among the binding parameters, the effect of kinetic rate constants is confined largely to the boundary shapes. Therefore, their best-fit estimates would be most susceptible to unrecognized polydispersity; however, polydispersity may also potentially be correlated with other parameters, such as diffusion constants or their implicit buoyant molar masses.

---

[13]Simultaneously, fitting of discrete species Lamm equation solutions to non-interacting was introduced, such as described in 1980 by Holladay [82].

[14]As outlined in Appendix A of Part II [2], computer algorithms for solving the Lamm equation should not blindly be accepted to produce correct concentration profiles. This is particularly true for the more complex and coupled partial differential equations of interacting systems. This problem provides additional motivation to achieve physical understanding and derive laws governing the boundary structure as a reference point for numerical solutions.

[15]A clear demonstration of this is the data discussed in [90] for the glycosylated NK receptor fragment, which in a single-species model produces a fit of moderate quality suggesting a monomeric state, whereas a detailed size-distribution model with $c(s)$ reveals the microheterogeneity and leads to the correct result of dimeric molecules. As discussed in Part I, Section 4.3.5 (Fig. 4.29) [1], a close inspection of the residuals allows the clear distinction between the two cases.

Multi-component interacting systems where each of the components can be studied separately can offer a simple and stringent test for the applicability of direct Lamm equation modeling of the interacting mixtures: Any of the interacting components that is considered a single species must provide a satisfactory fit to a discrete single-species non-interacting Lamm equation solution, over the entire concentration range used later for the interacting system. If the single-species model does not fit, then obviously a meaningful fit of the data from the interacting mixtures with a coupled system of Lamm equations cannot be expected. This test will exclude the majority of potential applications of direct Lamm equation modeling to heterogeneous interacting systems. Even ideal behavior of the individual components will not guarantee the success of interaction modeling.[16]

The great size-resolution and spectral discrimination of SV analytical ultracentrifugation carries the potential for multi-site and multi-component interacting systems to be studied. In view of such applications, however, the susceptibility to sample imperfections of direct Lamm equation modeling is a significant problem, increasingly so for systems with higher numbers of components.

Finally, even for perfectly homogeneous samples, if the interaction mode is not known the Lamm equation modeling approach would be increasingly cumbersome, requiring a trial-and-error approach of fitting different interaction models, each containing parameters potentially hard to initialize and/or poorly determined by the data, such as kinetic rate constants. For data such as shown in Fig. 1.9, this seems a daunting proposition, since the boundary structure gives few clues about the underlying sedimenting species and their relationship.

For these reasons, it is crucial to develop quantitative tools that describe the reaction/diffusion/sedimentation process on a lower level of detail with fewer parameters, and with greater robustness against experimental imperfections. Likewise, in order to make more complex systems tractable, it is critically important to develop approaches that can more directly reveal the reaction scheme and the stoichiometry of the complexes. Finally, it is indispensable for the experimenter to develop a deeper understanding of the phenomenology of sedimentation of interacting systems. An intuition of these processes can help us to recognize different factors affecting the practical data analysis, and guide us to a meaningful quantitative interpretation. Solving the Lamm equations is only the beginning!

Accordingly, the topic of the next chapters will be the development of robust quantitative analysis strategies that do not aim at a full description of all aspects of the reaction model and, correspondingly, also do not aim to interpret all aspects of the experimental data. This requires first a more comprehensive discussion of the phenomenology of reaction/diffusion/sedimentation processes, and a deeper understanding of the physical processes governing the sedimentation boundaries (Chapter 2). Toward this end, we will introduce a theoretical concept that treats the reacting

---

[16]Remaining problems may include aggregation of the complex(es), and/or chemical polydispersity, e.g., in phosphorylation, that will lead to heterogeneity in the binding properties [85].

system as a single effective particle, and outline the reasons for the polydispersity of apparent sedimentation coefficients even for perfectly pure samples.

We will then proceed with approximate solutions of the Lamm equation that focus on the sedimentation process with different levels of detail (Chapter 3). The set of tools developed in Part II for non-interacting polydisperse systems will come to bear [2]. This includes first the transport method for determining the weighted-average sedimentation coefficient s_w (Section 3.1). Like in the analysis of non-interacting species, s_w is only concerned about the overall mass balance irrespective of the boundary structure or detailed shape. Second, we will examine the quantitative aspects of the effective particle model (Section 3.2), which explains in a simple diffusion-free picture the multi-modal boundary structure, but not the boundary shape. Third, the Gilbert–Jenkins theory (Section 3.3) is still more detailed, in that it provides quantitative expressions for the polydispersity of the reaction boundary, though also limited to a diffusion-free approximation. The constant bath theory (Section 3.4) is orthogonal to the effective particle model and the Gilbert–Jenkins theory as it is focused on the diffusion properties of the sedimenting system, which finally highlights a deeper relationship with sedimentation coefficient distributions $c(s)$ of non-interacting particles (Chapter 4). These methods provide quantitative analysis tools in the form of composition-dependent isotherms that can provide estimates for the equilibrium binding constants of the reaction, as well as qualitative assessments of the time-scale of the reaction kinetics, while not requiring direct Lamm equation fitting.

CHAPTER 2

Phenomenology of Sedimentation of Interacting Systems

SOME of the salient aspects of the transport behavior of rapidly reacting systems were discovered in the 1930s, dating back to the early experiments examining the properties of proteins by electrophoresis, chromatography, and sedimentation [18,19]. From the work of Tiselius, it was known that in moving boundary transport experiments no resolution of species' boundaries will occur if the components are in a rapid equilibrium compared to the rate of migration [91]. Rather, in this case an average migration velocity would be expected, reflecting the time-average state of the sedimenting molecules, but sometimes more complicated and counter-intuitive behavior was observed. Biophysical and mathematical studies on the principles of transport of rapidly reacting systems were of great interest at the time and published by many groups in the following decades [43,92–98]. Approximate Lamm equation solutions, as they were developed, appear to have merely confirmed rather than clarified the complex behavior.

In the present chapter, we will systematically explore the phenomenology of reversibly reacting systems in the centrifugal field. The sedimentation patterns are fascinating and rich in detail, but can be confusing to the unprepared.

2.1   DISTINCTION OF SLOW AND FAST KINETICS

The principal distinction of sedimentation behavior of interacting systems is that between either hydrodynamically resolved sedimentation boundaries of long-lived chemical species or patterns of dynamically co-sedimenting components in rapid local exchange of species. The observation of a relatively sharp transition in rate constants exhibiting either type of behavior is similar for heterogeneous associations and for homogeneous (self-) associations. The transition occurs in the k_{off}-range of 10^{-4}/sec to 10^{-3}/sec, and it appears at slightly higher rate constants for larger

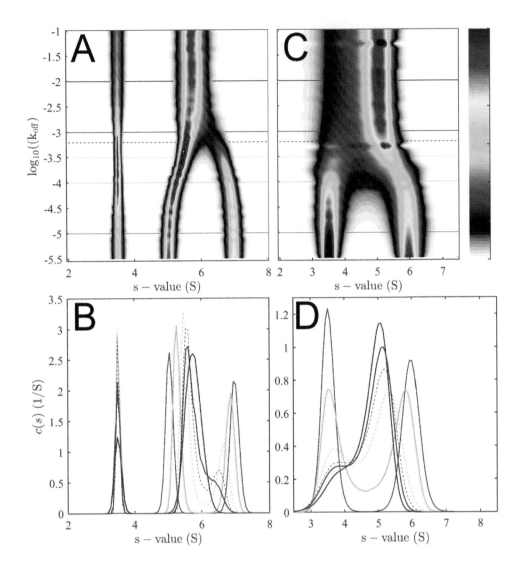

Figure 2.1 Sedimentation coefficient distributions for interacting systems as a function of kinetic rate constant k_{off}. *Panels A* and *B* show $c(s)$ distribution from the simulated sedimentation of the heterogeneous interacting system A + B ↔ AB, whereas *Panels C* and *D* are for a monomer-dimer self-association A + A ↔ A$_2$. Both are at component concentrations equal to K_D. The top panels depict arrays of $c(s)$ distributions as a surface plot with color temperature scaled linearly according to the $c(s)$ values. The bottom panels show select $c(s)$ distributions in the standard overlay plots at values of $log_{10}(k_{off}) = -5$ (magenta), -4 (green), -3.5 (dashed green), -3.2 (dashed blue), -3 (blue), and -2 (red). For reference, these values are highlighted as horizontal lines in the contour plots above. For all data, sedimentation profiles were simulated with coupled Lamm equation systems, ~1% normally distributed noise was added, and a standard $c(s)$ analysis was carried out with frictional ratio optimized separately for each data set. Simulation parameters for the heterogeneous system in *Panels A* and *B* were based on a 100 kDa, 5 S-species A binding a 60 kDa, 3.5 S-species B to form a 7 S complex, sedimenting in a standard 12 mm solution column at 50,000 rpm, recorded in 300-sec intervals for 254 min. The self-associating system was simulated analogously for a 60 kDa, 3.5 S monomer forming a 6 S dimer. Relative signal contributions were assumed to be proportional to species' masses.

molecules than for small molecules. The complex lifetime at the transition point is roughly comparable with the time it takes the interacting molecules to transverse an appreciable fraction of the solution column. Any reactions that take place on a time-scale much faster than that will be nearly indistinguishable from the limiting case of instantaneous chemical equilibrium at all radii and all points in time. In contrast, reactions that are slower than that have species that are sufficiently long-lived to be physically separated and hydrodynamically resolved, and the observed sedimentation patterns approach those of free and complex species sedimenting as if they were non-interacting.

To illustrate this in concrete examples, let us take a look at a bimolecular reaction $A + B \leftrightarrow AB$ and a monomer-dimer reaction $A + A \leftrightarrow A_2$, and examine the sedimentation profile for equimolar concentrations at K_D as a function of the kinetic off-rate constant k_{off}. Since the $c(s)$ distribution provided us with the most faithful representation of the distribution of sedimenting particles for non-interacting species, we will use this tool here, too, for deconvoluting the $\sqrt{t}$-dependent contributions of diffusion. (We will examine later why this still works well for interacting systems.) Fig. 2.1 shows the $c(s)$ distributions for systems with $-1 < \log_{10}(k_{\text{off}} \times \text{sec}) < -5$ side-by-side in a color contour map. The most striking feature is that the sedimentation patterns largely fall into two classes — one characteristic for fast sedimentation in the upper half of the plot, and the other for slow sedimentation in the lower half — and that there is only a narrow range of off-rate constants for which we see a transition between the two. Within each of the kinetic regimes, there are no obvious features in the SV concentration profiles that could provide further information on the reaction rate constant.[1]

Generally, macromolecules exhibit different binding kinetics in different solvent conditions or when locked in different conformational states [99–101]. This can provide the opportunity to experimentally observe sedimentation patterns characteristic of rapid and slow interconversion from the same molecules in different states. An example is the monomer-dimer interaction of N-cadherin domain NCAD12 [101], which shows a pattern of very slowly interconverting kinetically trapped monomer and dimer populations in the apo state, but rapidly exchanging and equilibrating monomer and dimer in the calcium saturated state. This is shown in Fig. 2.2. Similar changes in association kinetics have been observed for EBV protease in the presence or absence of glycerol [99], and for HVC core protein in the presence of different detergents [100].

Only in a narrow kinetic transition region is the AUC a suitable instrument to measure quantitatively the chemical kinetics. This intermediate case will be discussed in more detail below. More commonly, the chemical reaction kinetics can

---

[1]This is not an artificial effect of the deconvolution of diffusion in $c(s)$, which can be shown by inspecting corresponding ls-$g^*(s)$ traces [28] that show a similar pattern with regard to different k_{off}. A peculiarity of the application of $c(s)$ to data with different kinetics is a characteristic slight decrease in the size of the kinetic transition region, joining normal values similar to those that would be expected for the individual species in both slow and fast kinetic regimes, respectively. This will be discussed in more detail later.

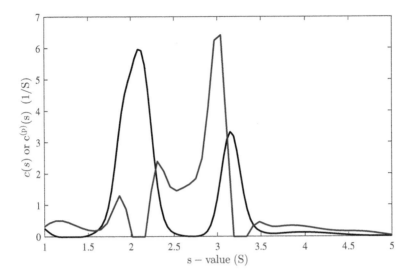

Figure 2.2 Change of monomer-dimer exchange kinetics of N-cadherin domain NCAD12 in the presence of calcium [101]. Shown are samples in the apo-form of NCAD12 at 1 mg/ml with (magenta) and without (blue) 1 mM calcium. In the absence of calcium (blue line), exchange between monomer and dimer is virtually absent, as indicated by the concentration-independent s-values of monomer ($\sim$2.10 S) and dimer ($\sim$3.27 S) in the $c(s)$ distribution. After addition of calcium (magenta line), exchange of monomer and dimer is promoted, and takes place on a time-scale that is fast relative to sedimentation. In order to enhance the difference in the sedimentation coefficient distributions, for the sample with calcium the Bayesian $c^{(p)}(s)$ distribution is shown with the Bayesian prior that the same s-values are populated as in the slowly exchanging sample. However, this prior distribution cannot be accommodated in the fit, for which populations of intermediate s-values are essential. (To partially accommodate the impostor prior, a substructure appears in the magenta peak at the edges toward the blue peaks.) In the presence of calcium, all samples assume equilibrium distributions with s_w following mass action law with $K_D \approx 24$ μM [101].

be readily classified as being either slow or fast on the time-scale of sedimentation, and to this limited extent usually we can still learn about the kinetics of the reacting system. On the other hand, a convenient consequence of the narrow transition regime is that we can capture most of the phenomenology by discussing what happens in the slow and in the fast regime. Therefore, we turn to the concentration-dependence of the sedimentation patterns in the two regimes. Fig. 2.3 shows a series of $c(s)$ distributions again as a color contour map, but now at fixed k_{off} and variable concentration. Two fundamentally different processes occur.

2.1.1 Recognizing Slow Exchange Patterns

The slow case is certainly the easier one to describe and intuitively understand with the tools we already have developed in Part II [2], since for any individual SV experiment the sedimentation process is similar to that of a mixture of populations of non-interacting species. In contrast to non-interacting mixtures, however, the relative contributions from the different species change, dependent on the total

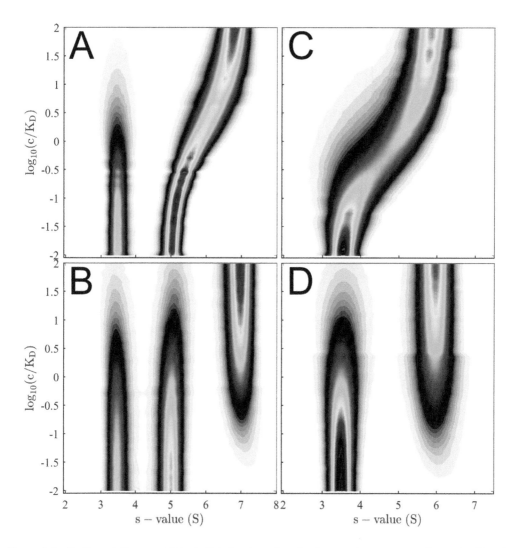

Figure 2.3 Sedimentation coefficient distributions for the interacting systems shown in Fig. 2.1 now as a function of concentration. *Panels A* and *B* show $c(s)$ distribution from the simulated sedimentation of the heterogeneous interacting system $A + B \leftrightarrow AB$, whereas *Panels C* and *D* are for a monomer-dimer self-association $A + A \leftrightarrow A_2$. The top *Panels A* and *C* show the systems with rapid equilibration ($k_{\text{off}} = -1$), whereas the bottom *Panels B* and *D* show the case of slow equilibration ($k_{\text{off}} = -5$). $c(s)$ traces are normalized by the total signal at each concentration, and shown in the same color temperature map as Fig. 2.1. Concentrations are equimolar for the hetero-association, and scaled logarithmically in units K_D, ranging from 0.01-fold to 100-fold K_D. Conventional overlay plots of $c(s)$ distributions at select concentrations are shown in Fig. 2.4.

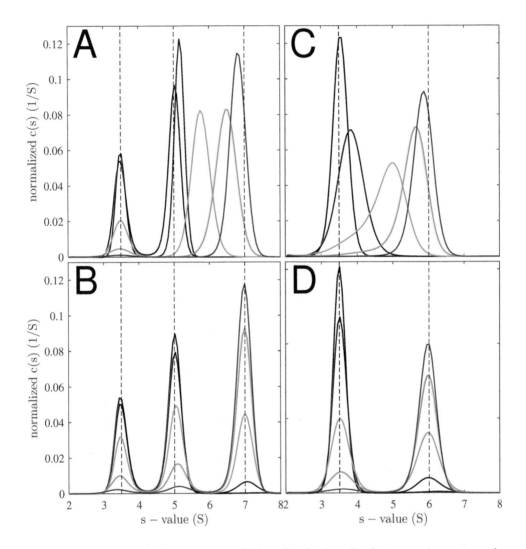

Figure 2.4 Overlay plot of sedimentation coefficient distributions for the interacting systems shown in Fig. 2.3 at (equimolar) concentrations of $0.01 \times K_D$ (black), $0.1 \times K_D$ (blue), K_D (cyan), $10 \times K_D$ (green), and $100 \times K_D$ (magenta). As in Fig. 2.3, *Panels A* and *B* show $c(s)$ distribution from the simulated sedimentation of the heterogeneous interacting system $A + B \leftrightarrow AB$, whereas *Panels C* and *D* are for a monomer-dimer self-association $A + A \leftrightarrow A_2$. The top *A* and *C* show the systems with rapid equilibration ($k_{off} = -1$), whereas the bottom *Panels B* and *D* show the case of slow equilibration ($k_{off} = -5$). Vertical dashed lines indicate s-values of the underlying sedimenting species. $c(s)$ traces are normalized by the total signal at each concentration. Simulation parameters are identical to those in Fig. 2.1.

loading concentrations (Fig. 2.3B/D and Fig. 2.4B/D). Assuming that the loading mixtures are in chemical equilibrium prior to sedimentation, we will observe the species' equilibrium populations in SV. As a consequence, for slowly interacting mixtures, we expect to observe as many boundaries or $c(s)$ peaks as there are species participating in the interaction (i.e., two for reaction schemes A + A ↔ A_2, three for $4A ↔ 2A_2 ↔ A_4$ and A + B ↔ AB, four for $2A + B ↔ A + AB ↔ A_2B$, etc.), although these may not always be resolved due to limitations in resolution and signal/noise.

A hallmark of the sedimentation of slowly reversible systems is that the s-values of the peaks stay virtually constant when comparing experiments at different loading concentrations. Accordingly, slow systems can be recognized from an overlay of $c(s)$ distributions obtained at different concentrations spanning a large range of concentrations in the vicinity of K_D. Fig. 2.4 illustrates such overlays for slow and fast systems. The $c(s)$ traces are shown normalized relative to the total loading concentration to permit their easier comparison. When assessing the peak positions we have to recognize the fact that regularization will tend to make the peaks broader at lower concentrations due to the lower signal/noise ratio, unless this is compensated in the data acquisition using more sensitive detection.[2]

It is noteworthy that, even though the s-values of the different species do not change much with loading concentration in slow systems, the peak positions for the complexes may still not be exactly identical to the s-values of the true underlying species. For systems such as those depicted in Figs. 2.1–2.4, with off-rate constants between 10^{-4}/sec and 10^{-5}/sec, the differences are sufficiently small to not interfere with the separation and identification of the species, as well as the classification of the interaction, but frequently the deviations are too large to permit the reliable interpretation of the peak s-values observed in $c(s)$ in terms of molecular hydrodynamic shapes. One can understand the origin of this deviation from the consideration that the complexes will not be absolutely stable during the SV experiment, since there is a finite probability for them to dissociate while on their way traversing the solution column: Once dissociated, these molecules will proceed to sediment with the lower s-values of their free components, until they eventually recombine. This will lead to a reduction in the s-value of the $c(s)$ peak of the complex that is dependent on the complex lifetime, as well as the loading concentrations. *Vice versa*, there is a finite probability for initially free molecules to combine to form a complex, slightly increasing the time-average s-value of these molecules. These effects will be less with lower k_{off}-values; $c(s)$ peaks of complexes

---

[2]In an attempt to eliminate this confounding factor, one can test whether or not the data is — within the noise of the data acquisition — consistent with the profile from a higher concentration. This is accomplished by taking the $c(s)$ profile from a higher concentration as a prior expectation $p(s)$ in the Bayesian $c^{(p)}(s)$ analysis of the data at lower concentrations (Part II, Section 5.7). The peak positions of the prior can generally be honored for slow systems, but may not be consistent in fast systems.

will be closer to complex s-values at higher loading concentrations, and free species will be more accurately represented in $c(s)$ at lower concentrations.[3]

For slow interactions, the chemical reaction and sedimentation are 'orthogonal' and independent: One can analyze the chemical equilibrium without interpreting the sedimentation, just by taking advantage of the empirical fact of the separation of the different species, through the analysis of their population as a function of the loading concentration. The equivalent loading signals concentrations a_i for each species may be determined by integration of respective $c(s)$ peaks. This is referred to in the following as a 'population isotherm,' abbreviated with the symbol $a_{\text{pop},i}(\{c_{k,\text{tot}}\})$, where each species simultaneously obeys mass action law and mass conservation for the given set of total component loading concentrations $\{c_{k,\text{tot}}\}$ inserted in the experimental sample. In more detail, given the free component concentrations $c_{k,1}$, the population of complexes can be written as:

$$a_{\text{pop},i}(\{c_{k,\text{tot}}\}) = \varepsilon_{i,\lambda} K_i \prod_{k=1}^{N_k} c_{k,1}(\{c_{k,\text{tot}}\})^{S_k^i}$$

$$c_{k,\text{tot}} = \sum_{i=1}^{N_i} S_k^i c_i \ , \tag{2.1}$$

with the second equation expressing mass conservation (including free species for which $K_k = 1$ and $S_k^i = \delta_{ik}$), which must hold for all components. It should be noted that the equivalent loading signal concentrations of species determined by integration of $c(s)$ do not suffer from the same imperfections as the species' peak s-values.

> SEDPHAT allows data sets of $a_{\text{pop},i}(\{c_{k,\text{tot}}\})$ to be loaded along with given total component loading concentrations. This defines the experiment type of a population isotherm, which can be part of a model with mass action law embedding a particular interaction scheme. For detailed data format and analysis strategy, see Section 6.5.2.2.

The analysis will be stronger, however, if we do take into account the additional information from the sedimentation process, in particular, the shift of the total

---

[3]In the measurement of the hydrodynamic shape of the complex (and, similarly, that of free species in self-associating systems) one may be able to circumvent this difficulty by deriving a better estimate for the s-value from modeling of the entire isotherm of the weighted-average sedimentation coefficients as a function of loading concentration, $s_w(\{c_{k,\text{tot}}\})$ with mass action law models (Section 6.5.1). These implicitly extrapolated s-values can be better suited for hydrodynamic modeling if the experimental concentration range is very large. On the other hand, if the concentration range is far from saturation (or complete dissociation, respectively) then the $c(s)$ peak locations may be the best estimate for the species s-values in the isotherm modeling.

signal weighted-average sedimentation coefficient, $s_{w,\lambda}(\{c_{k,\text{tot}}\})$. It usually follows

$$s_{w,\lambda}(\{c_{k,\text{tot}}\}) = \frac{\displaystyle\sum_{i=1}^{N_i} s_i \varepsilon_{i,\lambda} c_i(\{c_{k,\text{tot}}\})}{\displaystyle\sum_{i=1}^{N_i} \varepsilon_{i,\lambda} c_i(\{c_{k,\text{tot}}\})} , \tag{2.2}$$

where the species concentration follow mass action law in Eq. (2.1).[4] s_w can also be determined with high precision from the sedimentation data by integration of $c(s)$ across all peaks that reflect species participating in the interaction. As we have seen in Section 3.4 of Part II [2], the resulting values will depend neither on the ability to describe the sedimentation process with the correct physical model, nor on hydrodynamic separation of the different species. Rather, due to its foundation in the transport method, s_w-values only depend on a good fit with non-systematic residuals that ensures that the depletion of signal (i.e., the area under the measured profiles) is faithfully matched. Therefore, s_w-values present observations independent from $c(s)$ species peaks.

In fact, multiple isotherms $s_{w,\lambda}(\{c_{k,\text{tot}}\})$ and $a_{\text{pop},i}(\{c_{k,\text{tot}}\})$ may be available from multi-signal detection of heterogeneous mixtures, if we can take advantage of distinctive components' extinction coefficients $\varepsilon_{k,\lambda}$.[5] Furthermore, due to the stability and separation of their complexes, slow interactions also are particularly suitable for study by multi-signal $c_k(s)$ for the determination of stoichiometry and molar mass of complexes (Part II, Chapter 7 [2]) [38]. The practical application of these and other analysis strategies will be discussed in Chapters 4 and 6.

$s_{w,\lambda}(\{c_{k,\text{tot}}\})$ isotherms can also be loaded for modeling in SEDPHAT, either individually or in the context of global modeling, as described in more detail in Sections 6.5.1 and 6.5.2.

2.1.2 The Concentration-Dependence of Fast Exchange Patterns

A rapidly reacting mixture can generally not just sediment in a single boundary with a single average s-value. The inspection of an overlay of $c(s)$ distributions obtained from rapidly reacting systems in Fig. 2.4A and C at different concentrations underscores the conclusions drawn from the color temperature plots of Fig. 2.3: The

---

[4]Corrections are necessary for rapidly reversible systems with signal changes upon complex formation [102], as described in Section 3.2.2.

[5]As discussed in Section 4.3.2 of Part I [1], we often use the term 'extinction coefficient' loosely synonymous with a general molar signal increment or signal coefficient, such as absorbance extinction coefficients, molar fringe increments, count rates, etc., dependent on the optical detection system used. For convenience, we also contract the optical pathlength into the value of ε, unless noted otherwise.

number of peaks equals the number of components and the peak positions are concentration-dependent except for the slowest boundary in multi-component systems.

For two-component systems, the faster moving boundary — the one with the concentration-dependent peak position — is termed 'reaction boundary,' since it is formed by a mixture of all species undergoing chemical conversions. As we will support theoretically later, a set of interconverting molecules can be regarded as an equivalent 'effective particle' that sediments at an average s-value of the interconverting species, hence leading to the concentration dependence. It is crucially important to note that *generally, the set of interconverting species in the reaction boundary is not a combination of molecules in a kinetic steady-state of binding and rebinding at the concentration ratio equal to the complex stoichiometry.* Similarly, *the rate of coupled migration is generally not the weighted-average s-value of all species in solution.*[6] The slower boundary that usually appears in hetero-associating systems and exhibits concentration-independent s-values is termed 'undisturbed boundary,' since it is formed by a single component that sediments behind as if not participating in the chemical reactions. There are some special cases to these rules which will be discussed in detail below.

When inspecting $c(s)$ overlays from a series of experimental data acquired at a range of sample concentrations such as in Fig. 2.4A and C, again, the classification may be hampered at low concentrations by lack of resolution, and as a consequence of regularization applied to data of low signal/noise ratio, unless this is compensated by using a detection mode with higher sensitivity for the lower concentrations.[7]

Due to the strong concentration-dependence of the peaks, the interpretation of species s-values is non-trivial. Assuming it is known (from independent sources) that there is only one complex species with given mass and stoichiometry, for determining its hydrodynamic shape it is best to perform experiments at a series of concentrations and molar ratios, including experiments at high concentrations that saturate binding as much as possible. However, it would require even higher concentrations than for slow interacting systems ($> 1,000$-fold K_D) to shift the experimental value close enough to the s-value of the complex species to permit hydrodynamic modeling on the basis of the observed peak s-value. Instead, the resulting isotherm of s-values of the fast peak, $s_{\text{fast}}(\{c_{k,\text{tot}}\})$, can be fitted to arrive at an estimate for the true complex s-values. Empirical extrapolation should be avoided; appropriate models for the isotherm of the fast boundary component as a

---

[6]This arises directly from the unequal constituent velocities of the different components, if averaged over all molecules of a component in the entire sample as depicted in Fig. 1.2, i.e., $s_{\text{A,tot}} \neq s_{\text{B,tot}}$ in Eq. (1.34).

[7]In contrast to the case of slow interacting systems, the Bayesian expectation that the peak locations be invariant will not be honored for data with sufficient signal/noise ratio. An example for this is the data in Fig. 2.2. At low signal/noise ratios, however, the fit is more flexible and might comply with the Bayesian prior independent of whether the system is fast or slow.

function of loading composition require more detailed considerations of the physical nature of dynamic co-transport, and will be developed in Sections 2.4 and 3.2.

Similar to $s_{w,\lambda}(\{c_{k,\text{tot}}\})$, the isotherms of the fast peak $s_{\text{fast}}(\{c_{k,\text{tot}}\})$ can be be loaded for modeling in SEDPHAT. This is discussed in detail in Section 6.5.2.3.

2.2 BOUNDARY SHAPES OF SELF-ASSOCIATING SYSTEMS WITH FAST KINETICS

The overall characteristics of reaction boundaries of rapidly reversible self-associating systems can be readily understood from the idea that the molecules exhibit a sedimentation rate that reflects their time-average association state. Any molecule starting out as an oligomer will quickly dissociate before it has sedimented far, but then the monomers will also reassemble quickly again to oligomers. The time spent in the oligomeric state leads to short periods of faster sedimentation, whereas the time spent in the dissociated state is associated with slower sedimentation. For ergodic systems, the time-average state of a single molecule is equal to the population-average of all molecules. Therefore, we can relate the sedimentation coefficients with the local concentrations, as expressed in Eqs. (1.16)–(1.18). Due to the concentration gradients in the sedimentation boundary, driven in part by diffusional spread, a drawn-out boundary exhibiting a polydisperse range of sedimentation coefficients is produced for rapidly self-associating systems. Examples for different self-associating systems are shown in Fig. 2.5.

In a back-of-the-envelope calculation, at concentrations $c_0 \gg K_D$ (or $c_0 \gg K^{-1/(n-1)}$ for reversible n-mer formation, respectively), the molecules will be predominantly assembled, and therefore sediment rapidly, whereas at concentrations $c_0 \ll K_D$ the molecules will exist predominantly in the disassembled state and will sediment at a rate close to that of the monomer. Considering that experiments with loading concentrations far above K_D will still have low concentration regions in the trailing edge of the diffusionally broadened sedimentation boundary, we expect a broad, multi-modal distribution of sedimentation coefficients for self-associating systems at high loading concentrations. As observed first by Gilbert [71], this multi-modal feature will be increasingly more pronounced for monomer-n-mer systems with larger oligomer size n, essentially due to the steeper isotherm of oligomer assembly as a function of total concentration. This is visible, for example, in the sedimentation coefficient distribution from the highest concentration of the monomer-tetramer system in Fig. 2.5.

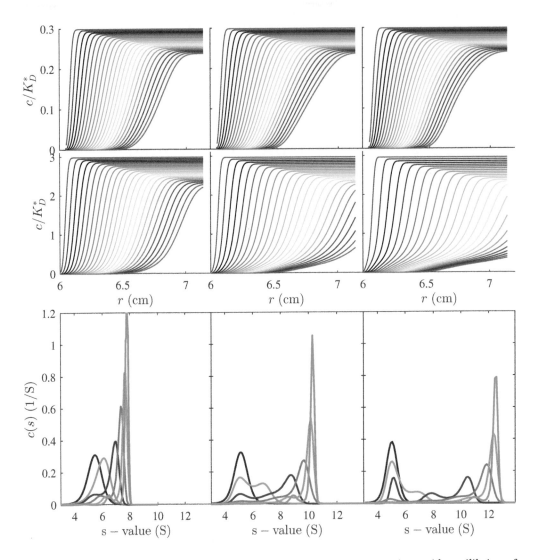

Figure 2.5 Boundary shapes of sedimenting monomer-n-mer systems in rapid equilibrium for different oligomer size: monomer-dimer (*Left Column*), monomer-trimer (*Middle Column*), and monomer-tetramer (*Right Column*). The boundary shapes are simulated assuming a monomer of 100 kDa monomer of 5 S, oligomerizing without significant change in the hydrodynamic frictional ratio, leading to an 8 S-dimer, 10.4 S-trimer, or a 12.6 S-tetramer, respectively. All panels show calculated profiles at the same equidistant time points. Total loading concentrations are 0.3-fold K_D^* (*Top Row*) and 3-fold K_D^* (*Middle Row*) (with $K_D^* = K^{-1/(n-1)}$ representing the monomer concentration where $c_1 = c_n$). Clearly, a multi-modal character of the boundaries appears for $n > 2$, a fact distilled even more clearly in the corresponding asymptotic boundaries of Gilbert theory shown in Fig. 3.12. It may also be discerned that the transition point from the slow trailing to the more rapidly sedimenting leading boundary portion occurs in these data at approximately the same concentration for the different self-association schemes. No multi-modal character is visible at low concentrations. The *Bottom Row* shows $c(s)$ profiles at concentrations 0.1 (gray), 0.3 (cyan), 1 (red), 3 (blue), 10 (orange), and 30 (green) fold K_D^*.

2.3 BOUNDARY SHAPES OF HETERO-ASSOCIATING SYSTEMS WITH FAST KINETICS

The boundary shapes of rapidly reversible hetero-associating systems are far more complex than those of self-associating systems [43, 103]. This arises from the fact that in hetero-associating systems the free species of each component may migrate at different velocities, and, moreover, that the time-average sedimentation coefficients of molecules from different components are different (compare Fig. 1.2). Another added complication for data analysis is that we cannot deduce from the local total signal alone the local species composition and the local average s-values, since this would require the local concentrations of all components.[8]

In contrast to the self-associating systems that showed a single boundary, though potentially stretched and multi-modal at high concentrations, for two-component hetero-associating systems we normally observe the undisturbed boundary separated from the reaction boundary by a plateau region. The existence of the separating plateau depends on diffusion and relative s-values, but principally not on concentrations. More generally, it can be shown that n-component mixtures will usually form n boundaries (Appendix B), all but the slowest one being reaction boundaries. The s-values, amplitudes, and composition of these boundaries all depend on the component concentrations of the loading mixture (in their totality also referred to as the 'concentration space'), on each species' s-value, and on the binding constants. They exhibit numerous properties that may appear counter-intuitive at first, but provide rich characteristics for qualitative and quantitative data analysis. In the following we will focus on two-component systems unless otherwise mentioned.

The reaction boundary sediments at rates between the s-value of the fastest sedimenting free species and the s-value of the complex species, or the largest complex species for multi-site binding, respectively.[9] In some cases the reaction boundary can appear polydisperse, covering a broad range of s-values, and can even be somewhat multi-modal; however, for simple bimolecular reactions in large regions of the concentration space it consists of boundaries that migrate and diffuse very similar to the sedimentation patterns of single, non-interacting species. Still, it is crucial that the reaction boundaries are not being confused with those of non-interacting species, and be identified rather as the sedimentation of an effective

---

[8]For some systems this information may be available through acquisition of data at multiple characteristic signals.

[9]For the purpose of this discussion we will assume throughout that the complexes sediment faster than the free species. This applies to the vast majority of interacting molecules described thus far in the literature. For macromolecules of similar density, slower sedimenting complexes could be caused by large conformational changes in complexes. Exceptions may be encountered more readily in the study of interactions between particles of very dissimilar density, such as metal nanoparticles and proteins, where the relative increase in buoyant mass from protein binding to the particle is small compared to the relative increase in Stokes radius [104].

particle representing the dynamically co-sedimenting system. This is depicted in Fig. 2.6.

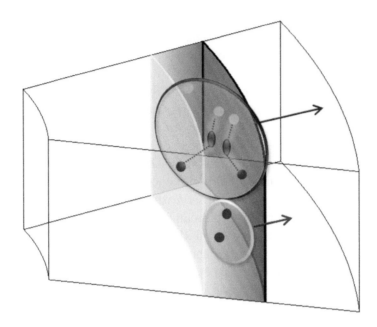

Figure 2.6   Cartoon of the sedimentation process for a rapid bimolecular reaction, with distinction between the reaction boundary and the undisturbed boundary, in contrast to the component-average sedimentation depicted in Fig. 1.2. Based on the implications from mass action law and the requirement of co-sedimentation we arrive at a view where one component is carried completely within the faster boundary, and the other component partitions into two boundaries: The reaction boundary (purple circle) sediments at an s-value that is identical to the component-average s-value of the blue component completely contained in the boundary. It also contains an appropriate fraction of the red component such that its time-average state leads to an s-value that precisely matches that of the blue component. The excess red component cannot be taken up into the fast reaction boundary, and constitutes the slower undisturbed boundary. In this way, the reacting mixture can co-sediment while obeying mass action law in the zones of co-existence. The quantitative relationships of this view are developed in the effective particle model.

The comparison of the sedimentation behavior of the system at different loading concentrations over a range covering K_D can give clues to their dynamic nature as a reaction boundary. As mentioned above, the number of the $c(s)$ peaks and their concentration-dependent shift permits identification as a rapidly reacting system. However, it is important to be aware of peculiar differences in phenomenology to the concentration-dependence of $c(s)$ peaks of self-associating systems: Although the (average) s-value of the reaction boundary of self-associating systems generally increases with increasing loading concentrations, this is not always true for hetero-associating systems. For example, when comparing the reaction boundaries from a series of experiments arranged as a titration series where the concentration of one component is kept constant while the other one is increased, the s-value of the reaction boundary may be virtually constant at low concentration of the titrant and then decrease at increasing concentrations of titrant! This is until a threshold concentration is exceeded, above which there is a sudden steep increase in the s-value. This non-monotonic behavior is dependent on the identity (s-value of the free species) and concentration of the component held constant in the titration series. By contrast, experiment series at constant molar ratios will usually lead to a monotonically increasing s-value of the reaction boundary.

It is important to realize that the reaction boundary is composed of a mixture of all complex(es) and all free species. This is a direct consequence of the coupled migration. For example, in interactions of a large macromolecule with a smaller molecule, at concentrations where the small molecule contributes the undisturbed boundary, some additional free small molecule species will always be part of the reaction boundary and co-sediment with the s-value of the reaction boundary. A direct demonstration of this is presented in Fig. 2.7, which shows simulated Lamm equation solutions for rapidly equilibrating bimolecular reaction for a case where only free A contributes to the signal (as would be recorded, for example, for a system with a single fluorescent component and complete quenching of signal in the complex). The presence of two boundaries of free A can be clearly discerned, the faster one being part of the reaction boundary.

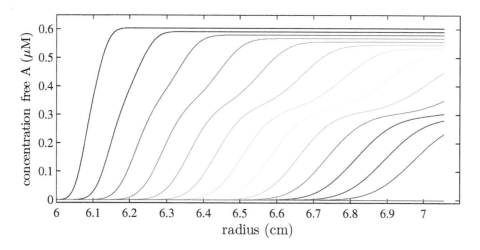

Figure 2.7 Demonstration of co-sedimenting free species in the reaction boundary predicted by Lamm equation solutions (Eq. 1.33) for instantaneous bimolecular reactions. Sedimentation velocity profiles are calculated based on coupled Lamm equations with instantaneous chemical equilibration, for a 150 kDa, 7 S component A sedimenting in the presence of an undetected 200 kDa, 9 S component B forming a rapidly reversible complex with 13 S that completely quenches the signal, such that only free A generates a signal. Loading concentrations are $c_{A,tot} = c_{B,tot} = K_D$, in a solution column of 12-mm length at a rotor speed of 50,000 rpm; signal profiles are shown in 10-min intervals. The slower boundary is the undisturbed boundary at the velocity of free A, the faster boundary is the reaction boundary.

However, despite the fact that all species are contributing to the reaction boundary, the molar ratio of the components in the reaction boundary does not directly represent that of the complex(es) or the stoichiometry of the reaction. Rather, it is additionally dependent on the loading concentrations, each species s-values, and the binding constants. In fact, the stoichiometry of the reaction boundary always exhibits an excess of the component with the faster-sedimenting free species.

The undisturbed boundary is mono-disperse and consists of only one of the components. Since the boundary migrates in the absence of any other components, short of self-association, the sedimentation pattern will be identical to that of one of the free species. We can imagine the component providing the undisturbed boundary to be in some form of excess relative to the concentration that the reaction boundary is able to carry along. The amplitude of this undisturbed boundary generally does not correspond to the equilibrium population of the same free species in a resting mixture, as predicted by mass action law: It is lower due to the fraction of free species co-sedimenting in the reaction boundary. Therefore, the amplitude of the undisturbed boundary is not only dependent on loading concentrations and the binding constants, but also each species' s-value.

The undisturbed boundary can even vanish, leaving the reaction boundary as the sole boundary of the sedimenting system. In a titration or dilution series of experiments, the nature — and therefore the s-value — of the undisturbed boundary can change. This switch of component does not occur where the loading concentrations are in a molar ratio equaling the complex stoichiometry. Instead, it will require higher concentrations of the species sedimenting faster in the free state. The s-value of the reaction boundary at this transition point will depend on concentrations and binding constants, and will fall between that of the faster sedimenting species and the complex.

These initially puzzling properties — most notably the existence of only two boundaries, the possibility of a decrease in the reaction boundary s-value with increasing concentration, the switch of undisturbed boundary at concentrations different from the molar ratio, and the possibility of the system exhibiting a single boundary that corresponds to none of the underlying species — will become clear if we look closer at the physics of the sedimenting system.

2.4 THE EFFECTIVE PARTICLE AS A PHYSICAL PICTURE FOR RAPIDLY REACTING SYSTEMS

The model of a 'sedimenting particle' in the context of solvation and its effect on buoyancy and hydrodynamic friction was already introduced in Chapter 2 of Part I [1]. In the present section, the idea that we can summarize the microscopic structural and dynamics aspects into a macroscopic particle equivalent for the purpose of sedimentation and diffusion will be further developed for rapidly reversible interacting systems. The 'effective particle of the reacting system' [41] leads to quantitative expressions that correctly capture the behavior of sedimenting interacting systems and fully predict their boundary structure; they will be derived mathematically within the theoretical framework of sedimentation in Section 3.2. First, we concentrate on the physical picture of the process.

2.4.1 The Effective Particle Picture

For simplicity, let us consider a system $A + B \leftrightarrow AB$ where the components A and B at total loading molar concentrations $c_{A,tot}$ and $c_{B,tot}$, and reversibly form a complex AB. At all times, the free and complex species concentrations c_A, c_B, and c_{AB} follow mass action law $c_{AB} = K c_A c_B$ with the equilibrium constant K, and they follow mass conservation $c_A + c_{AB} = c_{A,tot}$ and $c_B + c_{AB} = c_{B,tot}$, respectively. Without loss of generality we designate A and B such that their sedimentation coefficients follow $s_A \leq s_B$. We assume the complex to sediment faster than either free species. In order to highlight the salient features of the boundary pattern, we make the approximation that all boundaries be mono-disperse, and use the traditional simplification of diffusion-free sedimentation with constant force and rectangular geometry (see below).

The effective particle model is based on the recognition that there can be at most two distinct boundaries for rapidly reacting two-component systems, and that one of the components is completely migrating within the faster sedimenting reaction boundary. Only this can avoid the creation of zones where local species concentrations violate mass action law, while honoring the co-migration with different component-average s-value. The component completely contained in the reaction boundary is referred to as the 'dominant' component. It may be either A or B — both are possible modes of propagation in certain ranges of the concentration space. Depicting the concentration space as a phase diagram, we refer to them as 'phases.' We denote the phase in which B is dominant as $B\cdots(A)$, and the phase where A is dominant as $A\cdots(B)$. Since the dominant component is engulfed entirely in the reaction boundary, the latter must exhibit the sedimentation properties of the time-average state of that dominant component. The component that is not dominant is viewed as secondary, providing a ligand to the dominant component. Its concentration generally exceeds the fraction co-sedimenting with the dominant component. The secondary component therefore supplies the material for the undisturbed boundary, which sediments simply like the free species of that component. The key recognition is that *the population of the secondary component co-sedimenting in the reaction boundary is such that the time-average sedimentation coefficient of the co-sedimenting fraction of the secondary component matches the time-average sedimentation coefficient of the dominant component.* Otherwise no stable reaction boundary could form.

This simple principle is sufficient to explain most of the sedimentation behavior of rapidly reacting systems. Parallel to the formal derivations in Section 3.2, we can continue here in the physical picture to establish the essential relationships.

2.4.1.1 Phases and s-values of the Reaction Boundary

Let us first focus on $B\cdots(A)$ under the condition $c_{A,tot} \gg c_{B,tot}$, and consider the ensemble of molecules B in solution (upper plot of Fig. 2.8). Each molecule of B will spend on average a fractional time τ_{AB} occupied in the complex, out of a total $\tau_B + \tau_{AB}$:

$$\frac{\tau_{AB}}{\tau_B + \tau_{AB}} = \frac{c_{AB}}{c_B + c_{AB}} = \frac{1}{1 + (Kc_A)^{-1}}. \tag{2.3}$$

During this fractional time it will sediment with the velocity s_{AB}, and the remaining time it will sediment free with velocity s_B. Accordingly, we would expect the boundary to migrate with a velocity following the time-average sedimentation velocity of a transiently occupied molecule B [41]

$$s_{B\cdots(A)} = \frac{s_{AB}}{1 + (Kc_A)^{-1}} + s_B \left(1 - \frac{1}{1 + (Kc_A)^{-1}}\right) = \frac{s_B + s_{AB}Kc_A}{1 + Kc_A}. \tag{2.4}$$

This is closely related to the constituent s-value in Eq. (1.34).[10]

While B is unliganded, the copies of A previously in the complex are now free and unable to keep pace with their previous partner B. For the advancing B, however, another copy of A from the uniformly filled solution column ahead can be taken up to serve as the new binding partner.

What happens to A? From the pool of molecules of this component, a fraction will be transiently 'dragged along' in occasional binding events with B. During these encounters they will be transported ahead of where they would have been if they had been traveling constantly at the velocity of free A. These boosts in migration occur only in the zone of the reaction mixture, and they will serve to establish a concentration difference of A between the zone of the undisturbed boundary lagging behind and the zone of the reaction boundary. The elevation in concentration of A partitioned into the reaction boundary zone will travel along with the latter, and therefore appear macroscopically as a subpopulation of A that is part of the reaction boundary, although, of course, there is no physical distinction between those molecules of A in reaction and undisturbed boundaries.

In view of the undisturbed boundary, how big is the population of A left behind by the advancing front of the reaction boundary? This will depend on the relative velocities and the total loading concentrations. Let us consider two cases: If the free state of A were to migrate with the same velocity as the free state of B ($s_A = s_B$), then at equal loading concentration the time-average velocity of all of A would naturally match $s_{B\cdots(A)}$, and there would be no undisturbed boundary lagging behind: with time, all free molecules will encounter binding partners at sufficient rate to boost their average velocity. Similarly, even if free A would migrate slower ($s_A < s_B$), but there would be enough B in the reaction boundary to cause binding events at a sufficiently high rate, different copies of B could take turns boosting each A, all

---

[10]In the juxtaposition of Eqs. (1.34) and (2.4), the more subtle question arises: What exactly is c_A? We will come back to this later, but for now in the effective particle model of (2.4) it will be the concentration of free A of the equilibrium loading mixture prior to sedimentation, $c_{A,0}$, as may be calculated by mass action law *via* Eq. (1.32). This is the correct and only choice in this picture due to the absence of radial dilution and the neglect of diffusion. This is different from Eq. (1.34), where χ_A stands for the momentary local concentration $\chi_A(r,t)$.

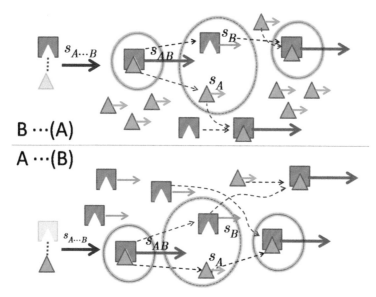

Figure 2.8 Cartoon of the effective particle A···B formed by the individual species A, B, and AB that are in rapid chemical equilibrium. A dynamic representation in the form of an animated cartoon can be found online [105]. Two phases are shown, B···(A) with A being in excess (top), A···(B) with B in excess (bottom). For each, the three circles depict the two states of the effective particle, the assembled state (left and right circle) with fast sedimentation in the complex (indicated by long arrows), and the dissociated state (center circle) where sedimentation is slow like the individual free species. Since nothing physically links the molecules in the dissociated state, other molecules of the same kind that happen to be localized nearby can exchange with the free constituent species. However, the relative number and relative velocities in the effective particle is such that it can form a self-sufficient network of particles co-sedimenting with the same time-average s-value. For example, for free A in B···(A) (top, middle circle) which sediments slower than B and therefore will fall behind in the dissociated state, another free B sedimenting behind its original partner will be available to recombine, and by doing so speed up both molecules to migrate ahead. The key is that the time-average velocity of A and B matches. Since the free state of A always lasts a shorter time than the free state of B, there are always fewer copies of A in the sedimenting system than B, leading to a stoichiometry of the virtual particle of total A to total B of less than unity. Therefore, the exact composition and the velocity of the effective particle depend on the sedimentation rates of all species as well as the total concentrations of both A and B.

of A could be dragged along. Generally, as much A can be taken up by the reaction boundary as can be made migrating with the time-average velocity match $s_{B···(A)}$; any higher number would not get boosts at a high enough rate to keep up. This will fix the co-sedimenting fraction of free A, and thus the amplitude of the undisturbed boundary.

Due to the high reaction rate constant (the effective particle model describes the limiting case of instantaneous equilibria), these exchanges of states happen very rapidly and over small spatial scales. From a macroscopic perspective it is not possible or meaningful to attribute any differences to distinguish copies of A since they are all equivalent, but by the same argument we have a choice of how we would count them on a microscopic scale. In particular, we can imagine them sorted into a subset of A that gets recycled for complexation with B and stays with the reaction boundary, having the same time-average velocity and taking part in a concerted migration pattern of binding and unbinding. This is illustrated in Figs. 2.8 and 2.9, and in animated cartoons [105].

The case of A⋯(B) is very similar (lower plot of Fig. 2.8). It occurs when a sufficiently large excess of B is present such that not only all of A can co-migrate in the reaction boundary, but that the over-abundance of B would leave many of them unoccupied and cause their time-average sedimentation to be less than that of A. In this case, the boundary pattern flips and the roles of dominant and secondary boundary are reversed, with some of B being left behind representing the undisturbed boundary (now sedimenting at s_B). Now the time-average state of A determines the velocity of A⋯(B) [41]:

$$s_{A\cdots(B)} = \frac{s_A + s_{AB}Kc_B}{1 + Kc_B} \tag{2.5}$$

Finally, it is possible that the species concentrations and species velocities are just right so that the time-average of A precisely matches that of B, i.e., $s_{A\cdots(B)} = s_{B\cdots(A)}$, in which there would be only a single boundary undergoing the concerted migration depicted in 2.9. In this case, there is only a single boundary, representing the 'effective particle.'

2.4.1.2 Component Populations in the Reaction and Undisturbed Boundary

As we have seen, the secondary component splits up into two fractions — the co-sedimenting fraction in the reaction boundary and the undisturbed fraction. The requirement that the co-sedimenting fraction must exhibit the same time-average sedimentation velocity as the dominant component allows us to predict quantitatively the population of both fractions. From this unfold all major characteristics of the boundaries. Specifically, the partitioning of the secondary component determines the population of the undisturbed boundary as well as the composition of the reaction boundary.

For the case B⋯(A), if we designate the concentration of the co-sedimenting A as $\widetilde{c}_A$, this requires

$$\frac{\widetilde{c}_A s_A + c_{AB}s_{AB}}{\widetilde{c}_A + c_{AB}} = \frac{c_B s_B + c_{AB}s_{AB}}{c_B + c_{AB}}, \tag{2.6}$$

leading to [41]

$$\widetilde{c}_A = \frac{Kc_A c_B (s_{AB} - s_B)}{(s_B - s_A) + Kc_A (s_{AB} - s_A)}. \tag{2.7}$$

As predicted above, this is always smaller than c_B. The total stoichiometry A:B of the effective particle B⋯(A) then follows as

$$S_{B\cdots(A)} = \frac{\widetilde{c}_A + Kc_A c_B}{c_B + Kc_A c_B} = 1 + \frac{1}{1 + Kc_A \dfrac{(s_{AB} - s_A)}{(s_B - s_A)}}. \tag{2.8}$$

For given binding constant and s-values, it depends only on the concentration of free A, and for very large c_A it approaches unity, i.e., for complete saturation of B the stoichiometry of the virtual particle is that of the true complex AB. The concentration of the remaining A neither in complex nor co-sedimenting in the effective particle as free A is

$$c_{A,u} = c_{A,\text{tot}} - \widetilde{c}_A - K c_A c_B \,, \tag{2.9}$$

and this is left to form the slow undisturbed boundary [41].

When B is in sufficient excess to form the phase $A\cdots(B)$, analogous considerations lead to the concentration of co-sedimenting B

$$\widetilde{c}_B = \frac{K c_A c_B \left(s_{AB} - s_A \right)}{K c_B \left(s_{AB} - s_B \right) - \left(s_B - s_A \right)} \,, \tag{2.10}$$

leaving

$$c_{B,u} = c_{B,\text{tot}} - \widetilde{c}_B - K c_A c_B \tag{2.11}$$

to sediment in the undisturbed boundary, and causing the effective particle to have the stoichiometry [41]

$$S_{A\cdots(B)} = 1 - \frac{\left(s_B - s_A \right)}{1 + K c_B \left(s_{AB} - s_B \right)} \,. \tag{2.12}$$

It should be noted that $A\cdots(B)$ still contains more B than A, just like the phase $B\cdots(A)$.[11]

2.4.1.3 The Phase Transition

Finally, we ask if we can arrive at a uniform picture of $A\cdots(B)$ and $B\cdots(A)$ and where the transition between the phases occurs in concentration space. Let us consider a point in the phase $B\cdots(A)$, where A is in large excess, and from there we draw a line along a trajectory of increasing $c_{B,\text{tot}}$ at constant $c_{A,\text{tot}}$ ending in the phase $A\cdots(B)$ under conditions where B is in large excess. The transition between phases will occur where the molar ratio of loading concentrations equals the stoichiometry of the effective particle. This transition point describes the condition where all of the free species are co-sedimenting within the effective particle, and so the undisturbed boundary vanishes.

For $B\cdots(A)$, the requirement $\widetilde{c}_A = c_A$ leads to an expression for the free concentration of B at the transition point

$$\dddot{c}_B(c_A) = \frac{K c_A \left(s_{AB} - s_A \right) + \left(s_B - s_A \right)}{K \left(s_{AB} - s_B \right)} \,, \tag{2.13}$$

---

[11] Although the stoichiometry for two cases can be written in a symmetric way, this was not done here in order to maintain positive differences of s-values (i.e., $s_B - s_A$ rather than $s_A - s_B$), highlighting the fact that we applied a designation of A and B that ensures $s_A < s_B$.

or, in terms of total loading concentrations [41],

$$\dddot{c}_{\mathrm{B,tot}}(c_{\mathrm{A,tot}}) = c_{\mathrm{A,tot}} + \frac{(s_{\mathrm{B}} - s_{\mathrm{A}})}{2K\,(s_{\mathrm{AB}} - s_{\mathrm{B}})}\left(1 + \sqrt{1 + \frac{4c_{\mathrm{A,tot}}K\,(s_{\mathrm{AB}} - s_{\mathrm{B}})}{(s_{\mathrm{AB}} - s_{\mathrm{A}})}}\right). \quad (2.14)$$

Thus, for any given loading concentration of **A** the phase **B**⋯(**A**) is supported as long as the loading concentration of **B** does not exceed $\dddot{c}_{\mathrm{B,tot}}$.

We may approach this question from the other side: For the transition point of the effective particle in **A**⋯(**B**) we expect $\widetilde{c}_{\mathrm{B}} = c_{\mathrm{B}}$ to be fulfilled, which leads to the expression for the free concentration of **A** at the transition point

$$\dddot{c}_{\mathrm{A}}(c_{\mathrm{B}}) = \frac{Kc_{\mathrm{B}}\,(s_{\mathrm{AB}} - s_{\mathrm{B}}) - (s_{\mathrm{B}} - s_{\mathrm{A}})}{K\,(s_{\mathrm{AB}} - s_{\mathrm{A}})}, \quad (2.15)$$

or

$$\dddot{c}_{\mathrm{A,tot}}(c_{\mathrm{B,tot}}) = c_{\mathrm{B,tot}} - \frac{(s_{\mathrm{B}} - s_{\mathrm{A}})}{2K\,(s_{\mathrm{AB}} - s_{\mathrm{A}})}\left(1 + \sqrt{1 + \frac{4c_{\mathrm{B,tot}}K\,(s_{\mathrm{AB}} - s_{\mathrm{A}})}{(s_{\mathrm{AB}} - s_{\mathrm{B}})}}\right). \quad (2.16)$$

Here, given any loading concentration of **B** we can support effective particles **A**⋯(**B**) where **B** is in excess as long as the total loading concentration of **A** does not exceed $\dddot{c}_{\mathrm{A,tot}}$.

Interestingly, since there are no negative concentrations, we can deduce from Eq. (2.15) that there can be no transition point, i.e., there is no **A**⋯(**B**) effective particle possible, for $c_{\mathrm{B}} < (s_{\mathrm{B}} - s_{\mathrm{A}})/\bigl(K\,(s_{\mathrm{AB}} - s_{\mathrm{B}})\bigr)$. This ensures that the reaction boundary of **A**⋯(**B**) is sedimenting faster than free **B**, which was a requirement that appeared above to make possible the physical picture depicted in the lower plot of Fig. 2.8. This asymmetry is an important consequence of the difference in the s-values of **A** and **B**, and we will come back to this point later.

So far, we have shown the concentration limits for the effective particles **B**⋯(**A**) to be sustainable. What happens beyond these concentration limits? Does the type of effective particle switch? The answer is yes, since we can show that the transition point predicted from **B**⋯(**A**) and that predicted from **A**⋯(**B**) is the same, i.e., where one type is impossible, the other type can exist. The identity of the concentration limits can be demonstrated most easily by showing that [41]

$$\dddot{c}_{\mathrm{A}}\bigl(\dddot{c}_{\mathrm{B}}(c_{\mathrm{A}})\bigr) = c_{\mathrm{A}} \quad (2.17)$$

is indeed fulfilled, based on Eqs. (2.13) and (2.15). Importantly, the s-value of the reaction boundary at the transition point is continuous across the branches, as follows from the identity [41]

$$s_{\mathrm{B}\cdots(\mathrm{A})}\bigl(\dddot{c}_{\mathrm{A}}(c_{\mathrm{B}})\bigr) = s_{\mathrm{A}\cdots(\mathrm{B})}(c_{\mathrm{B}}) \quad (2.18)$$

based on Eqs. (2.4) and (2.5), and, likewise, based on Eqs. (2.8) and (2.12) the

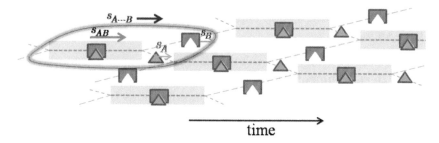

Figure 2.9 Cartoon of the effective particle A⋯B at the phase transition point, in the absence of excess A or B relative to the reaction boundary composition. Depicted is the fractional time spent by A and B in the different states, either in complex (grayed time interval) or free (white background). The fractional time spent by A in the free state is less than the fractional time spent by B in the free state. For this reason, the overall stoichiometry in the reaction boundary of A:B is less than unity. The difference in the fractional time of being free allows for the time-average sedimentation velocity of A and B to be equal. The system then sediments from a macroscopic point of view like an effective particle encircled in red, with the velocity $s_{A⋯B}$.

stoichiometry of the effective particle is continuous at the transition point between the two branches [41]

$$S_{A⋯(B)}\left(\overset{...}{c}_B(c_A)\right) = S_{B⋯(A)}(c_A). \tag{2.19}$$

The unambiguous and continuous transition between the effective particles A⋯(B) and B⋯(A) allows us now to restore the two into a single consistent picture A⋯B. It is apparent now that what initially seemed to be contradictory views are only two different branches that have a clearly delineated transition point, much like a first-order phase transition (with discontinuous derivatives). Exactly at the transition point, there will be no undisturbed boundary at all, and a depiction of the coupled sedimentation process is in Fig. 2.9. It is important to reiterate that for dissimilar-sized molecules this transition point is not at equimolar concentration. Rather, a lower concentration of the slower sedimenting component A co-exists with a higher concentration of the faster-sedimenting component B. This is a consequence of the requirement for a match of the time-average velocities, which permits only shorter time intervals in which A can stay unliganded as compared to B.

This concerted motion is best illustrated in an animation. The SEDPHAT effective particle explorer tool can be used to produce a movie, after the desired heterogeneous interaction model with equilibrium constants and species velocities has been set. Invoking the menu function Options ▷Interaction Calculator ▷Effective Particle Explorer produces a color temperature plot of various quantities in concentration space, i.e., as a function of total concentration A and B, respectively. After right-clicking at a particular point in the plot, an animated cartoon of the reaction boundary migration can be played for the particular conditions. See also Fig. 6.2 on p. 192.

2.4.2 Overview of the Sedimentation Patterns in Concentration Space for a System A + B ↔ AB

In summary, the essence of the reaction boundary can be explained by the need for both components to co-sediment. Due to the difference in sedimentation velocity of the free species, this must result in different concentrations of these species: Less of the slower-sedimenting free species than of the faster-sedimenting free species is necessary to achieve a time-average sedimentation velocity $s_{A\cdots B}$, which must lie between s_B and s_{AB}. The behavior of the effective particle A$\cdots$B and its two phases in the parameter space is illustrated in Figs. 2.10 and 2.11 for a system A + B ↔ AB. Some previously counter-intuitive properties of the reaction boundary now unfold naturally from the effective particle picture.

For example, decreasing s-values of reaction boundaries in titration series can now be easily explained. If we inspect $s_{A\cdots B}$ along trajectories of constant concentration of the secondary component (e.g., along horizontal lines in the upper left quadrant of *Panel B* of Fig. 2.10 showing the A$\cdots$(B) phase, or vertical lines in the lower region of the B$\cdots$(A) phase), we find a decrease of $s_{A\cdots B}$ prior to the phase transition line, followed by a steep increase after the phase transition. This is naturally explained by the time-average occupation state of the dominant component determining the sedimentation velocity of the effective particle: At concentrations below the phase transition, when increasing the concentration of the dominant component, more total complex is formed but the average time dominant molecules spend in the complex state decreases, since the additional complex population grows more slowly than the total of the dominant component. As a consequence, the sedimentation velocity of the reaction boundary decreases.[12] This happens despite the overall increase of s_w, which is driven by an increasing amplitude of the reaction boundary. In addition, increasing the concentration of the dominant component requires a larger fraction of the secondary component to co-sediment in the reaction boundary, until, at the transition point, there is no excess in the undisturbed boundary left and the secondary component will become the dominant.

If the concentration ratio of the loading mixture is equal to the stoichiometry of the effective particle, the singularity occurs that there is only the boundary from the effective particle and no undisturbed boundary. At this point there is a distinct phase transition. If there is more A in the loading mixture, the case B$\cdots$(A) will develop, and *vice versa*, the case A$\cdots$(B) will develop for excess B.

---

[12]This may be easily verified through inspection of Eq. (2.3), since an increase in $c_{B,tot}$ will lead to a greater saturation of A, hence lower concentration of its free form. Smaller c_A will increase the denominator and reduce τ_{AB}.

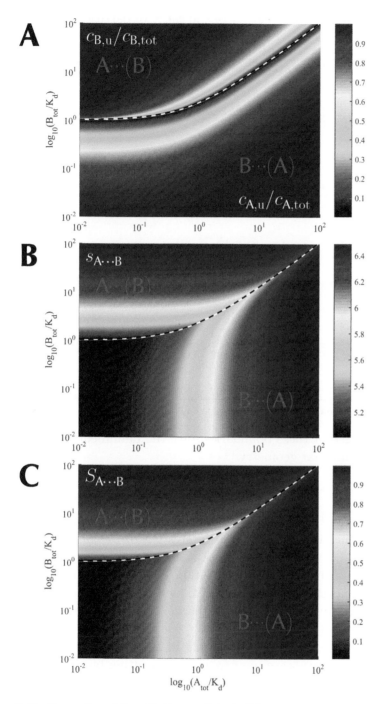

Figure 2.10 Properties of the effective particle A⋯B in concentration space, plotted as a function of $log_{10}(c_{A,tot}/K_d)$ and $log_{10}(c_{B,tot}/K_d)$ across the range from $1/100$ K_d to 100-fold K_d for the system $A + B \leftrightarrow AB$ with values $s_A = 3.5$ S, $s_B = 5.0$ S, and $s_{AB} = 6.5$ S. The phase transition line according to Eq. (2.16) is shown as dashed line. *Panel A*: The fraction of $c_{A,u}/c_{A,tot}$ or $c_{B,u}/c_{B,tot}$, respectively; *Panel B*: Sedimentation coefficient $s_{A\cdots B}$ following Eq. (2.20); *Panel C*: Stoichiometry $S_{A\cdots B}$ following Eq. (2.21). In all plots, the phase A⋯(B) is in the upper left quadrant, and the phase B⋯(A) everywhere else.

We can summarize the sedimentation velocity of the effective particle as

$$
s_{A\cdots B} = \begin{cases} \dfrac{(c_A s_A + c_A c_B K s_{AB})}{(c_A + c_A c_B K)} & \text{for } c_B > \ddot{c}_B(c_A) \\[2em] \dfrac{(c_B s_B + c_A c_B K s_{AB})}{(c_B + c_A c_B K)} & \text{else} \end{cases} , \qquad (2.20)
$$

and the stoichiometry as

$$
S_{A\cdots B} = \begin{cases} 1 - \dfrac{(s_B - s_A)}{K c_B (s_{AB} - s_B)} & \text{for } c_B > \ddot{c}_B(c_A) \\[2em] 1 - \left(1 + K c_A \dfrac{(s_{AB} - s_A)}{(s_B - c_A)}\right)^{-1} & \text{else} \end{cases} , \qquad (2.21)
$$

with $c_B > \ddot{c}_B(c_A)$ following Eq. (2.13). The amplitude of the undisturbed boundary is

$$
c_u = \begin{cases} c_{B,\text{tot}} - K c_A c_B - \dfrac{K c_A c_B (s_{AB} - s_A)}{K c_B (s_{AB} - s_A) - (s_B - s_A)} & \text{for } c_B > \ddot{c}_B(c_A) \\[2em] c_{A,\text{tot}} - K c_A c_B - \dfrac{K c_A c_B (s_{AB} - s_B)}{(s_B - s_A) + K c_A (s_{AB} - s_A)} & \text{else} \end{cases} . \qquad (2.22)
$$

We will show in Section 3.2 that these boundaries solve the Lamm equation in the absence of diffusion in rectangular geometry.[13]

2.4.3 Limiting Cases

The most intriguing feature of the effective particle is the phase transition. It is indicated by the dashed lines in Fig. 2.10. It is interesting to imagine a series of experiments at different $c_{A,\text{tot}}$ where we adjust $c_{B,\text{tot}}$ to $\ddot{c}_{B,\text{tot}}(c_{A,\text{tot}})$, i.e., explore the parameter space right along the phase transition line. The isotherms of the sedimentation coefficient and the stoichiometry of the effective particle are shown in Fig. 2.11. It is significantly broader than isotherms that would be measured in a titration along the line of constant $c_{B,\text{tot}}$, let's say in the phase B$\cdots$(A) with $c_{B,\text{tot}} < K_d$. The stretched shape of the isotherms in the vicinity of the phase transition arises from the relative scarcity of free co-sedimenting binding partner, allowing each molecule of the dominant component to be occupied only for shorter times.

Fig. 2.12 illustrates the dependence of the phase transition line on the relative size of the interacting molecules. An observation that naturally follows from

---

[13]This allows us to identify the local free concentrations c_A and c_B as those characterizing the initial equilibrium at the start of the experiment, $c_{0,A}$ and $c_{0,B}$.

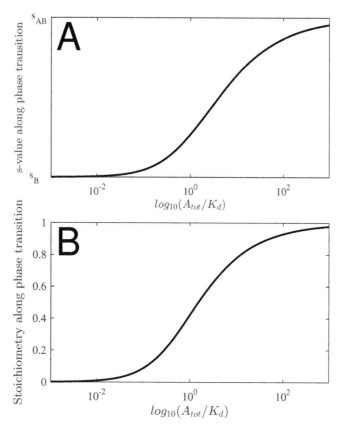

Figure 2.11 $s_{A\cdots B}$ (*Panel A*) and stoichiometry $S_{A\cdots B}$ of the effective particle along the trajectory of the phase transition line $\ddot{c}_{B,tot}(c_{A,tot})$ for the system $A + B \leftrightarrow AB$ with values $s_A = 3.5$ S, $s_B = 5.0$ S, and $s_{AB} = 6.5$ S as shown in Fig. 2.10.

the effective particle picture is that for **A** to be entirely in the reaction boundary, its fractional occupation must lead to an average sedimentation velocity at least matching that of free **B**. This leads to the perhaps unexpected result that for low concentrations of **A** and **B**, in principle even a very large excess in the molar ratio of **B** over **A** may not be able to cause **A** to be the dominant component. This ultimately gives rise to the asymmetric shape of the phase transition line in parameter space.

Indeed, in the lower limit of concentrations of **A**, the transition line Eq. (2.14) approaches a constant value of **B** given by [41]

$$\ddot{c}_{B,tot,min} = \frac{(s_B - s_A)}{K(s_{AB} - s_B)}. \tag{2.23}$$

At concentrations of $c_{B,tot}$ below $\ddot{c}_{B,tot,min}$, **B** cannot generate the undisturbed boundary, irrespective of even very large molar excess (compare Fig. 2.12). The dependence of the limiting concentration Eq. (2.23) on the s-values of the different species is noteworthy. From Eq. (2.23) we can also recognize that for small ligands where $s_B - s_A \gg s_{AB} - s_B$, the asymptotic transition point will be at $\ddot{c}_{B,tot,min} \gg K_D$

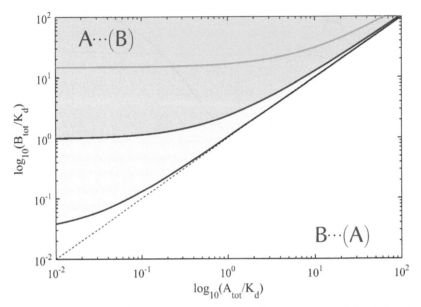

Figure 2.12 Phase transition in the parameter space according to Eq. (2.16) for different size ratios of A and B. Shown is the same effective particle A⋯B for the system A + B ↔ AB as in Fig. 2.10 with values $s_A = 3.5$ S, $s_B = 5.0$ S, and $s_{AB} = 6.5$ S (red line), for more similar-sized molecules with values $s_A = 4.9$ S, $s_B = 5.0$ S, and $s_{AB} = 8.5$ S (blue line), and for the case of a small ligand with $s_A = 0.5$ S, $s_B = 5.0$ S, and $s_{AB} = 5.3$ S (green line). Equimolar conditions are shown as dashed line. The colored patches indicate the region in the parameter space where A⋯(B) exists, all regions below the respective phase transition lines are B⋯(A).

(e.g., green line in Fig. 2.13), whereas for similar-sized molecules it will be at $\dddot{c}_{B,tot,min} \ll K_D$ (e.g.., blue line in Fig. 2.13). In fact in the limit that $s_A = s_B$ the phase transition line will not have an asymptotically limiting concentration of $\dddot{c}_{B,tot}$, and instead divide the parameter space along the diagonal line of the reaction stoichiometry. These observations can be helpful considerations for the experimental design (Section 6.1.4.1).

2.5 POLYDISPERSITY OF SEDIMENTATION COEFFICIENTS IN THE REACTION BOUNDARY

The effective particle model is appropriate to describe the overall principles of sedimentation of rapidly interacting systems, and to understand the resulting structure of the sedimentation boundaries. While the lack of detail about the precise boundary shapes is an attractive feature, allowing robust data analysis [41, 106], it is not entirely satisfactory in arriving at a complete physical picture of sedimentation.

A reaction boundary as we have considered it so far, with the effective particle exhibiting a concerted sedimentation uniformly propagating with $s_{A\cdots B}$, would only be meta-stable. Any disturbance that creates the slightest concentration gradients, such as invariably arising from diffusion or convection, will lead to increasing departure from ideal step-functions: Since the concentrations in the leading edge of the reaction boundary are higher in comparison with the concentrations in the trailing edge of the reaction boundary, the populations of the reacting species will be shifted more to the assembled state in the front and to the disassembled state in the back, thereby causing somewhat faster sedimentation in the leading edge and somewhat slower sedimentation in the trailing edge of the reaction boundary. This creates a dispersion of sedimentation velocities that will reinforce the concentration gradient.

Thus, we can refine our view of the reaction boundary by moving from an integral picture based on the time-average behavior of the whole reaction boundary to a more detailed differential picture of individual boundary slices caused by concentration gradients in the reaction boundary: This idea becomes clearer in a simple Gedankenexperiment where we imagine we could keep track of the sedimentation of complexes and free molecules, and assume them to persist through finite time intervals. The sketch in Fig. 2.13 depicts different stages of the sedimentation process:

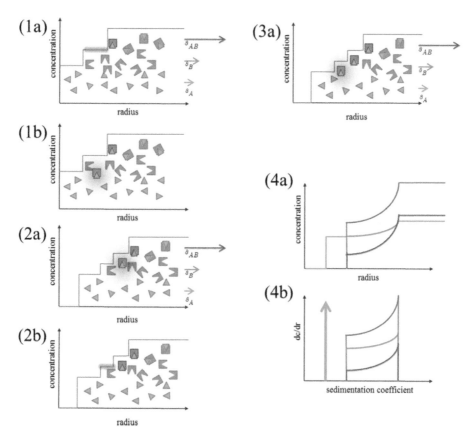

Figure 2.13 Illustration of a Gedankenexperiment considering the polydispersity in the reaction boundary. We imagine infinitesimal steps during which progressively more of the complex peels off the boundary to sediment ahead, leaving sample with lower concentration and correspondingly smaller velocity in the trailing part of the reaction boundary. For a detailed description see text.

(*Panel 1a*) After a small time-increment, one would initially expect a radial region to be depleted of complex.[14] (*Panel 1b*) However, since there is co-existing free A and free B, although at lower concentration than the initial loading concentration, they will reform complexes AB. (*Panel 2a*) After a further infinitesimal time-increment, these newly formed complexes should have sedimented, too. Importantly, we note that since the total concentration of A and B in this region is lower than the loading concentration in the first step, the *s*-values of the corresponding effective particles in this spatial region will be lower than those sedimenting ahead initially. (*Panel 2b*) The sedimentation of this second (infinitesimal) batch of complex has depleted a new spatial region of AB, and (*Panel 3a*) new complexes will form, sedimenting at a still lower rate, etc., continuing until there is no free B left. We notice that the last remaining material is the free A of the undisturbed boundary. *Panel 4a* shows the overall result — it appears that there is a continuous concentration gradient of A, B, and AB present in the reaction boundary, corresponding to a different composition of the effective particles sedimenting at the different rates even within the reaction boundary. As depicted in *Panel 4b*, this concentration gradient can be transformed into a differential sedimentation coefficient distribution (with the vertical arrow indicating a δ-function arising from the single velocity of the undisturbed boundary).

A quantitative description of this idea is achieved in the theory by Gilbert and Jenkins in the familiar case of non-diffusing particles migrating under constant force in rectangular geometry [43, 107, 108] (outlined in Section 3.3). From a phenomenological point of view, an important result is that we should expect the reaction boundary to exhibit a distribution of sedimentation coefficients, with the average described by the effective particle model.

As we will see in later examples of the computed sedimentation coefficient distributions, reaction boundaries can be quite narrow and essentially mono-disperse, or they can be broad and appear even multi-modal. This depends on the loading conditions and the phase transition point. The closer to the phase transition point, the higher the dispersion exhibited by the boundaries. *Vice versa*, the farther away from the transition point the more mono-disperse the asymptotic boundaries will become and merge with the predictions of the average properties from the effective particle model. This can be understood on the basis of the range of saturation of the dominant component in the broadened reaction boundary, which is determined by the amplitude of the secondary component in the undisturbed boundary.

Since the Gilbert–Jenkins theory is still based on the approximation where diffusion effects are neglected, the boundaries in real experiments are even broader than predicted by the velocity dispersion. The sedimentation coefficient distributions of the reaction boundaries are historically termed the asymptotic boundaries,

---

[14]In this simplified view we will not achieve a completely consistent picture, whether what is initially sedimenting is an effective particle, just the complexes having finite lifetimes (departing from the idea of an instantaneous reaction), or there is a pre-existing small concentration gradient causing the initial dispersion in velocity.

based on the equivalence of diffusion-free sedimentation and the extrapolation of diffusion-free boundaries to infinite time; the same limit can be approximated by deconvolution of diffusion in $c(s)$ [109] (Chapter 4).

2.6 INTERMEDIATE KINETICS

For slowly reacting sedimenting systems, we have observed concentration-independent sedimentation coefficients of boundary components that reflect the populations of the individual species, whereas rapidly reacting sedimenting systems exhibit only one undisturbed boundary reflecting species sedimentation and $N_k - 1$ reaction boundaries (if N_k is the number of components in the mixture), with the reaction boundaries showing concentration-dependent sedimentation coefficients. Although intermediate behavior occurs only in a narrow range of kinetic rate constants, what happens at the transition between these two kinetic regimes is important for understanding the complete picture of reacting systems in sedimentation.

Toward this end, in Fig. 2.14 we have simulated sedimentation data for an interacting hetero-association system $A + B \leftrightarrow AB$ at a range of equimolar concentrations, and assumed different kinetic rate constants under otherwise identical conditions [106]. The left column of plots shows the normalized $c(s)$ distributions, the middle column the weighted-average s-value of the reaction boundary from integration of the fast peaks, s_{fast}, and the right column shows the signal amplitudes from integration of $c(s)$ of the undisturbed and reaction boundary a_u and a_{fast}, or species boundaries a_A, a_B, and a_{AB}, respectively, dependent on interpretation of the kinetics. The lines are the best-fit isotherms, leading to K_D estimates indicated.

For this system, at a rate constant $k_{off} = 10^{-3}$/sec (second row) we still observe sedimentation patterns that are nearly identical to those of essentially infinitely fast reactions. However, in a more detailed inspection it seems that the bimodality of the reaction boundary that is usually observed only at high concentrations has become more pronounced as compared to the case of fast reactions ($k_{off} = 10^{-1}$/sec in the top row). Although the data are not quite identical to the limiting case of instantaneously equilibrating reactions, the isotherms of both the amplitude and the s-value of the reaction boundary can be modeled very well with the predictions for the effective particle model (Eqs. 2.20–2.22). As the rate constant becomes only little more than twofold slower, at $k_{off} = 4 \times 10^{-4}$/sec (middle row), we see clearly multi-modal reaction boundaries, which could already be interpreted as reflecting the populations of the individual species. Since both the slow and fast designation could be applied, the boundary amplitudes are shown for both integrations, across the range $9 - 14$ S in the fast reaction model, as well as across individual partial peaks in the slow reaction model, as shown in the split middle panel of the right column. Remarkably, both models fit the data reasonably well (blue and green lines, respectively). As the reaction becomes only slightly slower still, at $k_{off} = 10^{-4}$/sec (second row from bottom), we see clearly separated species peaks. Although they exhibit a slight concentration-dependence from the only slight influence of the finite

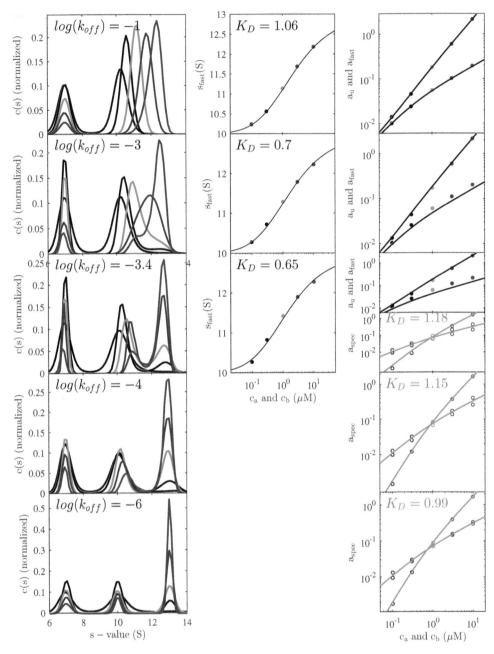

Figure 2.14 Simulation of a dilution series and the effect of intermediate kinetics on the boundary structure and its analysis with the limiting models of effective particles or mass action law. *Left Panels*: $c(s)$ traces (normalized by the integrated area) derived from simulated SV data based on numerical Lamm equation solutions for a 1:1 interaction (simulated for a 7 S, 100 kDa-protein reacting with a 10 S, 200 kDa-protein to form a 13 S complex), using equimolar loading concentration of $0.1 \times K_D$ (black), $0.3 \times K_D$ (blue), $1.0 \times K_D$ (green), $3 \times K_D$ (magenta), and $10 \times K_D$ (red). The different rows show the data from simulations with different kinetic rate constants, with $log_{10}(k_{off}) = -1, -3, -3.4, -4$, and -6 (*Top to Bottom*). *Middle Panels*: s_{fast} (solid circles). *Right Panels*: Signal amplitudes of the undisturbed and the reaction boundaries (top three panels, solid circles) or the species signal amplitudes (lower three panels, open circles), respectively, and global fit with effective particle model (blue lines) or mass action law model (green lines), respectively, resulting in the K_d-value indicated. Reproduced from [106].

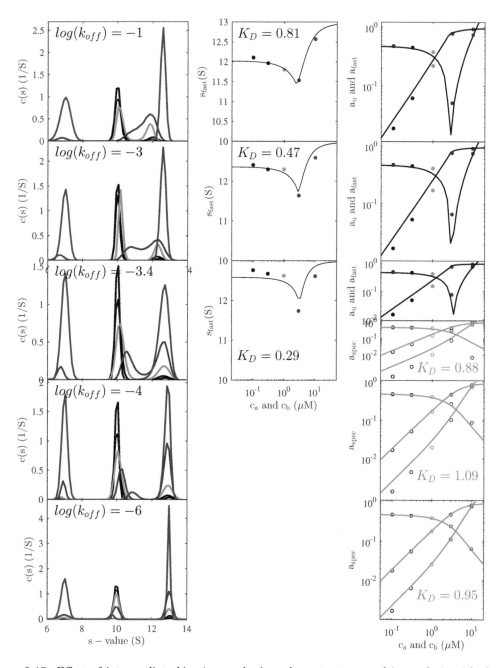

Figure 2.15 Effect of intermediate kinetics on the boundary structure and its analysis with the limiting models of effective particles or mass action law. Simulations and analyses are shown for the same system and in analogous presentation as in Fig. 2.14, but for a titration series of constant concentration of the larger molecule at 4-fold K_d with increasing concentration of the smaller concentration at $0.1 \times K_d$ (black), $0.3 \times K_d$ (blue), $1.0 \times K_d$ (green), $3 \times K_d$ (magenta), and $10 \times K_d$ (red).

reaction kinetics, they are already very similar to those at virtually infinitely slow reactions (bottom row). In either case, the quantitative analysis of the boundary amplitudes follows very well the mass action law, which we would expect to be exact in the limit of infinitely slow reactions relative to sedimentation. For reactions with $k_{off} = 10^{-6}$/sec, we observe the individual species' peaks fully developed.

It is of interest to examine the analogous simulation for the case of a titration of a constant concentration of the faster sedimenting species with increasing concentrations of the slower. This is shown in Fig. 2.15 with otherwise identical conditions and plots. This titration exhibits a phase transition, and for fast systems a decrease of the reaction boundary s-value prior to the phase transition. Again, it can be discerned that the $c(s)$ peaks that were bimodal for fast reactions will increasingly split up with decreasing reaction rate constants, and represent species populations for slow reactions. Further, in the transition region analyses with either models for slow or fast behavior lead to moderate errors in K_D.

The example of Figs. 2.14 and 2.15 have focused on the transition region of the kinetics, which is far from the assumptions of infinitely fast or infinitely slow reaction kinetics made in the effective particle and species analysis models, respectively. Under these conditions, the quantitative results for the estimated equilibrium binding constant are not exact and not quite satisfactory for the effective particle model, especially in the titration configuration with the phase transition. This can be understood considering the increased stability of the complex leading to higher s-values of the reaction boundary and to underestimates of the undisturbed boundary signal amplitude. This represents the limit of applicability of this model. Nevertheless, the qualitative results are striking and the quantitative results are sufficient to initialize parameter values of a more detailed full Lamm equation model.

In summary, importantly, the intermediate case does not exhibit entirely new behavior. We observe that not only is the transition region confined to a very small range of kinetic rate constants, but also the quantitative analysis models exhibit properties that morph from one to the other. We see the boundary shapes merge and reflect partial features of both limiting cases, with the quantitative models for the boundary positions and amplitudes being remarkably robust, in a sense that they show only modest deviations when the kinetic assumptions are not fully met.

2.7 THE ROLE OF DIFFUSION

We have introduced above the concept of an effective particle to describe the sedimentation properties of the rapidly interacting system in a diffusion-free scenario. We now ask how the system would diffuse [42]. In the simplest picture neglecting dispersion of the velocities, we can imagine a population of effective particles sedimenting at the velocity $s_{A\cdots B}$ in a bath of free species from the undisturbed boundary of the secondary component. Since the dominant component is sedimenting in its entirety in the reaction boundary, we may approximate the effective time-average diffusion coefficient of the effective particles via the constituent diffusion coefficient of the dominant component. This is indeed a quite reasonable approximation, as will be shown in more detail in Section 3.4.

The most important result from a phenomenological point of view is that the reaction boundary, in a first-order approximation, exhibits diffusional properties that are very similar to those of non-interacting species, with the ordinary $\sqrt{t}$-dependent boundary spreading, but with a composition-dependent diffusion coefficient $D_{A\cdots B}$ [42]. This is illustrated in Fig. 2.16, which shows the shape of the reaction boundary in excellent agreement with a single species model — though not exhibiting an easily interpreted M^*. This demonstrates that the boundary broadening in this particular case arises mainly from diffusion, with only a small influence from the polydispersity of the reaction boundary, and that the diffusion proceeds approximately normally.

An important consequence is that, since the diffusion contributions proceed at least approximately with the ordinary $\sqrt{t}$-dependence, it can be deconvoluted with the $c(s)$ distribution [28, 42]. This will be more closely examined in Chapter 4, but

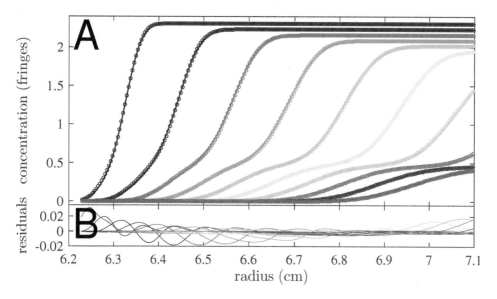

Figure 2.16 Demonstration of $\sqrt{t}$-dependent diffusion in the reaction boundary. *Panel A*: Total concentration profiles for a simulated sedimentation of the system A + B ↔ AB in the limit of instantaneous reaction at equimolar concentrations at $K_D = 10\ \mu M$, calculated with the full Lamm equation solutions for a component A with 100 kDa and 7 S, a component B with 200 kDa and 10 S, both forming a complex with 13 S, assuming molar signal increments of 100,000 for both components (circles). The profiles are fitted to a model with two non-interacting species, one accounting for the undisturbed boundary and one for the reaction boundary. The best-fit to this data (lines) was found with parameters $M^* = 88$ kDa, $s = 7.02$ S for the slow boundary, and $M^* = 191$ kDa, and $s = 11.1$ S for the fast boundary, respectively, with an rmsd of 0.0039 fringes. *Panel B*: Overlay of the residuals. Relative to the loading signal of 2.4 fringes, the residuals are small, and would not be recognizable at common experimental signal/noise levels. Similar data are shown in [28], and the same case was made by Urbanke, Witte, and Curth showing two-species fits to experimental data from interacting EcoSSB and χ protein [66].

it is already illustrated here in Fig. 2.17 for a system of interacting small molecules, where the non-deconvoluted boundary shapes (represented by their ls-$g^*(s)$ distributions) only show broad features for both fast and slow kinetics. Once diffusion is deconvoluted in $c(s)$, given sufficient signal/noise ratio, the underlying sedimentation coefficient distributions show the familiar concentration-dependent positions for fast reactions, and concentration-independent peaks for slow reactions. More quantitatively, the $c(s)$ distributions unravel boundary patterns with properties consistent with the predictions of the effective particle model (Fig. 2.17C). Furthermore, although the SV data usually do not have sufficient information content to display the fine structure of the sedimentation coefficients distributions of reaction boundaries well, a Bayesian analysis shows very good correspondence between $c(s)$ results and the asymptotic boundaries predicted from the Gilbert–Jenkins theory [109] (Fig. 4.5).

Generally, even though diffusion creates additional concentration gradients of free A and B, which complicate the analysis computationally — in a second-order approximation they would lead to the polydispersity of sedimentation velocities and diffusion coefficients — diffusion does not contribute significant new aspects. However, it can diminish the characteristic features we rely on to recognize the different regimes of reaction kinetics. This is true, in particular, for smaller particles, as illustrated in Fig. 2.18: In the comparison of boundary shapes for an A + B ↔ AB system composed of large *vs.* small particles, the stronger diffusion-broadened small particle boundaries in Fig. 2.18B show relatively shallow features at all rate constants. By contrast, the large particle boundaries in Fig. 2.18A exhibit multi-modal species boundaries characteristic of slow systems that may be readily discerned and distinguished from bimodal reaction boundaries of fast systems. Fortunately, even from strongly diffusion-broadened boundaries of small systems, the diffusion can be deconvoluted with the $c(s)$ distribution, as was illustrated in Fig. 2.17.

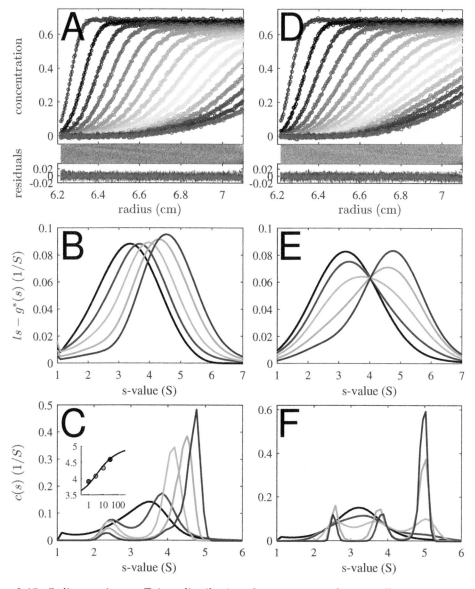

Figure 2.17 Sedimentation coefficient distributions from a system of two small proteins reversibly associating at different time-scales. Sedimentation was simulated for a protein of 25 kDa, 2.5 S binding to a 40 kDa, 3.5S species forming a 5 S complex with $K_d = 3$ μM for a fast reaction ($k_{off} = 10^{-2}$/sec; *Panels A, B, C*) and a slow reaction ($k_{off} = 3 \times 10^{-5}$/sec; *Panels D, E, F*). Data were simulated for an interference optical experiment at 50,000 rpm with typical signal-to-noise ratio, and typical molecular extinction coefficients, with 50 scans over a time interval of 250 min. *Panels A* and *D*: Example for simulated data (showing every 3rd data point of every 3rd scan) at equimolar concentrations of 3 μM with $c(s)$ fit. *Panels B* and *E*: Representation of the diffusion-broadened boundary shape in the best-fit ls-$g^*(s)$ sedimentation coefficient distribution for equimolar concentrations of 0.3 μM (black), 1.0 μM (red), 3.0 μM (blue), 10 μM (green), and 30 μM (magenta). The plot shows the distributions normalized to equal area. *Panels C* and *F*: Corresponding diffusion-deconvoluted sedimentation coefficient distribution $c(s)$, exhibiting the characteristic fast and slow patterns (where signal/noise permits resolution) using standard regularization. The inset in *Panel C* are signal-averaged s_{fast}-values where the peak structure is recognizable, and best-fit isotherm of $s_{A \cdots B}$ leading to K_d estimate of 2.7 μM. See also [28].

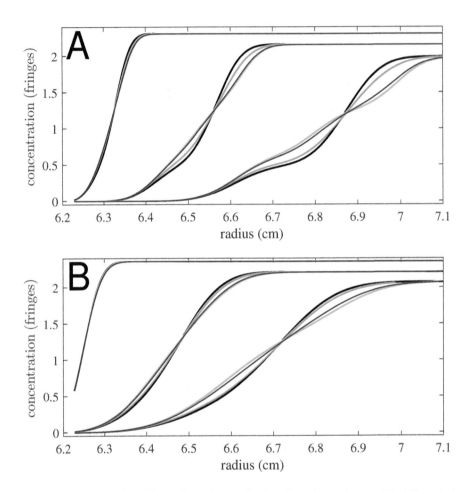

Figure 2.18 Diffusion broadened boundary shapes for A + B ↔ AB systems with different size at a range of kinetic rate constants. Shown are representative sedimentation boundaries from simulated Lamm equation solutions with $k_{\text{off}} = 10^{-2}$/sec (black), $k_{\text{off}} = 10^{-3}$/sec (blue), $k_{\text{off}} = 10^{-4}$/sec (magenta), and $k_{\text{off}} = 10^{-5}$/sec (green). *Panel A*: Representative behavior of large macromolecules with a 7 S, 100 kDa-protein binding a 10 S, 200 kDa-protein to form a 13 S complex; shown at times 850 sec, 2,050 sec, and 3,550 sec during sedimentation at 50,000 rpm at equimolar concentration at K_D. *Panel B*: Example for stronger diffusion in smaller systems, showing a 3 S, 30 kDa-protein binding a 4 S, 50 kDa-protein to form a 6 S complex; shown at times 850 sec, 3,550 sec, and 6,250 sec during sedimentation at 50,000 rpm at equimolar concentration at K_D.

CHAPTER 3

Approximate Solutions for the Coupled Lamm Equations of Reacting Systems

H AVING examined the phenomenology of interacting systems in SV, we are in a position to turn our attention back to the quantitative aspects, in order to complement the strict computational solutions of the Lamm equations for reacting systems with insightful approximate solutions. Our goal is to develop simpler approaches that are not as demanding in the application to experimental data with regard to the ideality of the experiment and purity of the samples. This can be achieved with models that are less detailed in the boundary modeling, yet still capture the salient features of SV and allow us to reliably quantitatively interpret the sedimentation boundaries of interacting systems.

As in the previous chapter, we proceed from general to more detailed aspects: The description of the weight-average s_w from the transport method as an overall measure of the sedimentation of the entire system is followed by analysis of the boundary structure in the effective particle model in the rectangular constant-force approximation. These quantities can be predicted for the entire concentration space at once, and along experimental trajectories therein. Gilbert and Gilbert–Jenkins theories provide, for specific given loading concentrations, a detailed prediction of the polydispersity of sedimentation coefficients within the boundaries. Finally, this is followed by predictions of the diffusive properties of the reaction boundary in the framework of the constant bath theory and the effective particle model.

3.1 THE TRANSPORT METHOD AND WEIGHTED-AVERAGE SEDIMENTATION COEFFICIENTS

The use of the weighted-average and constituent sedimentation coefficients has a long history in the analysis of SV of interacting systems [19, 20, 95, 110–112]. To begin, let us recapitulate the transport method for non-interacting mixtures as discussed in Part II, Section 2.3 [2]. This approach is concerned with the mass balance of material that is above[1] a certain reference radius, chosen arbitrarily in the plateau region of the concentration profile. Essentially, we count the number of particles that cross the reference radius per unit time, weighted with the signal that they generate in the given optical detection system. The weighted-average sedimentation coefficient was defined in Part II as the rate of change in the radial integral under the boundaries, taken to the limit to zero time [2]

$$
s_w =: \lim_{t \to 0} \left[\frac{1}{a(r_p, t)\omega^2 r_p^2} \left(-\frac{d}{dt} \int_m^{r_p} a(r, t) r \, dr \right) \right], \tag{3.1}
$$

where r_p is the reference radius in the plateau region at which point we measure the signal $a(r_p, t)$. The reason for choosing a reference radius in the plateau region was to ensure that diffusion fluxes will not contribute to the mass balance, since they are absent if there is no gradient. In this way, the change in the total mass above the reference radius is determined by the sedimentation fluxes only. The extrapolation to the start of sedimentation was necessary in order to obtain an unambiguous value characterizing a mixture of particles, free of the time-dependence introduced by their different rates of radial dilution (see Figure 2.15 of Part II [2]).

Further, we have shown in Section 3.4 of Part II that s_w is naturally consistent with sedimentation coefficient distributions based on direct boundary modeling, such as $c(s)$ [2]:

$$
s_w = \frac{\int c(s)s\,ds}{\int c(s)\,ds} \tag{3.2}
$$

From a practical perspective, one of the most important features of s_w is its complete independence of the boundary shape. Moreover, any model that is able to fit the experimental data faithfully with regard to the integral $\int_m^{r_p} a(r, t) r \, dr$ will be sufficient to determine s_w.

In fact, it is possible to test to what extent residuals of the fit $\delta a(r, t)$ from boundary modeling constitute unaccounted mass transport that leads to an error δs_w of the weighted-average sedimentation coefficient. Splitting the signal into the fit model and residuals, the additivity of the parenthesis of Eq. (3.1) leads to an

---

[1]'Above' means at a smaller radii in the centrifugal field, analogous to the height in the gravitational field.

expression for error propagation

$$\delta s_w =: \lim_{t \to 0} \left[\frac{1}{a(r_p, t)\omega^2 r_p^2} \left(-\frac{d}{dt} \int_m^{r_p} \delta a(r, t) r \, dr \right) \right]. \tag{3.3}$$

In practice, this δs_w can be evaluated very quickly, though only for a subset of the fitted data that exhibit a solution plateau. δs_w is not to be confused with the statistical error of s_w, but considered a measure to what extent a misfit of the model introduces bias in s_w.

3.1.1 Isotherms of s_w for Reacting Systems

We will discuss in the following how this idea can be applied consistently to the sedimentation of reacting systems, and how it relates to the quantities determining the chemical reaction.[2] Let us assume there are N_i species present in the mixture of N_k chemically interconverting components. Their local reaction fluxes q_i couple the sedimentation of species in this mixture, following the general Lamm equations for reacting systems (Eq. 1.2). The signal observed in the sedimentation profiles is the sum of those from all species, weighted by their respective signal increment, $a_\lambda(r, t) = \sum_i \varepsilon_{i,\lambda} \chi_i(r, t)$.[3,4] Inserted in the definition of s_w in Eq. (3.1), we obtain

$$\begin{aligned} s_{w,\lambda} &= \lim_{t \to 0} \left[-\frac{1}{a_\lambda(r_p, t)\omega^2 r_p^2} \left(\int_m^{r_p} \sum_i \varepsilon_{i,\lambda} \frac{d\chi_i(r, t)}{dt} r \, dr \right) \right] \\ &= \lim_{t \to 0} \left[-\frac{1}{a_\lambda(r_p, t)\omega^2 r_p^2} \left(\int_m^{r_p} \sum_i \varepsilon_{i,\lambda} \left[q_i - \frac{1}{r} \frac{\partial}{\partial r} \left(\chi_i s_i \omega^2 r^2 - D_i \frac{\partial \chi_i}{\partial r} \right) \right] r \, dr \right) \right]. \end{aligned} \tag{3.4}$$

With the index λ of we emphasize that the weighted-average sedimentation coefficient $s_{w,\lambda}$ will be weighted by the species' signal coefficients.[5]

---

[2]Historically, several alternative formulations of weighted-average sedimentation coefficients have been derived in the context of the analysis of interacting systems, for example, accommodating properties of Schlieren data [113]. For more detailed information, see [47].

[3]Note that we are using molar concentration units, whereas s_w is dependent on signal units.

[4]The signal coefficient is considered temporally constant.

[5]In a shorthand, s_w used to be referred to as weight-average sedimentation coefficient, which stems from considerations of refractive index-based optical data where the signal increments are considered proportional to the molecular weight of the species. This is mostly correct for self-associations, usually a reasonable approximation for hetero-associations [114], but breaks down for absorbance and fluorescence data where signal increments correlate little with molecular weight. A detailed discussion of signal increments for different macromolecules in different detection systems can be found in Part I, Section 4.3 [1].

Contributions to $s_{w,\lambda}$ consist of two terms,

$$s_{w,\lambda} = \lim_{t \to 0} \left[-\frac{1}{a_\lambda(r_p,t)\omega^2 r_p^2} \left(\int_m^{r_p} \sum_i \varepsilon_{i,\lambda} q_i r dr \right) \right]$$
$$+ \lim_{t \to 0} \left[\frac{1}{a_\lambda(r_p,t)\omega^2 r_p^2} \left(\int_m^{r_p} \sum_i \varepsilon_{i,\lambda} \left[\frac{1}{r} \frac{\partial}{\partial r} \left(\chi_i s_i \omega^2 r^2 - D_i \frac{\partial \chi_i}{\partial r} \right) \right] r dr \right) \right],$$

(3.5)

the first of which vanishes with mass and signal conservation

$$\sum_i \varepsilon_{i,\lambda} q_i = 0.$$

(3.6)

For this to hold true, either the chemical reaction fluxes have to be slow on the time-scale of sedimentation, or there cannot be hyper- or hypochromicity or changes in fluorescence quantum yield in conjunction with the chemical reaction event [102].[6] Under this condition, we proceed to resolve the second term, as previously in Part II, Section 2.3.1 [2]. Briefly, after executing the trivial integration, we can evaluate the terms in parenthesis recognizing that the diffusion fluxes vanish in the plateau region where $\partial\chi/\partial r = 0$ holds, and that the sedimentation flux vanishes at the meniscus. After rearrangements and expansion of the plateau signal, this results in

$$s_{w,\lambda} = \lim_{t \to 0} \left[\frac{\sum_i \varepsilon_{i,\lambda} s_i \chi_i(r_p,t)}{\sum_i \varepsilon_{i,\lambda} \chi_i(r_p,t)} \right] = \frac{\sum_i \varepsilon_{i,\lambda} s_i \chi_{i,0}}{\sum_i \varepsilon_{i,\lambda} \chi_{i,0}},$$

(3.7)

with the index 0 in the second identity referring to each species' initial equilibrium concentration in the loading mixture. The case of fast reactions with signal changes where Eq. (3.6) is invalid requires an extended theory incorporating the effects from coupled co-transport on the time-average signals [102], as described in Section 3.2.2.

This shows that the transport method can be applied to reacting mixtures just as well as to non-reacting mixtures, independent of the reaction type or kinetics. As long as signal is conserved in the reaction, $s_{w,\lambda}$ will reflect the signal-average sedimentation coefficient from the population of each species in the initial loading mixture. Therefore, the dependence of $s_{w,\lambda}$ on the loading composition $s_{w,\lambda}(\{c_{k,\text{tot}}\})$ is

---

[6]Signal changes accompanying the binding reaction may be detected, for example, by monitoring the total signal as a function of loading concentration. The problem arising for fast systems without signal conservation can be illustrated with the extreme example of a heterogeneous reaction in which only one component contributes to the signal and the complex signal is completely quenched. Due to the coupled sedimentation, all species are present in the reaction boundary, including the free species that can be detected. This makes contributions to the overall migration of the measured s_w, and is the meaning of the first term in Eq. (3.5). By contrast, a naïve average of signals from equilibrium populations would remain constant at the s-value of the free species since the complex is silent [102].

an isotherm that reports on the equilibrium thermodynamics of the reaction, and can be modeled with the mass action law of the chemical reaction.[7]

For example, we can study a self-associating system with signal conservation in a series of SV experiments at different loading concentrations, determine s_w at each concentration and model the resulting isotherm as

$$s_w(c_{\text{tot}}) = \frac{1}{c_{\text{tot}}} \sum_i i K_{1,i} c_1^i s_i \,, \tag{3.8}$$

with i enumerating the different oligomers, c_1 the free molar monomer concentration, and $K_{1,i}$ denoting the monomer-oligomer association constant with $K_{1,1} = 1$. In this equation, we have adopted molar protomer concentration units for c_{tot}, which allows the extinction coefficients in Eq. (3.7) to cancel out. See Fig. 3.3 for an example of such an isotherm for a monomer-dimer system.

Similarly, for a bimolecular hetero-association reaction $A + B \leftrightarrow AB$ that is slow on the time-scale of sedimentation Eq. (3.6) holds and the first term of Eq. (3.5) disappears, such that the signal-weighted average sedimentation coefficient is

$$s_w(c_{\text{A,tot}}, c_{\text{B,tot}}) = \frac{\varepsilon_A c_A s_A + \varepsilon_B c_B s_B + (\varepsilon_A + \varepsilon_B + \Delta\varepsilon_{AB}) K c_A c_B s_{AB}}{\varepsilon_A c_A + \varepsilon_B c_B + (\varepsilon_A + \varepsilon_B + \Delta\varepsilon_{AB}) K c_A c_B} \,. \tag{3.9}$$

This allows for a potential molar extinction coefficient change $\Delta\varepsilon_{AB}$ in complex AB due to hypo- or hyperchromicity or changes in fluorescence quantum yield.[8] If there is no such signal change and $\Delta\varepsilon_{AB} = 0$, then Eq. (3.9) is valid also for systems that exhibit rapidly interconverting species on the time-scale of sedimentation. Fig. 3.1 shows the isotherm of s_w as a function of total loading concentration of A and B. The maximum ridge is quite sharp, and modulated in position slightly by the relative extinction coefficients of A and B. Importantly, s_w does not reach the complex value s_{AB} short of extremely high equimolar concentration (relative to K_D). Even for equimolar loading concentrations at 100-fold K_D the measured s_w-value is still significantly lower than the sedimentation coefficient of the complex.

---

[7]The key as to why SV, despite being a transport method, reveals equilibrium thermodynamics is that the sample is considered to be in chemical equilibrium prior to the run, and that we take the limit $t \to 0$. Alternatively, in the section on Gilbert–Jenkins theory (Section 3.3) we will be concerned with asymptotic boundaries in the limit $t \to \infty$, which will also lead to thermodynamic equilibrium information.

[8]It is interesting to note that in Eq. (3.9) an unaccounted spectral change $\Delta\varepsilon_{AB}$ would be absorbed into an apparent binding constant $K' = K(\varepsilon_A + \varepsilon_B + \Delta\varepsilon_{AB})/(\varepsilon_A + \varepsilon_B)$. The significance of this deviation from K may be assessed for any particular case by comparison with common concentration errors, the accuracy of the extinction coefficients, as well as statistical errors in the determination of K.

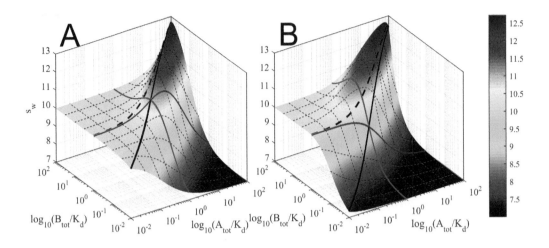

Figure 3.1 s_w isotherms for a bimolecular hetero-association reaction A + B $\leftrightarrow$ AB, with $s_A = 7$ S, $s_B = 10$ S, $s_{AB} = 13$ S. To illustrate the dependence on ε_A and ε_B, *Panel A* is for $\varepsilon_A = 0.1\varepsilon_B$, whereas *Panel B* is for $\varepsilon_A = 10\varepsilon_B$. The red and pink lines indicate titration series of constant B with varying A and constant A with varying B, respectively; the black solid line highlights the isotherm for equimolar experiments, and the black dashed line indicates the s_w-values at the phase transition line of the effective particle, $\ddot{c}_{B,tot}(c_{A,tot})$ according to Eq. (2.14).

For given sedimentation and extinction parameters, the isotherm $s_w(c_{A,tot}, c_{B,tot})$ can be displayed in SEDPHAT as a color contour plot in the Effective Particle Explorer function of the Options ▷Interaction Calculator submenu. Prior to this, the interaction model must be selected, and parameters be specified in the Global Parameters box.

3.1.2 The Effect of Radial Dilution on s_w

At this point, the practical question arises how to determine $s_{w,\lambda}$ in the limit of zero time, as required in Eq. (3.7). In principle, one could attempt using Eq. (3.1) directly as a recipe to integrate the experimental concentration profiles, and extrapolate the results to zero time. However, when applied to experimental data the result would suffer from unacceptable statistical errors due to the effect of unknown baseline offsets, in particular when using a few scans only to represent a small time-increment for extrapolation, and it would be very susceptible to amplification of statistical noise in particular for data at low loading concentrations.[9,10]

---

[9] Any RI noise would cause errors in the time-derivative of the integral, and all forms of noise including ubiquitous constant offsets and TI noise offsets would cause errors in the determination of the plateau concentration $a(r_p, t)$.

[10] Also, the time-derivative method for $g^*(s)$ (Part II, Section 6.2 [2]) provides no obvious relationship useful in the evaluation of Eq. (3.1): Even though only a small fraction of the experimental

There are two causes for $\chi_i(r_p, t)$ in Eq. (3.7) to vary with time. First, the fact that radial dilution will diminish the concentrations of each species at a rate that depends on each species sedimentation coefficient, $\sim e^{-2\omega^2 s_i t}$, changes their relative proportions and their contributions toward s_w^t. Second, as a result of this radial dilution, the chemical equilibria between the species will readjust, and cause reaction fluxes even in the plateau region [18,21,47]. The first point is automatically accounted for when directly modeling the observed sedimentation boundaries with sedimentation coefficient distributions rooted in Lamm equation solutions, such as $c(s)$, that propagate species loading concentrations with time. Unfortunately, however, the second point strictly violates an assumption inherent in all distribution methods developed thus far, that the particle size distribution remains constant throughout the experiment.

Let us examine the impact of re-adjustment of species populations due to the radial dilution. During sedimentation across a standard solution column extending from 6.0 cm to 7.2 cm, the square dilution law (Eq. (1.9) of Part I [1]) predicts the dilution to be $\sim 30\%$. Based on mass action law, rapidly equilibrating systems may adjust the relative species concentration ratio by as much or more during the experiment. However, what is measured in standard boundary modeling is not the instantaneous velocity of the boundary, but rather the distance from the meniscus $\bar{r} - m$ traveled by the boundary after a certain time t_{scan}.[11] This defines a time-average velocity during the sedimentation process up to that point (Fig. 3.2). Accordingly, more relevant than the final species concentrations is the time-average of the species concentrations during the sedimentation up to the time t_{scan} [47]:

$$
\frac{\bar{\chi}_i(t_{\text{scan}})}{\chi_{i,0}} = \frac{1}{t_{\text{scan}}} \int_0^{t_{\text{scan}}} e^{-2\omega^2 s_i t'} dt' = \frac{1}{2\omega^2 s_i t_{\text{scan}}} \left(1 - e^{-2\omega^2 s_i t_{\text{scan}}} \right)
$$
$$
\approx 1 - \frac{1}{2} \left(2\omega^2 s_i t_{\text{scan}} \right) = 1 - \frac{1}{2} \log \left(\frac{\bar{r}_i^2}{m^2} \right),
$$
(3.10)

where $\bar{r}_i$ is the second moment boundary position of species i. (In a first approximation, one may use s_w to determine average positions $\bar{r}_i$.) Thus, the time-average dilution is approximately half that of the instantaneous boundary position, or maximally half that at the end of the solution column. Further, in the SV analysis usually a whole set of scans is considered, uniformly representing the entire time-course of the sedimentation process. This will give additional weight toward the conditions at early times. If we were sampling the concentration profiles continuously, we could

---

scans are evaluated, the theoretical framework of this method invokes the entire sedimentation process starting at $t = 0$.

[11]To the extent that we do not fix the meniscus position, it may be replaced in this consideration by the radial position of the earliest scan included in the analysis. For standard configurations including the entire available time-range of scans reporting on sedimentation, as described in Chapter 8 of Part II [2], the distinction will not have an impact on the present conclusions.

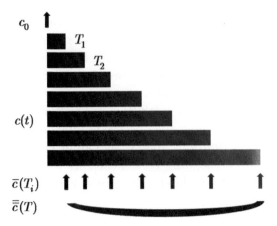

Figure 3.2 Illustration of the radial dilution and its representation in the boundary analysis. This is relevant for interacting systems to understand the extent of chemical redistribution of species during the SV run caused by the radial dilution. Concentrations are indicated as color temperatures, and the radial dimension is from left to right. The initial loading concentration c_0 is depicted as the dark blue arrow on top. The history underlying the sedimentation in each scan, indicated by the horizontal bar, shows macromolecules being increasingly diluted as the boundary migrates to higher radii, as depicted by the color gradients (with more dilute regions showing as purple and red colors). Experimental scans at times T_i are snapshots of the current boundary position, which reflects the entire history of the sedimentation process between $t = 0$ and $t = T_i$, not only the migration at the time T_i. The average conditions during the run up until $c(T_i)$ are indicated by the set of colored arrows below the bars, as in Eq. (3.10). The analysis of the entirety of all scans will reflect an average condition from all these scans, as in Eq. (3.11), leading to a further shift toward the initial conditions. Thus, s_w-values from boundary modeling do not reflect equilibria at the diluted conditions observed at times T_i, but are skewed toward the initial condition.

describe this approximately as

$$\frac{\bar{\bar{\chi}}_i}{\chi_{i,0}} = \frac{1}{t_{\max}} \int\limits_0^{t_{\max}} \frac{\bar{\chi}_i(t')}{\chi_{i,0}} dt' \approx 1 - \frac{1}{4} \log\left(\frac{\bar{r}_i^{\,2}}{m^2}\right), \qquad (3.11)$$

which further reduces the radial dilution by another factor of two (Fig. 3.2). This reduces the effect of radial dilution on the s_w-value to less than 10% in concentration if the samples traverse the entire solution column. Thus, in most circumstances[12] the effect of the chemical re-equilibration seems small considering the concentration scale of mass action law, common concentration errors, and the typically obtainable precision of binding constants.

---

[12] A counter-example could be theoretically constructed as a highly cooperative assembly, or phase transition, where the starting concentration is less than 10% above a critical concentration.

The SEDFIT function Options ▷Calculator ▷Calculate Radial Dilution reports, for a species of interest with given s-value, the final instantaneous radial dilution at t_{scan}, the average radial dilution $\bar{\chi}_i(t_{\text{scan}})/\chi_{i,0}$ for the species experienced during the experiment up to the time of the last scan (according to Eq. (3.10)), as well as the average radial dilution at conditions reported on in all scan time-points $(\bar{\bar{\chi}}_i(t_{\text{scan}})/\chi_{i,0}$ from Eq. (3.11)) for the loaded data. The latter factor may be used to correct species concentration in the context of binding isotherms.

3.1.3   Determining s_w with Sedimentation Coefficient Distributions $c(s)$

If a sedimentation coefficient distribution reflects the true distribution of species in solution, then the integration of the sedimentation coefficient distribution in Eq. (3.2) seems a natural way to calculate a weighted-average s-value. However, sedimentation coefficient distributions are defined as distributions of non-interacting species of different s-values, and — even though we have seen how rapidly interacting systems can be approximated as pauci-disperse distribution of effective particles — the relationship between distributions of non-interacting species and species sedimenting with rapid chemical equilibration is not obvious.[13] After all, for example, diffusion in the boundary causes major concentration gradients and concomitant chemical transformations of species during the SV run. For all but slow interactions on the time-scale of sedimentation, this makes it necessary to distinguish the macroscopic 'apparent' or 'effective' species from true microscopic macromolecular entities whose system properties we want to capture in s_w.

The elegance and power of the transport method arises from the fact that it solely rests on the overall mass balance, in a way that is independent of all chemical re-arrangements in the boundary, and usually affected little by radial dilution, as we have seen above. Thus, a link between a rigorously defined s_w reflecting the properties of the interacting system on one hand, and the sedimentation coefficient distribution of non-interacting particles on the other hand, can be founded on the consistency of the distribution with these mass balance considerations: If the distribution leads to a description of the sedimentation boundary that correctly models the total signal above a reference radius in the plateau at all times, then the integration of the distribution Eq. (3.2) is equivalent to the definition of s_w in Eq. (3.1) (Part II, Section 3.4). This is the case for $c(s)$ or ls-$g^*(s)$ distributions that are directly fitting the sedimentation boundary — as long as the best-fit model is faithful to the area (or radial integral) described by the boundary. This can be guaranteed simply by a good fit. Since this connection to s_w is independent of the physical motivation and interpretation of the model, the conceptual questions of macroscopic 'apparent' or 'effective' species vs. true microscopic chemical species

---

[13]In other words, it is possible to formally insert any distribution function *ad hoc* into Eq. (3.2) to mathematically calculate a weighted-average quantity of this distribution. However, this is not meaningful *per se* for analysis of interacting systems without establishing a firm connection between the form of Eq. (3.2) for that distribution and the definition of s_w in the transport method of sedimentation.

is rendered irrelevant in the context of s_w. Thus, it is imperative and sufficient to achieve a good fit to the raw data with $c(s)$.[14]

This is generally the case, and illustrated in the example of Fig. 3.3. It shows simulated noise-free data of a monomer-dimer system in rapid equilibrium. At most concentrations the fit quality is ~0.002–0.003-fold the loading concentration, which is commensurate with the experimental signal/noise achieved in excellent detection conditions (Part I, Section 4.3.5 [1]). On the basis of Eq. (3.3), for the data shown, the residuals from the first half of the scans (those that still exhibit solution plateaus) propagate into a systematic error in s_w of 0.04 S or 0.5%.

With regard to the effect of radial dilution, if we carry out an isotherm experiment measuring $s_w(\{c_{k,\text{tot}}\})$ as a function of concentration, the shifts in the total concentration during the experiment should be very similar throughout (assuming we maintain equivalent experimental conditions for the SV experiments at different loading concentrations). Therefore, as a first-order correction for fast reactions we could apply a uniform correction factor based on Eq. (3.11), using $c_{k,\text{tot}} \times \bar{\bar{\chi}}_i/\chi_{i,0}$ instead of $c_{k,\text{tot}}$ in the isotherm analysis. This accounts for the average redistribution of species slightly changing the s_w-value obtained with the direct boundary distribution model. The improvement provided by this correction for the example of a monomer-dimer system is shown in *Panel B* of Fig. 3.3 (black *vs.* magenta line). In practice, this correction is rarely applied because the shift of populations during the SV run would translate in an error in the estimated binding constant on the order of 10%, which is usually negligible considering alone the common problems in determining the precise experimental concentration of active protein, as well as statistical errors in determining K_d. Further, 10% differences in binding constants of bimolecular reactions of biological macromolecules are usually inconsequential in the context of interpreting the biological function.

In summary, the transport method provides a rigorous theoretical background for interpreting the isotherms $s_w(\{c_{k,\text{tot}}\})$ with models of mass action law and equilibrium thermodynamics. Although some approximations are necessary to estimate s_w from SV data of rapidly interacting systems (negligible for slowly interacting systems), this is one of the most widely used approaches of analyzing interacting systems in SV. The theoretical appeal of this method is that it does not rely in any way on the available information of the actual shape of the sedimentation boundary. At the same time, the lack of utilizing any boundary shape information also represents the limitation of this method.

---

[14]Equivalence between the integration of the sedimentation coefficient distribution and the radial integrals underlying the definition of s_w can also be shown theoretically for apparent sedimentation coefficient distributions $g^*(s)$ that are based on the time-derivative dc/dt of the boundaries, as shown in Section 6.2.1 of Part II [2]. However, this is not the case for distributions $\tilde{g}^*(s, \Delta t)$ that arise from the estimation of $g^*(s)$ *via* finite time-difference $\Delta c/\Delta t$. See Section 4.2.1 for a more detailed discussion.

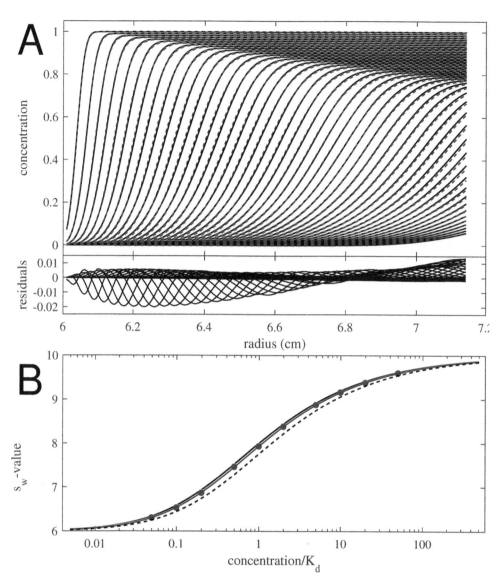

Figure 3.3 Example for the $c(s)$ fit of data from a rapid interacting system. *Panel A* shows the radial sedimentation profiles for a hypothetical monomer with 6 S in rapid equilibrium with a 10 S dimer sedimenting at 50,000 rpm in a 12 mm solution column at a loading concentration equal to K_d (black lines, plotted normalized relative to the loading concentration). The profiles were calculated using Eq. (1.33), and then fitted with the $c(s)$ model (red dashed lines). The residuals with rmsd of 0.0028 are shown in the lower plot. *Panel B*: The resulting peaks in $c(s)$ were integrated to determine s_w for several different loading concentrations, forming the $s_w(c_{tot})$ isotherm data (magenta circles). The solid black line shows the theoretically expected isotherm based on Eq. (3.8) using the correct s-values and K_d but not considering radial dilution. The magenta line is the isotherm after correctly accounting for time-average radial dilution following Eq. (3.11). If unaccounted, this leads to a small error in K_d of $\sim9\%$. By contrast, the naïve assumption that the instantaneous radial dilution of latest scans determines radial dilution would imply an overly large decrease in concentrations and s_w (dashed line), that would suggest an incorrect shift in K_d after isotherm analysis by $\sim40\%$.

3.2 EFFECTIVE PARTICLE THEORY

We have established in Section 2.4 the physical foundation of the effective particle model and explored some of its consequences. Here, we take a different approach and re-derive the basic equations for simple two-component systems directly from the Lamm equation. Appendix B presents a general theory for any number of components.

3.2.1 Basic Principle of Solving Coupled Lamm Equations for Single-Site Hetero-Associations

As we have seen from the previous discussion, the effects of radial dilution are rather small in the context of binding analysis. As a consequence, historically early on in the study of reaction/diffusion/sedimentation problems the approximation was adopted of sedimentation in linear geometry with a constant force (such as in electrophoresis). This is a very powerful approximation, because in this geometry the mathematical description is much more tractable [115].

A second time-honored simplification is the neglect of diffusion [43,103,116].[15] We have already noted above that, phenomenologically, diffusion in interacting systems does not evolve much differently from the familiar $\sqrt{t}$-dependent broadening behavior. We cannot expect the results of a diffusion-free picture to produce an accurate account of the actual boundary shapes observed in the experiment (except for very large particles), but we should expect the results to reflect the patterns of boundaries with regard to the number of boundaries, their velocities or velocity ranges, and their amplitudes. In this fashion a diffusion-free picture is able to highlight the essential features of the boundary patterns.

The neglect of diffusion allows us to rewrite the Lamm equation for a bimolecular reaction Eq. (1.31) on p. 16 as

$$\frac{\partial \chi_A}{\partial t} + v_A \frac{\partial \chi_A}{\partial x} = q$$
$$\frac{\partial \chi_B}{\partial t} + v_B \frac{\partial \chi_B}{\partial x} = q \qquad (3.12)$$
$$\frac{\partial \chi_{AB}}{\partial t} + v_{AB} \frac{\partial \chi_{AB}}{\partial x} = -q,$$

where we have used the symbols x and v to indicate the spatial coordinate and linear velocity in the linear geometry approximation (replacing r and s in radial coordinates, respectively), and with $q = -(k_{off}/K_D)\chi_A\chi_B + k_{off}\chi_{AB}$ abbreviating the chemical reaction flux.

As discussed above, the goal in effective particle theory is to arrive at average

---

[15]Even though Gilbert has not neglected diffusion outright, the focus on 'asymptotic boundaries' obtained after extrapolation to infinite time amounts to the same [103]. The diffusion-free picture was also used early in the interpretation of electrophoresis and chromatography of interacting mixtures [116–118].

boundary properties of the reaction and undisturbed boundaries, without considering any boundary spread, be it from diffusion or from differential sedimentation distributions. Specifically, we seek to predict the weighted-average s_w of the reaction boundary, since it is experimentally easy to determine (see Chapter 4). As discussed above, the origin of the weighted-average s_w value is the transport method, which considers the time-dependent depletion of the number of particles 'above' a reference radius. This mass balance is independent of boundary shapes, and — as long as the reference radius is in a plateau region — contributions from diffusion terms vanish. This justifies the expectation that the mass balance from the solution of Eq. (3.12) will reflect the correct mass balance, equal to that from the full set of Lamm equations including the diffusion terms.

The task to determine the weighted-average s_w value of the reaction boundary can be accomplished by an approach similar to the second moment analysis (Part II, Section 2.3.3 [2]), which aimed at determining the effective boundary position $\bar{r}$ that would correspond to a particle with the correct s_w^t-value. Accordingly, the starting point will be the attempt to write the concentration profiles for each species with the help of step-functions corresponding to mono-disperse (non-diffusing) boundaries.[16] The step-functions directly express the mass balance.[17]

A few additional properties of the Lamm equation solutions can be embedded outright. The solution column is initially uniformly loaded with the reaction mixture, and as a consequence, there can be only a zone reflecting the loading mixture and a zone depleted of one of the components. Correspondingly, there is an undisturbed boundary migrating at the velocity of a free species, which may be either $\mathbf{A}$ or $\mathbf{B}$, and a reaction boundary migrating at a velocity we will refer to as $v_{\mathsf{A}\cdots\mathsf{B}}$. We also know that some free A and free B will be co-migrating in the reaction boundary; consistent with the terminology in Section 2.4 this will be $\tilde{c}_\mathsf{A}$ and $\tilde{c}_\mathsf{B}$. The remaining portions of free $\mathbf{A}$ or free $\mathbf{B}$ will form the amplitude of the undisturbed boundary, here abbreviated $c_{\mathsf{A},\mathrm{u}}$ and $c_{\mathsf{B},\mathrm{u}}$, respectively. Finally, we know that since $v_{\mathsf{A}\cdots\mathsf{B}} > v_\mathsf{A}$ and $v_{\mathsf{A}\cdots\mathsf{B}} > v_\mathsf{B}$, the reaction boundary is completely immersed in the undisturbed boundary, and therefore the amplitude of the complex is determined by mass action law from the total free concentrations in this zone, $c_{\mathsf{AB}} = K(c_{\mathsf{A},\mathrm{u}} + \tilde{c}_\mathsf{A})(c_{\mathsf{B},\mathrm{u}} + \tilde{c}_\mathsf{B})$. This leads to the ansatz [41]

$$c_\mathsf{A}(x,t) = c_{\mathsf{A},\mathrm{u}} H(x - v_\mathsf{A}t) + \tilde{c}_\mathsf{A} H(x - v_{\mathsf{A}\cdots\mathsf{B}}t)$$
$$c_\mathsf{B}(x,t) = c_{\mathsf{B},\mathrm{u}} H(x - v_\mathsf{B}t) + \tilde{c}_\mathsf{B} H(x - v_{\mathsf{A}\cdots\mathsf{B}}t) \qquad (3.13)$$
$$c_{\mathsf{AB}}(x,t) = K(c_{\mathsf{A},\mathrm{u}} + \tilde{c}_\mathsf{A})(c_{\mathsf{B},\mathrm{u}} + \tilde{c}_\mathsf{B}) H(x - v_{\mathsf{A}\cdots\mathsf{B}}t) \, .$$

---

[16]Historically, it is interesting to note that the mathematical theory of distributions or generalized functions, such as step-functions and δ-functions, was not available until the 1950s [119], and likely not widely known yet (outside the field of quantum mechanics) when Gilbert theory and Gilbert–Jenkins theory were first conceived.

[17]In rectangular geometry, a boundary of the form $\chi(x,t) = \chi_0 H(x - vt)$ embeds the mass balance $\dfrac{dm}{dt} = \displaystyle\int_m^{r'} \dfrac{\partial}{\partial t}\chi(x,t)dx = -\chi_0 v \, .$

Since there is only one undisturbed boundary either $c_{A,u}$ or $c_{B,u}$ must be zero.

The spatial and temporal derivatives can be readily executed with the help of Dirac δ-functions [119]:

$$\frac{\partial c_A}{\partial t} = -v_A c_{A,u}\delta(x - v_A t) - v_{A\cdots B}\tilde{c}_A\delta(x - v_{A\cdots B}t)$$

$$v_A\frac{\partial c_A}{\partial x} = v_A c_{A,u}\delta(x - v_A t) + v_A \tilde{c}_A\delta(x - v_{A\cdots B}t) \ ,$$

$$\frac{\partial c_B}{\partial t} = -v_B c_{B,u}\delta(x - v_B t) - v_{A\cdots B}\tilde{c}_B\delta(x - v_{A\cdots B}t)$$

$$v_B\frac{\partial c_B}{\partial x} = v_B c_{B,u}\delta(x - v_B t) + v_B \tilde{c}_B\delta(x - v_{A\cdots B}t) \ ,$$

$$\frac{\partial c_{AB}}{\partial t} = -v_{A\cdots B}K(c_{A,u} + \tilde{c}_A)(c_{B,u} + \tilde{c}_B)\delta(x - v_{A\cdots B}t)$$

$$v_{AB}\frac{\partial c_{AB}}{\partial x} = v_{AB}K(c_{A,u} + \tilde{c}_A)(c_{B,u} + \tilde{c}_B)\delta(x - v_{A\cdots B}t) \ ,$$

(3.14)

Their insertion in Eq. (3.12), followed by ordering according to δ-function terms, leads from a partial differential equation system to a simpler algebraic equation system [41]. If we sum over the first equation for A and the third equation for AB, we find

$$0 = \delta(x - v_{A\cdots B}t)\big[- v_{A\cdots B}\tilde{c}_A + v_A\tilde{c}_A$$
$$- v_{A\cdots B}K(c_{A,u} + \tilde{c}_A)(c_{B,u} + \tilde{c}_B) + v_{AB}K(c_{A,u} + \tilde{c}_A)(c_{B,u} + \tilde{c}_B)\big] \ .$$

(3.15)

This is fulfilled if

$$v_{A\cdots B} = \frac{v_A\tilde{c}_A + v_{AB}K(c_{A,u} + \tilde{c}_A)(c_{B,u} + \tilde{c}_B)}{\tilde{c}_A + K(c_{A,u} + \tilde{c}_A)(c_{B,u} + \tilde{c}_B)} \ .$$

(3.16)

For the case A$\cdots$(B) when A is dominant, it is $c_{A,u} = 0$, $\tilde{c}_A = c_A$, and $c_{B,u} + \tilde{c}_B = c_B$, therefore

$$v_{A\cdots(B)} = \frac{v_A c_A + v_{AB}K c_A c_B}{c_A + K c_A c_B} \ .$$

(3.17)

This is identical to (2.5), which we previously had derived based on the physical consideration of time-average sedimentation coefficients. On the other hand, for the case B$\cdots$(A), it is $c_{B,u} = 0$, $\tilde{c}_B = c_B$, and $c_{A,u} + \tilde{c}_A = c_A$, therefore

$$v_{B\cdots(A)} = \frac{v_A\tilde{c}_A + v_{AB}K c_A c_B}{\tilde{c}_A + K c_A c_B} \ ,$$

(3.18)

which is an equation for $v_{B\cdots(A)}$ cast in terms of the co-sedimenting fraction of the secondary component.

More can be found if we insert the derivatives Eq. (3.14) into Eq. (3.12) and take the sum of the second equation for B and the third equation for the complex AB. In symmetry to Eq. (3.17) and Eq. (3.18), we find for the phase B$\cdots$(A)

$$v_{B\cdots(A)} = \frac{v_B c_B + v_{AB} K c_A c_B}{c_B + K c_A c_B} ,$$

$$(3.19)$$

which is equivalent to Eq. (2.4), and for A$\cdots$(B) an equation

$$v_{A\cdots(B)} = \frac{v_B \widetilde{c}_B + v_{AB} K c_A c_B}{\widetilde{c}_B + K c_A c_B} ,$$

$$(3.20)$$

cast, again, in terms of the co-sedimenting fraction of the secondary component. The combination of Eq. (3.18) with Eq. (3.19) allows us to solve for the amplitudes of the co-sedimenting fraction $\widetilde{c}_A$ for B$\cdots$(A) as

$$\widetilde{c}_A = \frac{K c_A c_B (v_B - v_{AB})}{(v_A - v_B) + K c_A (v_A - v_{AB})} ,$$

$$(3.21)$$

equivalent to Eq. (2.7). Analogously, the combination of Eq. (3.17) with Eq. (3.20) leads to an expression for $\widetilde{c}_B$ for A$\cdots$(B) equivalent with Eq. (2.10).[18] From this follows the stoichiometry of the reaction boundary as was described in detail in Section 2.4.2.

This recapitulates from a mathematical perspective the quantitative aspects of the effective particle model that we introduced previously based on physical arguments, corroborating both the quantitative expressions as well as their physical interpretation. In summary, the effective particle model rests on a linear Lamm equation in the absence of diffusion, and arrives at a quantitative description of the boundary pattern for their s-values, amplitudes, and composition, but not the exact boundary shapes. As shown in Section 2.4.2, this allows us to survey the phenomenology of interacting systems with analytical expressions over the entire parameter space, as illustrated in Figs. 2.10, 2.11, and 2.12. It will also be extremely useful for the robust interpretation of experimental data in conjunction with (diffusion deconvoluted) sedimentation coefficient distributions in Chapter 4.

3.2.2 Signal-Weighted Average Sedimentation Coefficients

The effective particle theory offers a natural extension of the interpretation of the isotherm of signal-weighted sedimentation coefficients of the entire mixture. In this approximation one can take the view that sedimentation proceeds as if there are two independently sedimenting species — the undisturbed species and the effective particle. We know their sedimentation velocity and we can determine their signal coefficients from their composition. Therefore, we can rephrase the signal-weighted

---

[18]It is interesting to note the special case of equivalent molecules A and B with $v_A = v_B$, in which case symmetry of the free species in the reaction boundary is obtained if $\widetilde{c}_A = c_B$. Therefore for systems with $v_A = v_B$ at equimolar loading concentrations the undisturbed boundary disappears.

average sedimentation coefficient from Eq. (3.9) on p. 69 as

$$s_w^{(EPT)} = \frac{\varepsilon_Y \left(c_Y - \tilde{c}_Y\right) s_{Y,u} + \left(\varepsilon_Y \tilde{c}_Y + \varepsilon_X c_X + \varepsilon_{AB} c_{AB}\right) s_{A\cdots B}}{\varepsilon_Y \left(c_Y - \tilde{c}_Y\right) + \left(\varepsilon_Y \tilde{c}_Y + \varepsilon_X c_X + \varepsilon_{AB} c_{AB}\right)}, \quad (3.22)$$

using a nomenclature referring to the dominant component as X and the secondary component as Y. Satisfactorily, after inserting the amplitudes and velocities followed by lengthy rearrangement, for both cases A$\cdots$(B) and B$\cdots$(A) the r.h.s. of Eq. (3.22) is exactly identical to Eq. (3.9) provided that signals are additive in the complex, i.e., $\varepsilon_{AB} = \varepsilon_A + \varepsilon_B$ [102]! However, now we can allow for the possibility that complexes may have a signal different by $\Delta\varepsilon_{AB} = \varepsilon_{AB} - \varepsilon_A - \varepsilon_B$ from their constituents. With this terminology, we can rephrase $s_w^{(EPT)}$ as

$$s_w^{(EPT)} = s_w^{\Delta_0} - \frac{\Delta\varepsilon_{AB}\, c_{AB}}{\varepsilon_Y c_{Y,tot} + \varepsilon_X c_{X,tot} + \Delta\varepsilon_{AB} c_{AB}} \times \frac{c_X}{c_X + c_{AB}} \times \left(s_{AB} - s_X\right) \quad (3.23)$$

where s_w stands for the value from Eq. (3.9). What is the origin of the extra terms for the $\Delta\varepsilon_{AB}$ in Eq. (3.23)? We recall the derivation of s_w through the transport method above, where we found that the conventional form of Eq. (3.9) does not apply for the case of rapidly reacting systems when signal changes are associated with the reaction and Eq. (3.6) is not fulfilled. This was because the signal-weighted average transport based on equilibrium species population does not account for the additional signal co-transport in the reaction boundary.[19] In short, in the picture of individual molecular species, the time-average state is the same as the population-average, but the time-average signal transport is not the population-average signal transport. $s_w^{(EPT)}$ provides a correction for this in the second term of Eq. (3.23). This is possible since in the picture of effective particle theory, we have subdivided the sedimenting species into *effective species* (i.e., the undisturbed species and the species constituted by the reaction boundary mixture) such that the transport state of the *effective* species is constant. Therefore, the time-average signal transport becomes identical with the population-average signal transport in the effective particle picture.

To examine more closely the physical meaning of the second term on the r.h.s. of Eq. (3.23), we may consider a situation when only X contributes to the signal and Y is silent with $\varepsilon_Y = 0$ (e.g., Y may not be carrying a fluorescent label). We envision the extreme case that the signal of X is completely quenched in the complex ($\Delta\varepsilon_{AB} = -\varepsilon_X$). Now we find the additional term of Eq. (3.23) reduces to the product $(c_{AB}/(c_X + c_{AB})) \times (s_{AB} - s_X)$, where the first factor describes the fractional time that X is in the complex and silent, which is multiplied with the additional migration that X will experience during this time. Together it amounts to the migration that will be missed in the pure population based consideration of s_w of Eq. (3.9).

Since $s_w^{(EPT)} = s_w$ in the absence of signal changes, we may consider $s_w^{(EPT)}$ a

---

[19]This co-transport is highlighted in Fig. 2.7 on p. 41. In the case that $\Delta\varepsilon_{AB} \neq 0$ it follows that $\sum_i \varepsilon_{i,\lambda} q_i \neq 0$ such that the first term in the r.h.s. of Eq. (3.5) on p. 68 does not vanish.

generalization of the signal-weighted-average sedimentation coefficient s_w extending its applicability to rapidly reversible systems with any signal changes upon complex formation. It cannot be applied, however, in the case of slowly reversible systems with signal changes, where the effect of coupled transport is absent. In that case, the traditional s_w of Eq. (3.9) will hold for any $\Delta\varepsilon_{AB}$.[20]

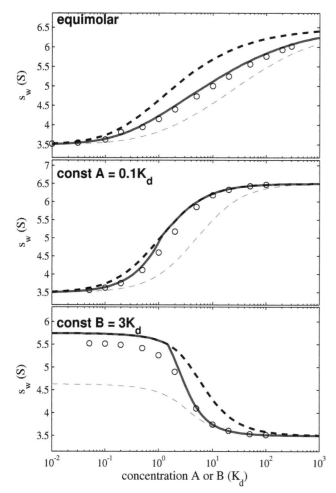

Figure 3.4 Effect of signal quenching on isotherms of rapidly reversible systems. Comparison of s_w-values extracted from coupled Lamm equation solutions (Eq. (1.33) followed by $c(s)$ analysis and integration, circles) with those from the standard population-based expression s_w (Eq. (3.9), gray dashed line), and the isotherm from effective particle theory $s_w^{(EPT)}$ (Eq. (3.23), red solid line). Simulations were for a system with a 45 kDa, 3.5 S component A sedimenting in the presence of a silent 75 kDa, 5 S component B forming a rapidly reversible complex with 6 S in a solution column of 12 mm length spinning at a rotor speed of 50,000 rpm. The signal of the complex was quenched to 0.2-fold that of free A ($\Delta\varepsilon_{AB} = -0.8\varepsilon_A$ and $\varepsilon_B = 0$). For comparison, the s_w isotherm in the absence of quenching is also shown (Eq. (3.9) with $\Delta\varepsilon_{AB} = 0$, blue dashed line). Concentrations were taken as equimolar dilution series (*Top Panel*), titration of constant concentration of A at 0.1-fold K_D (*Middle Panel*), and the titration of a constant concentration of B at 3-fold K_D (*Lower Panel*). Reproduced from [102].

The magnitude of this effect can be quite significant, as illustrated in the example of Fig. 3.4 for a system with signal only from component A, in the presence of strong quenching such that the complex signal is reduced to 0.2-fold that of free form of A. The circles represent the observed s_w from simulated Lamm equation solutions along dilution and titration series. Quenching causes the equimolar dilution series to exhibit a more shallow isotherm, and the titration series to be

[20]The necessity for a correction to the population-based s_w for rapidly reacting systems with signal changes was only recently discovered [102]. It adds a caveat to the discussion of hypo- and hyper-chromicity and changes in fluorescence quantum yield in Part I p. 151/152 [1], which is correct only for slowly interacting systems.

significantly steeper. As may be expected, not accounting at all for signal quenching (blue dashed line) fails to describe the data since the signal-transport contributions from the complex are over-estimated. When accounting for quenching of the complex in a conventional population-based isotherm, without considering the coupled co-transport as in Eq. (3.9), as the performance is even worse because the impact of quenching is overestimated (thin dashed lines). As a result, this population-based isotherm significantly underestimates s_w under all conditions except those that lead to complete saturation or to the absence of any complex. Compared to both, the isotherm accounting for co-transport, $s_w^{(EPT)}$ of Eq. (3.23) in the red line, is much more consistent with the data throughout.

An imperfect approximation can be discerned for conditions where **A** is predicted by effective particle theory to be engulfed entirely in the reaction boundary, mediated by an excess of **B**. However, the concentration gradients from diffusion allow for a small population of free **A** to escape and trail the reaction boundary, thereby lowering the observed s_w. This highlights a limitation of the diffusion-free

Figure 3.5 Effect of signal enhancement in the complex on s_w isotherms, analogous to Fig. 3.4. Using otherwise identical molecular parameters and concentrations, Lamm equation simulations were carried out to determine for a system where the complex contributes 3-fold the signal of the free species ($\Delta\varepsilon_{AB} = 2\varepsilon_A$ and $\varepsilon_B = 0$). As in Fig. 3.4, shown are s_w-values extracted from coupled Lamm equation solutions (circles), isotherms using a naïve population-based expression for s_w (gray dashed lines), isotherm from effective particle theory $s_w^{(EPT)}$ (red solid line), and isotherms in the absence of signal change (blue dashed lines). Concentrations were taken as equimolar dilution series (*Top Panel*), titration of constant concentration of **A** at 0.1-fold K_D (*Middle Panel*), and the titration of a constant concentration of **B** at 3-fold K_D (*Lower Panel*). There are no adjusted parameters in these plots. Reproduced from [102].

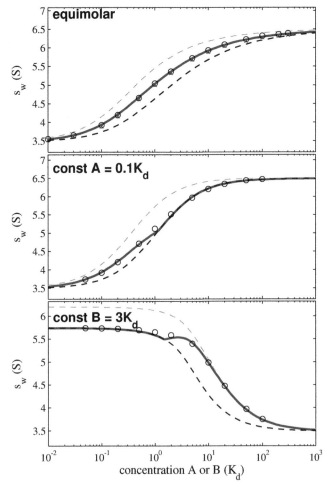

approximation of effective particle theory, though the effect is exacerbated by the strong quenching of the complex (see below). Nevertheless, despite this error $s_w^{(\mathrm{EPT})}$ of Eq. (3.23) still performs much better than the conventional isotherm s_w of Eq. (3.9) [102].

It should be noted that the same framework can be equally applied to any changes in the complex signal, including, for example, enhancement of fluorescence quantum yield of the complex. Fig. 3.5 shows such a case, where the complex generates 3-fold the signal of the free fluorescent species, in otherwise analogous representation to Fig. 3.4. Again, $s_w^{(\mathrm{EPT})}$ matches the sedimentation behavior of the system very well, here even the case where A should be driven entirely into the reaction boundary, since the diffusionally escaped population of A now contributes relatively little to the signal due to the large signal enhancement of the complex.

In summary, the anomaly of s_w measured by the transport theory departing from the naïvely assumed species population-average properties[21] is yet another consequence of the coupled transport process of rapidly reacting species, and naturally explained in the effective particle theory. It can be of practical importance, for example, when using fluorescence data acquisition where static or dynamic quenching, energy transfer, or any other change in fluorescence quantum yield occurs in reversibly formed complexes. Other mechanisms, such as hypo- and hyperchromicity in absorbance data of protein-protein or protein-nucleic interactions, would likewise be experimental situations where $s_w^{(\mathrm{EPT})}$ would apply if the interaction is rapidly reversible [102].

3.2.3 Two-Component Multi-Site Systems

It is straightforward to apply the same effective particle approach to more complex reactions with higher stoichiometry, based on more general sets of coupled Lamm equations than Eq. (3.12). For example, for the case of multiple complexes AB, A_2B, ... , A_NB in rapid equilibrium linked by equilibrium constants K_i, the phase transition is at [41]

$$\dddot{c}_\mathrm{B}(c_\mathrm{A}) = \frac{\sum_{i=1}^N K_i c_\mathrm{A}^i \left(v_{\mathrm{A}_i\mathrm{B}} - v_\mathrm{A}\right) + (v_\mathrm{B} - v_\mathrm{A})}{\sum_{i=1}^N i K_i c_\mathrm{A}^{i-1} \left(v_{\mathrm{A}_i\mathrm{B}} - v_\mathrm{B}\right) + \sum_{i,j=1}^N i \left(v_{\mathrm{A}_i\mathrm{B}} - v_{\mathrm{A}_j\mathrm{B}}\right) K_i K_j c_\mathrm{A}^{i+j-1}}, \tag{3.24}$$

and the reaction boundary exhibits the average velocity of

$$v_{\mathrm{A}\cdots\mathrm{B}} = \begin{cases} \left(v_\mathrm{A} c_\mathrm{A} + \sum_{i=1}^N i v_{\mathrm{A}_i\mathrm{B}} K_i c_\mathrm{B} c_\mathrm{A}^i\right) \Big/ \left(c_\mathrm{A} + \sum_{i=1}^N i K_i c_\mathrm{B} c_\mathrm{A}^i\right) & \text{for } c_\mathrm{B} > \dddot{c}_\mathrm{B}(c_\mathrm{A}) \\[3mm] \left(v_\mathrm{B} c_\mathrm{B} + \sum_{i=1}^N v_{\mathrm{A}_i\mathrm{B}} K_i c_\mathrm{B} c_\mathrm{A}^i\right) \Big/ \left(c_\mathrm{B} + \sum_{i=1}^N K_i c_\mathrm{B} c_\mathrm{A}^i\right) & \text{else} \end{cases} .$$

$$\tag{3.25}$$

---

[21]The assumption that s_w should reflect population-average s-values of interacting systems dates back many decades [19] and has been unquestioned until recently [102].

The co-sedimenting fractions of the secondary component are

$$
\tilde{c} = \begin{cases}
\dfrac{\displaystyle\sum_{i=1}^{N} K_i c_B c_A^i \left(v_A - v_{A_iB}\right) - \sum_{i,j=1}^{N} i(v_{A_jB} - v_{A_iB})K_i K_j c_B^2 c_A^{i+j-1}}{\left(v_B - v_A\right) + \displaystyle\sum_{i=1}^{N} i K_i c_B c_A^{i-1}\left(v_B - v_{A_iB}\right)} & \text{for } c_B > \dddot{c}_B(c_A) \\[4ex]
\dfrac{\displaystyle\sum_{i=1}^{N} i K_i c_B c_A^i \left(v_B - v_{A_iB}\right) - \sum_{i,j=1}^{N} i(v_{A_jB} - v_{A_iB})K_i K_j c_B c_A^{i+j}}{\left(v_A - v_B\right) + \displaystyle\sum_{i=1}^{N} i K_i c_A^i \left(v_A - v_{A_iB}\right)} & \text{else}
\end{cases}
\tag{3.26}
$$

from which the stoichiometry A:B of the reaction boundary follows with

$$
S_{\text{A}\cdots\text{B}} = \begin{cases}
\dfrac{c_{A,\text{tot}}}{c_{B,\text{tot}} - c_B + \tilde{c}_B} & \text{for } c_B > \dddot{c}_B(c_A) \\[3ex]
\dfrac{c_{A,\text{tot}} - c_A + \tilde{c}_A}{c_{B,\text{tot}}} & \text{else}
\end{cases}
\tag{3.27}
$$

These formulas apply to cases AB, A_2B, ... , A_NB, i.e., where the larger macro-molecule has multiple sites for the smaller. The opposite case of multiple copies of the larger macromolecule binding to multiple sites on the smaller, AB, AB_2, ... , AB_N, can be derived from symmetry operations on Eqs. (3.24) to (3.27).

An example for the greater complexity of multi-site binding with regard to the sedimentation boundary pattern exhibited is shown in Fig. 3.6. It depicts the sedimentation velocity of the reaction boundary of a two-site model with 10-fold (macroscopic) positive cooperativity of formation of double-occupied complexes ($K_2 = 10K_1$). At the present time, the resulting features have not been system-atically explored. Interestingly, in this case the phase transition line is not mono-tonically increasing, but exhibits a minimum between K_2^{-1} and K_1^{-1}. The limiting value of the phase transition at very low concentrations of A is governed by K_1.

3.2.4   Three-Component and General Multi-Component Systems

It is possible to generalize effective particle theory to K-component interacting systems, including E complexes created by reversible self-association and/or hetero-associations. This is shown in detail in Appendix B. The assumptions are:

1. All reactions are governed by mass action law and equilibration is considered quasi-instantaneous on the time-scale of sedimentation.

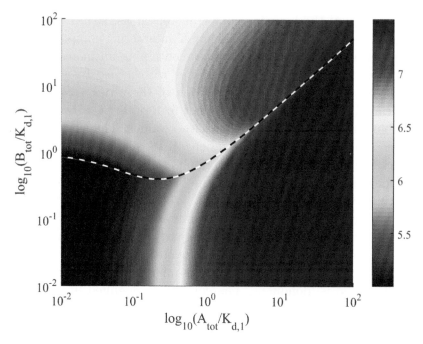

Figure 3.6 Properties of the effective particle A⋯B as a function of loading concentration $c_{A,tot}$ and $c_{B,tot}$, calculated for a two-site binding model $2A+B \leftrightarrow A+AB \leftrightarrow A_2B$ with macroscopic binding constants $K_1 = 1$ and $K_2 = 10$, with $s_A = 3.5$ S, $s_B = 5.0$ S, $s_{AB} = 6.5$ S, and s_{AAB} 7.5 S. Shown is a contour plot of the s-value of the reaction boundary. The line of transition points is plotted as a black-and-white dashed line.

2. All species sediment, and complexes sediment faster than their free constituent components.[22]

3. Diffusion is neglected. As in the two-component case, this will result in boundary s-values consistent with the transport method.

4. The geometry is simplified, again, by considering a constant force acting on molecules in rectangular, initially uniformly loaded solution column.

These assumptions are sufficient to predict the boundary patterns as a function of all components' loading concentrations. Although quantitative expressions for the boundary patterns can not generally be expressed in closed form, this is mainly due to the implicit relationship between species concentrations and total component concentrations imposed by mass action law and mass conservation. But the general principles can still be clearly recognized, and all quantities arise from straightforward algebraic relationships that can be computed efficiently for the entire concentration space.

---

[22]This will usually be true, for example, for macromolecules with similar partial-specific volume, but does not necessarily hold for pathological cases such as, for example, complexes with large ligand-induced unfolding, or with significant density differences among reactants.

The salient features of the boundary patterns appear as natural extensions of the two-component case. First, sedimentation of K interacting components in general generates a solution stratified into K zones containing an increasing number of components. They are separated by boundaries with a velocity that increases with the number of components. The composition of the zones, as well as velocities of the associated boundaries, are dependent in a non-trivial way on loading concentrations, binding constants, and all species s-values. The slowest boundary is an 'undisturbed' boundary of a single component. Its velocity represents that of a free single species, unless this component exhibits self-association which leads to a concentration-dependence. The zone with the fastest boundary is composed as the loading mixture, containing all components.

All components present in a given zone equilibrate to populate free and complex species conforming to mass action law. The particular component that under these conditions exhibits the highest population-average s-value (i.e., molecules of this component exhibit the highest time-average migration) sets the pace of the associated boundary and is 'dominant,' i.e., completely engulfed in this zone and absent from all lower, slower sedimenting zones. Other non-dominant components, however, will not be completely transported along. The fraction lagging behind participates in the mixture from the next adjacent slower-moving zone. The difference in concentration in the neighboring zones — the 'co-sedimenting' fraction — defines a subset of molecules of non-dominant components of the higher zone that has the same time-average s-value as the dominant component. Thus, the amplitude of the migrating boundary (which is the concentration differences between the adjacent zones) reflects a combination of molecules from all components that interconvert and combine transiently to states of different velocity, yet all jointly co-sediment at the same velocity as one system. We have discussed this process in detail for the two-component case above.

We have also already encountered phase transitions in boundary patterns of two-component mixtures. In multi-component mixtures they present far richer phenomenology. Not only the fastest zone, but any of the zones (except the last), can exhibit the special case that two components have the same time-average sedimentation that is larger than any other remaining components. Precisely at this condition of loading concentration the boundary pattern is degenerate and shows one fewer boundary. Viewed in concentration space, this is at the transition between conditions where either one of these components is dominant. At these points, the 'phase transitions,' the overall structure of the boundary pattern changes, since the composition and velocity of all the lower boundaries, including the undisturbed boundary, will exhibit a discontinuity. These phase transitions are therefore highly recognizable in isotherm experiments carried out in concentration series, as discontinuities between parts of isotherms where properties change smoothly with concentration. In multi-component systems the phase transitions describe hyper-surfaces in concentration space. The hyper-surfaces meet in points (on lower-dimensional surfaces) where three components have the same average sedimentation in the loading mixture. The conditions where *all* components have the same average sedimentation in the loading mixture describes a single line in concentration space.

These simple rules lead to complicated behavior. For example, Figs. 3.7–3.10 show the s-values of the fastest boundary and the phase transitions for a three-component system $2A+B+C \leftrightarrow AC+AB \leftrightarrow A+ABC$, where B does not interact with C but both bind to A, such as a bi-specific antibody interacting with its two independent ligands. Sedimentation profiles for this system were already considered in Section 1.3.5 but could not be further interpreted. With effective particle theory the origin of the different patterns becomes clear. As illustrated in Fig. 1.9 on p. 21, for such a system we will see up to three boundaries, including one or two undisturbed boundaries that are sedimenting at the free species (but are subject to discontinuous concentration-dependent jumps), and one or two reaction boundaries that show concentration-dependent composition and s-values.[23] The latter do not correspond to any distinct physical species. These also exhibit discontinuous jumps at the phase transitions. Sedimentation coefficients of the two fastest boundaries are shown in Figs. 3.8 and 3.9, respectively. Conditions where only two boundaries exist include the phase transition for the fastest reaction boundary (Figs. 3.7 and white lines in Fig. 3.8), as well as the concentrations for the phase transition for the intermediate reaction boundary (white lines in Fig. 3.9). The line where all phase transition surfaces meet describes concentrations where the entire system assumes a single boundary. This is highlighted in Fig. 3.7 using a color temperature scale reflecting the s-value of that boundary. Finally, Fig. 3.10 validates the predictions against full Lamm equation simulations followed by $c(s)$ analysis of the boundary pattern. This simulation highlights how rich the observed behavior of boundary patterns can be along a simple titration series, in Fig. 3.10B crossing several phase transitions: It can be discerned how with increasing concentration the fastest reaction boundary is decreasing in s-value, then increasing, and then decreasing again. Remarkably, the simple rules of effective particle theory can explain this behavior well.

---

[23] As a special case, in some concentration range where A sediments in the fastest boundary, both slower boundaries are the mutually non-interacting free species.

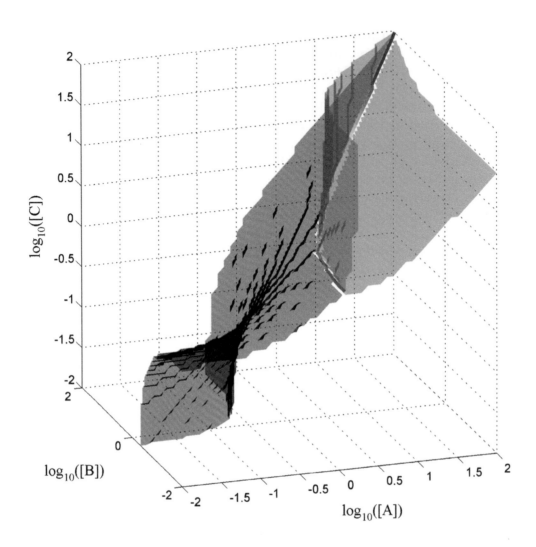

Figure 3.7 Phase transitions of the boundary pattern of a three-component system $2A+B+C \leftrightarrow AC+AB \leftrightarrow A+ABC$, where B does not interact with C but both bind to A. This system may be imagined as a bi-specific antibody interacting with its two ligands. All binding constants are equal, reflecting non-cooperative binding of equal strength, and concentrations are plotted relative to the K_D. Species s-values are $s_{Afree} = 7$ S, $s_{Bfree} = 5.5$ S, $s_{Cfree} = 8$ S, $s_{AB} = 10$ S, $s_{AC} = 11.5$ S, and $s_{ABC} = 15$ S. Loading concentrations $\{\chi_{A,load}, \chi_{B,load}, \chi_{C,load}\}$ that separate the conditions where either A or B are solely in the fastest boundary, dominating its s-value, are shown in basil; those separating A or C in this role are maroon; and those separating B and C are shown in olive. At these transition points, molecules of these two respective components have the same time-average s-value in the loading mixture, which leads to a degenerate boundary pattern where only one single-component boundary sediments behind. The surfaces described by these transitions meet in a single line that describes the concentrations at which all components have the same time-average s-value, and the entire system can sediment jointly in a single boundary. The s-value of this boundary is represented by color temperature, in the same color scale as Fig. 3.8

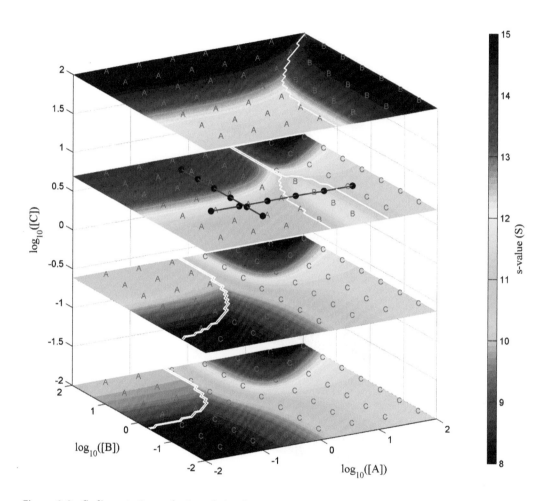

Figure 3.8 Sedimentation velocity of the fastest reaction boundary of the same three-component system 2A+B+C↔AC+AB↔A+ABC shown Fig. 3.7. Concentrations are plotted relative to K_D, which is equal for all interactions. The s-value is a function of all loading concentrations $\{\chi_{A,load}, \chi_{B,load}, \chi_{C,load}\}$, and only slices are represented as color contour plots for values of constant $\chi_{C,load}$ at 0.01, 0.24, 4.9, and 100 K_D. Letters indicate the component that is 'dominant,' i.e., solely present in the fastest boundary and governing its s-value. The white lines indicate the cross-section of the phase transitions shown in Fig. 3.7. The black circles and lines represent the trajectories for the Lamm equation simulations in Fig. 3.10.

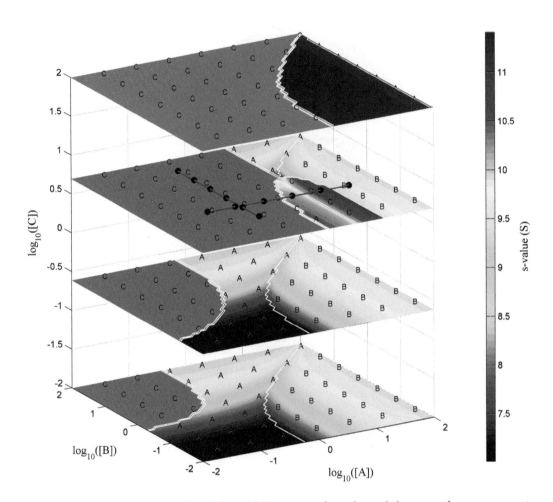

Figure 3.9 Sedimentation velocity of the middle reaction boundary of the same three-component system 2A+B+C↔AC+AB↔A+ABC shown Figs. 3.7 and 3.8. The color contour maps are presented at the same cross-sections of concentration space as in Fig. 3.8. The letters indicate the component that is dominant for the middle boundary, i.e., also contributing to the fastest reaction boundary but being excluded from the undisturbed boundary. The phase transition lines separating regions where different components are dominant in the middle boundary are shown in white. The black circles and lines represent the trajectories for the Lamm equation simulations in Fig. 3.10.

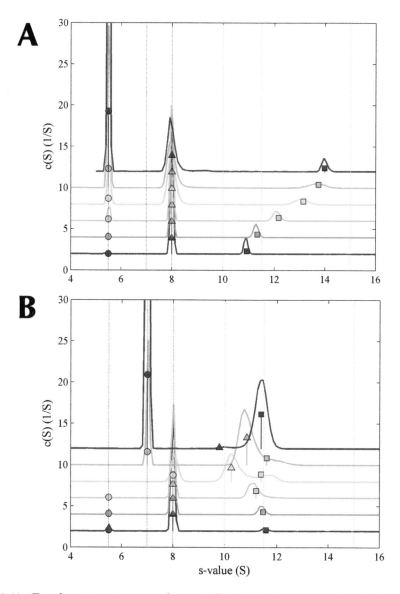

Figure 3.10 For the same system as shown in Figs. 3.7–3.9, a comparison between boundary s-values and amplitudes obtained from a full simulation of sedimentation with exact Lamm equations followed by $c(s)$ analysis (solid lines, arbitrarily offset for clarity) and the predictions from effective particle theory for the undisturbed boundary (circles), the middle reaction boundary (triangles), and the fastest reaction boundary (squares). Vertical gray dotted lines depict s-values of free and complex species. In more detail, Lamm equation solutions were calculated based on Eq. (1.2) for fast reactions, assuming molar masses of 150 kDa and extinction coefficients of 412,500 fringes $\times$ M^{-1}cm^{-1} for A and C and 100 kDa with 275,000 fringes $\times$ M^{-1}cm^{-1} for B sedimenting at 50,000 rpm, recorded in 300-sec intervals. Fig. 1.9 on p. 21 shows two representative examples for the sedimentation boundaries. *Panel A*: Total loading concentrations were along a titration series of constant 0.36 μM A and constant 5.0 μM C with increasing B (indicated by increasing color temperature). For reference, the K_D was 1 μM for both sites. *Panel B*: Titration series of constant 0.36 μM B and constant 5.0 μM C with increasing A. Both titration series are indicated with black circles in Figs. 3.8 and 3.9. Component concentrations in each boundary predicted by effective particle theory were multiplied with their extinction coefficient, and summed up to provide a prediction of the total signal in each boundary.

3.3 GILBERT THEORY AND GILBERT–JENKINS THEORY

In addition to predicting the boundary pattern, the Gilbert–Jenkins theory aims at describing the boundary shapes of the reaction boundaries. Historically it preceded the ability to numerically solve the Lamm equation for reacting systems, and the restriction to a linear geometry with constant force and a diffusion-free model was essential. The neglect of diffusion leads to boundary shapes equivalent to those in the presence of diffusion but in the limit of infinite time. Accordingly, the boundaries are termed 'asymptotic boundaries.' They are expressed in the form of differential sedimentation coefficient distributions dc/dv. The neglect of diffusion significantly limits the ability to directly compare the theoretically predicted boundary shapes with experimental boundary shapes. However, the neglect of diffusion is also an advantage in that it focuses the treatment to the essential characteristic features of the reaction boundaries.

A few approaches to experimentally observe the sedimentation properties of such diffusion-free boundaries have been proposed. A differential sedimentation method for self-associating systems was developed by Hersh and Schachman [120], where two solutions at slightly different composition are overlaid in a synthetic boundary cell. In the limit of vanishing concentration difference, the velocity of the differential boundary can be determined, and from a series of such experiments with different composition one can reconstruct the shape of the diffusion-free boundary [120,121]. Later, Winzor and coworkers [122,123] have proposed the extrapolation to infinite time of boundary slices (reminiscent of boundary divisions introduced by Baldwin [124], and in many aspects anticipating what was later introduced as the van Holde–Weischet method for polydisperse systems [125], see Part II, Section 6.1 [2]) in order to construct an asymptotic boundary. More recently, we have shown [109] how diffusion-deconvoluted $c(s)$ distributions with Bayesian regularization can be used to compare experimental data with theoretical asymptotic boundaries.

The work by Gilbert and Jenkins has been highly influential in many fields. It helped to achieve a general understanding of migrating systems of chemically reacting species in transport methods, including electrophoresis, chromatography, and sedimentation velocity. This was an important milestone in physical biochemistry. It was later generalized to the transport of systems with any physical interaction [126]. Restricting the discussion to reversible interacting systems, we will first review the theory for sedimentation of self-associating systems by Gilbert, and then discuss the more complex theory for sedimentation of two-component systems with hetero-associations by Gilbert and Jenkins.

3.3.1 Self-Associations

The asymptotic boundary shapes of self-associating systems rest on the fact that for self-associations, at any given point, the species populations are unambiguously determined from the total local concentration of a single component. As we will see, this leads to a close relationship between the asymptotic boundary shapes and the isotherm of weighted-average s-values $s_w(c_\text{tot})$. The following presentation is guided by the excellent review by Gilbert and Gilbert [108], simplified by the neglect of terms for hydrodynamic nonideality and the adoption of molar concentration units.

First, it is very useful to introduce the concept of a 'differential boundary.' Let us imagine a sedimentation boundary subdivided in many slices of infinitesimal height (Fig. 3.11). Due to the diffusion-free picture in rectangular geometry with constant force, and therefore the absence of radial dilution, we may introduce imaginary plateau regions to separate all steps without altering the sedimentation behavior of the steps. This clarifies the analysis of the fluxes into and out of the volume element occupied by the infinitesimal boundary step, and illustrates the relationship with the experimental technique of differential sedimentation [120].

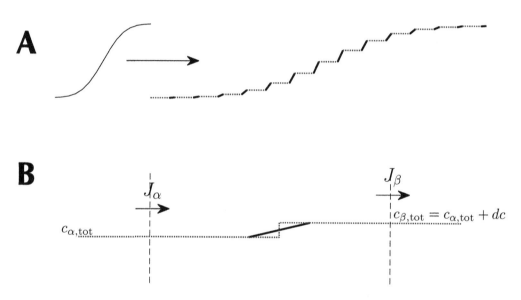

Figure 3.11 Concept of the differential boundary: *Panel A*: A boundary is divided in infinitesimal steps and, in a Gedankenexperiment, all steps are separated by artificial plateaus (red dotted lines). This will not change the sedimentation of the individual boundaries without diffusion in the model of linear geometry and constant force. (In the absence of diffusion, of course, the boundary shape would not be expected to resemble an error function, and the precise shape is to be established; however, the shape is irrelevant for the principle shown here.) *Panel B*: We may focus on a single step between a concentration $c_{\alpha,\text{tot}}$ and an infinitesimally larger concentration $c_{\beta,\text{tot}}$. The rate of movement of the boundary slice between $c_{\alpha,\text{tot}}$ and $c_{\beta,\text{tot}}$ will be determined by the fluxes J_α and J_β across imaginary planes in the plateau region, accounting for the material entering and leaving the boundary region, respectively. A real experiment with a finite step can be performed using a synthetic boundary cell, overlaying a solution column with a second solution that has a slightly lower concentration of macrosolute, as described by Hersh and Schachman [120].

In general, fluxes can be evaluated from velocities and concentrations with $J = vc$. As previously, in linear geometry we will not substitute $v = s\omega^2 r$, but instead describe the sedimentation velocity of a species i as the linear velocity v_i. For single-component self-associating systems, we are interested in the total flux of this component, rather than individual species, and for this the weighted-average velocity v_w is required. In the linear geometry, mass balance considerations for the weighted-average sedimentation velocity are analogous to those introduced above for the sector-shaped centrifugal geometry, leading to the weighted-average sedimentation coefficient for self-associating systems equivalent to Eq. (3.8)

$$v_w(c_{\text{tot}}) = \frac{1}{c_{\text{tot}}} \sum_i i K_{1i} c_1^i v_i \tag{3.28}$$

as a function of total molar protomer concentration,

$$c_{\text{tot}} = \sum_i i K_{1i} c_1^i \tag{3.29}$$

with c_1 representing the molar monomer concentration and K_{1i} the molar association constants ($K_{11} = 1$).[24] As depicted in Fig. 3.11B, we consider the material transported across planes in the plateaus on either side of a concentration step. With the total concentration $c_{\alpha,\text{tot}}$ at the left plane and the total concentration at the right plane $c_{\beta,\text{tot}}$, we can write the corresponding fluxes as $J_{\alpha,\text{tot}} = c_{\alpha,\text{tot}} v_w(c_{\alpha,\text{tot}})$ and $J_{\beta,\text{tot}} = c_{\beta,\text{tot}} v_w(c_{\beta,\text{tot}})$. The difference in the fluxes must represent the balance of material caused by the velocity v of the concentration step (indicated by the red dotted line in Fig. 3.11B), or

$$v \approx \frac{1}{c_{\beta,\text{tot}} - c_{\alpha,\text{tot}}} \left(c_{\beta,\text{tot}} v_w(c_{\beta,\text{tot}}) - c_{\alpha,\text{tot}} v_w(c_{\alpha,\text{tot}}) \right). \tag{3.30}$$

The approximation arises from the finite difference, but we can write in the limit of infinitesimal concentration steps

$$v = \frac{d(c_{\text{tot}} v_w(c_{\text{tot}}))}{dc_{\text{tot}}} = \frac{d}{dc_{\text{tot}}} \sum_i i K_{1i} c_1^i v_i. \tag{3.31}$$

For the differentiation with respect to c_{tot} we use the chain rule

$$d/dc_{\text{tot}} = (dc_{\text{tot}}/dc_1)^{-1} (d/dc_1) \tag{3.32}$$

to find

$$v = \frac{\displaystyle\sum_i i^2 K_{1i} c_1^i v_i}{\displaystyle\sum_i i^2 K_{1i} c_1^i}. \tag{3.33}$$

Though derived in a slightly different way, this is equivalent to the results in [71,123]. Going further, it is most convenient to arrive at a differential velocity distribution,

---

[24]The protomer concentration would be determined, for example, by dividing the mg/ml concentration by the monomer mass, or dividing the measured absorbance by the monomer molar extinction coefficient. The molar concentration units and binding constants used here are different from the weight-based units in [121, 123].

analogous to the familiar sedimentation coefficient distributions. To this end, we take the derivative of Eq. (3.33) with respect to the total concentration, using the relationship Eq. (3.32), which leads to

$$\frac{dv}{dc_{\text{tot}}} = \frac{\sum_i i^2 K_{1i} c_1^i \sum_i i^3 K_{1i} c_1^i v_i - \sum_i i^3 K_{1i} c_1^i \sum_i i^2 K_{1i} c_1^i v_i}{\left(\sum_i i^2 K_{1i} c_1^i\right)^3}. \qquad (3.34)$$

Finally, we will adopt a new notation by dropping the subscript 'tot' and adding a caret to indicate the diffusion-free approximation in rectangular geometry with constant force of Gilbert theory. The desired differential velocity distribution $d\hat{c}/dv$ now relates to Eq. (3.34) as $d\hat{c}/dv = (dv/dc_{\text{tot}})^{-1}$, and this leads us to the result

$$\frac{d\hat{c}}{dv} = \frac{\left(\sum_i i^2 K_{1i} c_1^i\right)^3}{\sum_i i^2 K_{1i} c_1^i \sum_i i^3 K_{1i} c_1^i v_i - \sum_i i^3 K_{1i} c_1^i \sum_i i^2 K_{1i} c_1^i v_i}. \qquad (3.35)$$

This general equation can be evaluated for any oligomeric system once we have determined the free monomer concentration as a function of total concentration, $c_1(c_{\text{tot}})$, for the range of concentrations from zero up to the loading concentration of the experiment.

Families of sedimentation boundaries at different concentrations for a monomer-dimer, a monomer-trimer, and a monomer-tetramer system are shown in Fig. 3.12. As is custom, the asymptotic boundaries are represented by colored bars with top edges following $d\hat{c}/dv$. The traces in this figure were normalized to loading concentration — without normalization the asymptotic boundaries at different concentrations would completely overlap and show different segments of the same curve. This is because the velocity of a boundary slice is dependent on the oligomeric populations, which is unambiguously determined by the local total concentration.

Further, as we have seen, the asymptotic boundary shapes of self-associating systems are entirely based on the local weighted-average s-value $s_w(c_{\text{tot}})$ — it is $d\hat{c}/dv = (d^2 v_w/dc_{\text{tot}}^2)^{-1}$. Therefore we should not expect to learn anything new about the self-association that could not be learned also, in principle, from the isotherm of weighted-average s-values $s_w(c_{\text{tot}})$. On the other hand, if we could extract the asymptotic velocity profile from experimental data, in principle, we could learn from a single experiment about the shape of the isotherm for all concentrations smaller than the loading concentration. Unfortunately, this is very difficult to accomplish due to the presence of diffusion and noise in real experimental data. For example, the diffusing boundaries and $c(s)$ distributions corresponding to systems in Fig. 3.12 can be found in Fig. 2.5 on p. 38; beyond their demonstration of consistency with Bayesian regularization [109, 127], no method has been established yet to exploit the correspondence for quantitative data analysis.

However, qualitative interpretations of $d\hat{c}/dv$ are possible. As discussed by

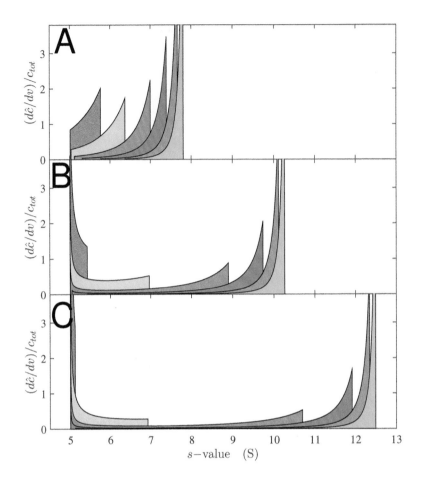

Figure 3.12 Shapes of asymptotic boundaries $d\hat{c}/dv$ as a function of loading concentration for different monomer-n-mer systems. In all cases, the monomer s-value is 5 S, the equilibrium association, constant is $K = 1$, and concentrations shown are 0.1 (gray), 0.3 (cyan), 1 (red), 3 (blue), 10 (magenta), and 30 (green). *Panel A*: Monomer-dimer system with dimer s-value of 8 S; *Panel B*: Monomer-trimer system with trimer s-value of 10.4 S; *Panel C*: Monomer-tetramer system with tetramer s-value of 12.6 S. For the same systems, simulated sedimentation boundaries and corresponding $c(s)$ traces are shown in Fig. 2.5.

Gilbert [71], the $d\hat{c}/dv$ profiles for monomer-n-mer systems with $n > 2$ exhibit two maxima: The first occurs at the monomer s-value, arising from dilution and dissociation to monomer of the oligomers in the trailing end of the boundary. The second is located at the leading edge of the velocity distribution at the highest concentration, which shows the largest population of oligomers. This is the origin of the peak structure obtained in $c(s)$ analyses of rapidly reversible self-associating systems at concentrations greater than K_D (Fig. 2.5, lower panels), which we had already observed in Section 2.2. The concentration at which the minimum s-value is achieved can be related approximately to the binding constant [116]. This corresponds to the concentration separating the two modes of the sedimentation boundaries in Fig. 2.5. We can also discern that the maxima and minima are more pronounced for higher oligomerization order n, as theoretically examined by Gilbert [71]. This is a reflection of the steeper binding isotherms in such systems.

In contrast, for a monomer-dimer system ($n = 2$) there is only a single maximum. This, along with other features of monomer-n-mer systems discussed by Gilbert [71], may be an aid in distinguishing a monomer-dimer system from higher oligomerization systems. However, it may not hold true completely in the presence of diffusion causing additional boundary spread and dilution.

In summary, Gilbert theory predicts the asymptotic boundary shapes of self-associating systems to be largely redundant to the information of a weighted-average isotherm $s_w(c_{tot})$, but more difficult to obtain experimentally. Importantly, it provides a theoretical basis for the broad, multi-modal sedimentation coefficient distributions that can be observed in $c(s)$ for self-associating systems. At present, the practical application seems limited to qualitative discrimination of the interaction scheme. This is much different in heterogeneous associations of two-component mixtures, where very information-rich boundary patterns occur. Therefore, applications of the Gilbert–Jenkins theory to the interpretation of hetero-associations are not restricted to the comparison of boundary shapes, but also involve the analysis of amplitudes and s-values of distinct boundaries, as will be described in the subsequent sections.

3.3.2 Hetero-Associations – Analytical Approach

We will limit the following only to the introduction of the main concepts, again assuming an instantaneous bimolecular reaction between component A and B reversibly forming a faster-sedimenting complex AB following mass action law. There are different ways to arrive at the asymptotic sedimentation coefficient distributions of interacting hetero-associating systems.[25] A formal approach presented by Gilbert and Jenkins [103] starts out at the linear Lamm equation for an infinitely long rectangular solution column with a constant force. As in the introduction of the effective particle theory, we start by substituting the radial coordinate r with the spatial coordinate x, and the species sedimentation coefficients s with the linear velocities v to indicate the rectangular geometry. This leads to the linear Lamm equation

$$\frac{\partial \chi_k}{\partial t} + v_k \frac{\partial \chi_k}{\partial x} - D_k \frac{\partial^2 \chi_k}{\partial x^2} = q_k , \tag{3.36}$$

for all species k with the reaction fluxes $q_A = q_B = -q_{AB} \equiv q$ as imposed by mass conservation. Rather than neglecting diffusion upfront, Gilbert and Jenkins introduce a transformation from space-time coordinates (x, t) to coordinates of velocity and inverse time (v, w) with $v = x/t$ and $w = 1/t$. We will label quantities expressed in these coordinates with a caret. The motivation is that this coordinate system will lend itself naturally to express the solutions in the space of velocity distributions, and to perform the limit $t \to \infty$ as $w \to 0$. Following the chain

---

[25]For more detailed reading, see [43, 103, 126, 128, 129].

rule, the differentials in the old and new coordinate system transform as $\partial/\partial t = -vw(\partial/\partial v) - w^2(\partial/\partial w)$ and $\partial/\partial x = w(\partial/\partial v)$, leading to [103]

$$(v - v_k)\frac{\partial \hat{\chi}_k}{\partial v} + w\left(\frac{\partial \hat{\chi}_k}{\partial w} + D_k\frac{\partial^2 \chi_k}{\partial v^2}\right) = \frac{q_k}{w}. \tag{3.37}$$

We note that the second term of the l.h.s. will vanish at $w \to 0$, i.e., diffusion disappears for $t \to \infty$. Thus, in this particular coordinate transform in this limit we are left with velocity coordinates and species concentrations only. Taking the limit $w \to 0$ is problematic for the r.h.s., but Gilbert and Jenkins argued that in the limit of infinite time, one would expect the absence of reaction fluxes between any species, i.e., $q_k \to 0$ with $w \to 0$, such that q_k/w remains finite. (This means that the reaction must achieve some form of steady-state.) As a result, using mass conservation we can equate the r.h.s. of (3.37) of all species and, in the limit $w \to 0$ obtain

$$(v - v_A)\frac{\partial \hat{\chi}_A}{\partial v} = (v - v_B)\frac{\partial \hat{\chi}_B}{\partial v} = -(v - v_{AB})\frac{\partial \hat{\chi}_{AB}}{\partial v}, \tag{3.38}$$

i.e., a relationship for the asymptotic differential velocity distributions $\partial\hat{\chi}_A/\partial v$, $\partial\hat{\chi}_B/\partial v$, and $\partial\hat{\chi}_{AB}/\partial v$. As shown by Gilbert and Jenkins [103], the definition of quantities $\mu = d\hat{\chi}_A/d\hat{\chi}_B = (d\hat{\chi}_A/dv)/(d\hat{\chi}_B/dv)$ and $\lambda = (2v_{AB} - v_A - v_B)/(v_A - v_B)$ leads to the ordinary differential equation

$$(1 - \lambda)b\mu^2 + \left[(1 - \lambda)\hat{\chi}_A + (1 + \lambda)\hat{\chi}_B + 2K_D\right]\mu + (1 + \lambda)\hat{\chi}_A = 0, \tag{3.39}$$

which can be solved analytically [103].[26]

In a second approach, leading to consistent results, Gilbert and Jenkins [43] also reported a part of the analytical solutions for the boundary shapes in the original space-time variables in the absence of diffusion, based directly on (3.12), which is equivalent to the asymptotic solution at infinite time in the presence of diffusion.

3.3.3   Hetero-Associations — Numerical Approach

A third approach to arrive at the same asymptotic boundaries was described by Gilbert and Gilbert [108] and is numerical in nature. This makes the Gilbert-Jenkins theory more generally applicable to systems for which no analytical solutions can be easily found, but importantly, also provides additional physical insight.

In the phenomenological considerations of reaction boundary polydispersity in Section 2.5 (p. 54), we have already conceptually drawn the conclusion that there must be a continuous distribution of sedimentation velocities supported by the local concentration mixture, exhibiting monotonically decreasing velocities for the differential boundary slices at lower concentrations toward the trailing edge of the reaction boundary (Fig. 2.13). To evaluate this quantitatively in more detail, let us

---

[26]However, the solution involves integration constants for which no explicit analytical expression can be given [103].

consider one slice of such a boundary, for example, following the propagation of a certain incremental boundary percentile as depicted in Fig. 3.13. Keeping in mind the absence of diffusion, the absence of radial dilution, and the constant force in the rectangular geometry, we recapitulate that a given boundary percentile will migrate synchronously with concentration fractions at a constant sedimentation velocity, let us say v'. The velocities of both components co-sedimenting in this boundary slice must each be equal to v'.

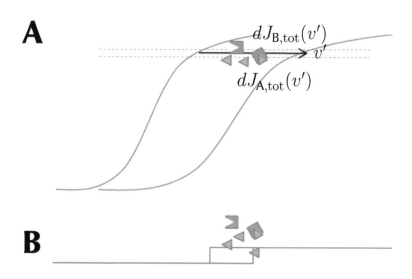

Figure 3.13 Schematics of the differential fluxes of A and B migrating with a velocity v', which deplete the boundary by a small amount. Due to the chemical reaction, the fluxes $dJ_{A,tot}(v')$ and $dJ_{B,tot}(v')$ are coupled, and this is the focus of the numerical approach of the Gilbert–Jenkins theory. In *Panel A* is shown a particular boundary slice from within an ordinary SV experiment, with boundaries shown at an earlier (blue) and later (green) time. *Panel B* depicts the situation of a differential boundary experiment, where in a synthetic boundary experiment a solution at slightly lower concentrations is initially layered on top of a solution with slightly higher concentration, creating a small concentration step whose evolution can be followed with time. As illustrated in Fig. 3.11, we can obtain the configuration of *Panel B* from that in *Panel A* in a Gedankenexperiment where we introduce hypothetical plateau regions between the boundary slices. Due to the absence of diffusion and the linear geometry, such plateaus would not alter the sedimentation properties of the individual slices.

In contrast to the effective particle theory introduced above, we are now not concerned with the time-average state of all of A or B, but ask much different questions: For this infinitesimal slice to be able to migrate ahead with a certain velocity v', how much of A and B will be needed? How are the amounts of co-sedimenting A and B coupled? And how much will be left for the trailing boundary?

As in Gilbert theory, we exploit the relationship between the velocity and the concentration of a species and the flux density. In this case, we distinguish the total flux of components A and B, $J_{A,tot}$ and $J_{B,tot}$. We note that the total flux for each component does include contributions to the transport arising from all species that contain that component, each weighted with the particular species velocity. Using this strategy of focusing on the differentials of the total flux, we will be able to determine how much of total A or B will migrate in the differential boundary slice. The goal is then to calculate how much of total A and total B is left behind to support the sedimentation of the next slower sedimenting slice.

The starting point of the calculation will be to establish the highest sedimentation velocity v' possible for the fastest boundary slice at given total loading composition. Knowing how much this slice depletes the total concentrations of A and B, we can repeat the same question and determine the highest sedimentation velocity v'' of the next slice, etc. In this manner we aim to iteratively establish the entire functions $\hat{\chi}_{A,tot}(v)$ and $\hat{\chi}_{B,tot}(v)$, i.e., the velocity distributions and boundary shapes.

This plan can be implemented in the following way:[27] In the infinitesimal slices $dJ_{A,tot}$ and $dJ_{B,tot}$ of the component fluxes, the material sedimenting with v' is $d\chi_{A,tot} = dJ_{A,tot}/v'$ and $d\chi_{B,tot} = dJ_{B,tot}/v'$, respectively. It is useful to write out the differentials

$$
\begin{aligned}
dJ_{A,tot} &= v_A d\chi_A + v_{AB} d\chi_{AB} = v_A d\chi_A + v_{AB} \left(K\chi_A d\chi_B + K\chi_B d\chi_A \right) \\
d\chi_{A,tot} &= d\chi_A + K\chi_A d\chi_B + K\chi_B d\chi_A ,
\end{aligned}
\tag{3.40}
$$

and analogous for B. Also, we can write the velocity as

$$
v' = \frac{dJ_{A,tot}}{d\chi_A} \frac{d\chi_A}{d\chi_{A,tot}} ,
\tag{3.41}
$$

and analogous for B. If we abbreviate the ratio of free species sedimenting in the boundary slice as

$$
\mu = d\chi_A / d\chi_B ,
\tag{3.42}
$$

we can rewrite Eq. (3.41) with the help of the differentials Eq. (3.40) and multiplication with Eq. (3.42) as

$$
v' = \frac{v_A \mu + v_{AB} K(\chi_A + \mu\chi_B)}{\mu + K(\chi_A + \mu\chi_B)} ,
\tag{3.43}
$$

and for B we arrive at [108]

$$
v' = \frac{v_B + v_{AB} K(\chi_A + \mu\chi_B)}{1 + K(\chi_A + \mu\chi_B)} .
\tag{3.44}
$$

Equations (3.43) and (3.44) are two equations with the two unknowns v' and μ. The system is quadratic in μ, but has only one physically meaningful root. It is straightforward to calculate *via* automated symbolic math, which can be embedded directly in computational algorithms, or, alternatively, be simplified for presentation

---

[27]This follows Gilbert and Gilbert [108] who carry out these calculations for a two-site system, reduced here for clarity to a single-site system.

to show the underlying structure of relationships. For the case predominant in sedimentation, $v_A < v_B < v_{AB}$ (i.e., the complex sediments faster than either free component), the solution is

$$v' =$$

$$\frac{(v_A + K\chi_B v_{AB})(\chi_A \Delta v_A - \chi_B \Delta v_B) + (\chi_B v_{AB} + \frac{v_A}{K})(\Delta v_{AB} - \sqrt{\Psi}) - 2K\chi_A \chi_B \Delta v_A v_{AB}}{(1 + K\chi_B)(\chi_A \Delta v_A - \chi_B \Delta v_B) + (\chi_B + \frac{1}{K})(\Delta v_{AB} - \sqrt{\Psi}) - 2K\chi_A \chi_B \Delta v_A}.$$

$$(3.45)$$

and

$$\frac{d\chi_A}{d\chi_B} = \mu = \frac{K(\chi_B \Delta v_B - \chi_A \Delta v_A) - \Delta v_{AB} + \sqrt{\Psi}}{2K\chi_B \Delta v_A}, \tag{3.46}$$

with the abbreviations $\Delta v_A = v_{AB} - v_A$, $\Delta v_B = v_{AB} - v_B$, $\Delta v_{AB} = v_B - v_A$, and

$$\Psi = K^2 (\chi_A \Delta v_A - \chi_B \Delta v_B)^2 + 2K v_{AB}(\chi_B \Delta v_B - \chi_A \Delta v_A) + \Delta v_{AB}^2. \tag{3.47}$$

For a given total concentration of **A** and **B**, we can calculate the free concentrations χ_A and χ_B using mass action law and then predict the velocity v_1' following Eq. (3.45). This is the highest velocity for the leading boundary slice that is supported by the solution composition. To find out how much of total **A** and total **B** is carried along in this leading boundary slice, we note that

$$\frac{d\chi_{A,tot}}{d\chi_{B,tot}} = \frac{d\chi_A + K\chi_A d\chi_B + K\chi_B d\chi_A}{d\chi_B + K\chi_A d\chi_B + K\chi_B d\chi_A} = \frac{\mu + K(\chi_A + \chi_B \mu)}{1 + K(\chi_A + \chi_B \mu)}. \tag{3.48}$$

As a consequence, a small boundary slice will carry along $\delta\chi_{B,tot}$ of **B** and $\delta\chi_{A,tot} = (d\chi_{A,tot}/d\chi_{B,tot})\delta\chi_{B,tot}$ of **A**, where the derivative is evaluated via Eq. (3.48) using μ from Eq. (3.46). This leaves behind a total concentration of $\chi_{B,tot,2} = \chi_{B,tot} - \delta\chi_{B,tot}$ and $\chi_{A,tot,2} = \chi_{A,tot} - \delta\chi_{A,tot}$. At these lower concentrations, another boundary slice with lower velocity v_2' will peel off the boundary, leading to a further decrease in the remaining total concentrations, and so on. We can write the iterative scheme for the asymptotic boundary as

$$\chi_{B,tot,i} = \chi_{B,tot,i-1} - \delta\chi_{B,tot}$$

$$\chi_{A,tot,i} = \chi_{A,tot,i-1} - \delta\chi_{B,tot}\frac{d\chi_{A,tot}}{d\chi_{B,tot}}(\chi_{A,i-1}, \chi_{B,i-1}) \tag{3.49}$$

$$v_i' = v'(\chi_{A,i-1}, \chi_{B,i-1}),$$

where $\chi_{A,i-1} = \chi_A(\chi_{A,tot,i-1}, \chi_{B,tot,i-1})$ and $\chi_{B,i-1} = \chi_B(\chi_{A,tot,i-1}, \chi_{B,tot,i-1})$. This scheme will be perpetuated until either $\chi_{A,tot,i} = 0$ or $\chi_{B,tot,i} = 0$. Whichever component is still populated at this point will constitute the undisturbed boundary, at the concentration remaining. With a suitable choice of $\delta\chi_{B,tot}$ (for example, 10^{-3} to 10^{-4}-fold the loading concentration of **B**), this algorithm allows us to iteratively construct the asymptotic differential velocity distributions $d\hat{\chi}_A/dv$, $d\hat{\chi}_B/dv$, and $d\hat{\chi}_{AB}/dv$, as

$$\frac{d\hat{\chi}_A}{dv} \approx \frac{\chi_{A,i} - \chi_{A,i-1}}{v'_i - v'_{i-1}}$$
$$\frac{d\hat{\chi}_B}{dv} \approx \frac{\chi_{B,i} - \chi_{B,i-1}}{v'_i - v'_{i-1}} \qquad (3.50)$$
$$\frac{d\hat{\chi}_{AB}}{dv} \approx \frac{\chi_{AB,i} - \chi_{AB,i-1}}{v'_i - v'_{i-1}},$$

and to determine the type and concentration of the undisturbed boundary, much as illustrated in the cartoon in Fig. 2.13 on p. 55. From the molar concentration scale of $d\hat{\chi}_A/dv$, $d\hat{\chi}_B/dv$, and $d\hat{\chi}_{AB}/dv$, we can predict the asymptotic boundary as it will be detected in any particular signal as

$$\frac{d\hat{\chi}}{dv} = \varepsilon_A d\frac{d\hat{\chi}_A}{dv} + \varepsilon_B d\frac{d\hat{\chi}_B}{dv} + \varepsilon_{AB} d\frac{d\hat{\chi}_{AB}}{dv}. \qquad (3.51)$$

Finally, after carrying out these calculations we can verify as a numerical control that integration of the asymptotic boundaries jointly with the undisturbed boundary does lead to the correct s_w as based on the transport method (Eq. 3.9 on p. 69 if $\Delta\varepsilon_{AB} = 0$).

3.3.4 Examples

What do these asymptotic boundaries look like? Fig. 3.14 shows two typical examples. Following a convention introduced by Gilbert and Jenkins, the 'spread' reaction boundaries are shown as filled or shaded bars, as opposed to the undisturbed boundaries, which in the differential distribution are infinitely sharp peaks (in the following represented as symbols with dropped lines). The asymptotic boundaries start at the highest supported velocity, and, dependent on the conditions, they may span almost the entire range from s_B to s_{AB}, or may be very narrow (see below). For two-component interacting systems, they are concave-shaped distributions. If the v-range includes regions very close to s_B the asymptotic boundaries will often have a maximum near s_B as well as a second maximum at the high-velocity end. In this case, the dominant component's distribution vanishes at the low-velocity end while the secondary component's distribution will remain finite and show a smaller gradient.

Obviously, since diffusion is not accounted for in Gilbert–Jenkins theory, asymptotic boundaries are of limited resemblance with experimentally observed boundaries, and even additional polydispersity in the s-values of reaction boundaries should be expected due to diffusionally-driven exacerbation of concentration gradients. Therefore, we may question to what extent the predicted asymptotic boundary shapes really reflect the sedimentation coefficient distributions produced from approximation-free Lamm equation solutions. At minimum, Gilbert–Jenkins

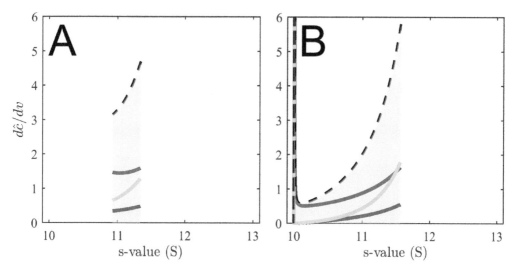

Figure 3.14 Asymptotic boundary profiles $d\hat{c}/dv$ calculated for a 7 S species A reacting with a 10 S species B to form a 13 S complex AB with an equilibrium constant $K = 1$. Shown are $d\hat{\chi}_A/dv$ (red), $d\hat{\chi}_B/dv$ (blue), $d\hat{\chi}_{AB}/dv$ (cyan), and the sum assuming equal signal increments of both species (dashed line). The range of s-values populated in the reaction boundary is shaded gray; the undisturbed boundary is not shown. *Panel A*: At concentrations of $\chi_{A,tot}$ and $\chi_{B,tot}$ equimolar at K_D, the system is of type B⋯(A) and the undisturbed boundary is composed of component A. These conditions correspond to the boundary profiles shown in 2.16 on p. 61. *Panel B*: At the same concentration of A at $\chi_{A,tot} = K_D$ but now with a sufficient excess of B, at $\chi_{B,tot} = 3K_D$, to cause the sedimenting system to be in the state A⋯(B) with dominant A such that the undisturbed boundary is composed of component B.

theory should provide a useful approximation for systems with small diffusion coefficients. This is examined in Fig. 3.15. It is apparent that the ls-$g^*(s)$ distributions derived from the evolution of the radial concentration profiles of interacting systems with low diffusion coefficients indeed approach the asymptotic boundary shapes, also including the correct species and amplitude of the undisturbed boundary [109]. Further, the overall quantitative aspects of the reaction boundaries are very consistent with those from the effective particle model. We will establish different facets of this in the following, using the effective particle model as a guide through the phenomenology of the asymptotic boundaries in parameter space.

To better understand the asymptotic boundaries, it is interesting to compare their shapes as they evolve in concentration series. Fig. 3.16A shows series of asymptotic boundaries for titrations at equimolar concentrations from 0.1 K_D to $10K_D$. Here, as we have already seen from the effective particle picture, the undisturbed boundary is always provided by the slower component, and its relative contribution decreases with increasing total concentrations. As can be expected, the reaction boundary increases in velocity, but we also observe an increase in width. In the titration series with constant $\chi_{B,tot}$ with increasing $\chi_{A,tot}$ over the range from 0.1 K_D to $10K_D$ (*Panel B*) a similar picture emerges, but with the reaction boundary covering a wider range of velocities in the course of the titration, and with

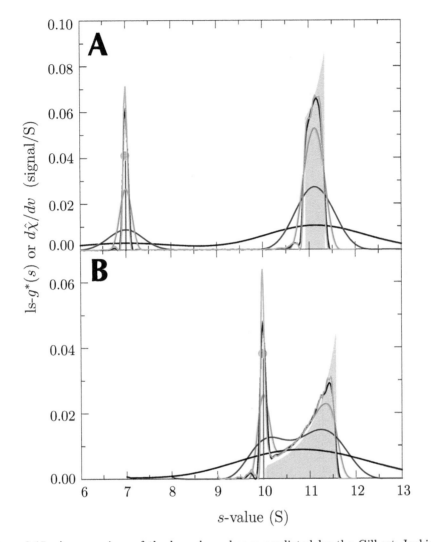

Figure 3.15 A comparison of the boundary shapes predicted by the Gilbert–Jenkins theory for rectangular cells at infinite time with solutions of the exact Lamm equation (Eq. 1.31) in the limit of instantaneous chemical equilibrium and very small diffusion coefficient. Lamm equations were solved for a 7 S species A reacting with a 10 S species B to form a 13 S AB complex. All diffusion coefficients were reduced from the physically realistic values (black) by a factor 10 (magenta), 100 (green), 1,000 (blue), and 10,000 (cyan). In each case, sedimentation profiles were calculated with Eq. (1.31) and transformed into sedimentation coefficient distributions using ls-$g^*(s)$ [130] as a representation of the boundary shapes (Part II, Section 4.2 [2]). For comparison, the profiles $d\hat{\chi}/dv$ from Gilbert–Jenkins theory are shown as gray bars and circles for the reaction and undisturbed boundary, respectively. The profiles were normalized to the same area, and calculated for the same species extinction coefficients as in the Lamm equation solution. Corresponding to the conditions in Fig. 3.14, *Panel A* shows the comparison with equimolar loading concentration of A and B equal to K_D, whereas *Panel B* is calculated with a 3-fold molar excess of B over A (such that B constitutes the undisturbed boundary). Reproduced from [109].

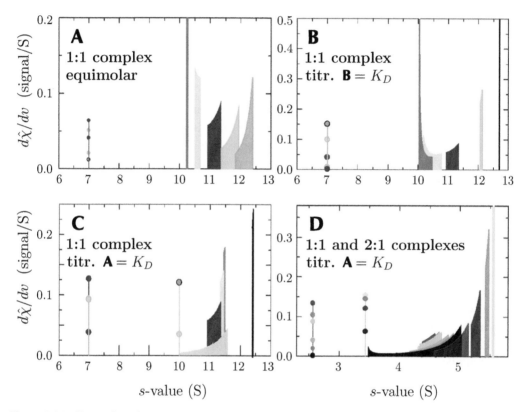

Figure 3.16 Examples of asymptotic boundary profiles $d\hat{\chi}/dv$ at different loading concentrations. *Panels A–C*: Calculated asymptotic boundary profiles for the same A+B↔AB system as shown in Fig. 3.15. $d\hat{\chi}/dv$ traces are normalized relative to the total concentration. Three titration series are shown: equimolar $\chi_{A,tot}$ and $\chi_{B,tot}$ (*Panel A*), constant $\chi_{B,tot} = K_D$ (*Panel B*), and constant $\chi_{A,tot} = K_D$ (*Panel C*). In all plots these, the variable concentration is $0.1 \times K_D$ (blue), $0.3 \times K_D$ (cyan), $1.0 \times K_D$ (red), $3 \times K_D$ (green), and $10 \times K_D$ (gray). The undisturbed boundary is indicated by the circle with dropped line. *Panel D*: Asymptotic boundary profiles for a reaction A+B+B↔AB+B↔BAB where a molecule A (2.66 S) has two identical non-cooperative sites for a larger ligand molecule B (3.56 S), resulting in 4.96 S and 6.11 S complexes, with a microscopic equilibrium dissociation constant of 1.7 μM. Concentrations were constant 4.96 μM for A, and for B 1.24 μM (magenta), 2.05 μM (light gray), 2.44 μM (light green), 4.20 μM (light blue), 6.04 μM (blue), 9.14 μM (black), 12.4 μM (violet), 17.7 μM (red), 23.3 μM (orange), and 28.7 μM (yellow). This system will be examined in more detail in Figs. 6.8 and 6.9 of Chapter 6. Reproduced from [109].

the velocity spread decreasing at higher concentrations. In contrast, a dramatically different behavior can be discerned in the titration series of constant $\chi_{A,tot}$ with increasing $\chi_{B,tot}$ (*Panel C*). Here, the undisturbed boundary is made of A at the lower concentrations but of B at the higher concentrations. While A is still the undisturbed boundary, increasing $\chi_{B,tot}$ leads to a slight reduction of the velocity and strong broadening of the reaction boundary. Once the switch has occurred, the asymptotic boundary quickly increases in velocity and sharpness. Similar patterns are observed for other two-component reactions (*Panel D*).

It is useful to put these observations into relation with the phase diagram of effective particle theory shown in Fig. 2.12 on p. 54. For the system simulated here, the red line in the phase diagram applies. For the first equimolar series in Fig. 3.16A, we are entirely within the B⋯(A) phase, but along the diagonal getting closer to the transition line at higher concentration. For the second titration in Fig. 3.16B, we start out close to the transition line but move further away at higher concentrations of A. Finally, in the third titration in Fig. 3.16C with constant A, we start in the phase B⋯(A) and with higher B move up through the transition point into the phase A⋯(B). As a common feature, we can discern that with greater distance from the transition point the polydispersity of the reaction boundary decreases.

In order to examine this in more detail, let us look at Fig. 3.17, which displays the shape of the reaction boundaries in a condensed presentation for different conditions. The abscissa is still showing the sedimentation coefficient range, but we now use the ordinate to plot the concentration along a certain trajectory of a titration or dilution, respectively, and apply a color temperature to indicate the relative heights of the asymptotic boundaries (much like the plot used in Fig. 2.3 on p. 31). This allows us to inspect more comprehensively the asymptotic boundaries as a function of loading concentration, crossing the parameter space $\{\chi_{A,tot}, \chi_{B,tot}\}$ along different lines. Without much loss of generality for the conclusions drawn, we select $s_A = 3.5$ S, $s_B = 5.0$ S, and $s_{AB} = 6.5$ S. As we already observed from the effective particle model, the key landmark to understand the character of the reaction boundary is the transition point where it branches between the region where A is in the undisturbed boundary (regions with white background) and where B plays this role (areas shaded gray).[28] Not shown here is the amplitude of the undisturbed boundary, which continuously decreases toward the transition point and completely vanishes at the singularity. Importantly, we find again that this transition point is not simply where the molar ratio of loading concentration equals the complex stoichiometry.

Let us move along the titration series and observe the change in the asymptotic boundaries, for example, the titrations of constant A shown in the *Middle Column*. Focussing on the top plot, at small values of B near the bottom of the plot (indicated by arrow #1), A is in excess and also provides the undisturbed component; thus, the asymptotic boundary is located in the plot area with white background. Far from

---

[28] For the same system, the concentration space in effective particle theory is shown in red in Fig. 2.12 on p. 54.

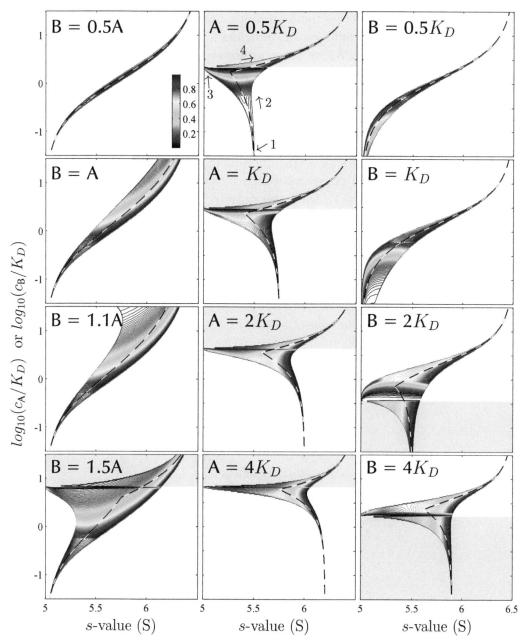

Figure 3.17 Asymptotic boundaries for the A+B↔AB system calculated with $s_A = 3.5$ S, $s_B = 5.0$ S, and $s_{AB} = 6.5$ S (the same model system as in Fig. 2.3A) as a function of loading concentration. The color temperature of the contour lines represents the shape of the reaction boundary, normalized at each concentration. *Left Column*: Dilution series with $\chi_{A,tot}$ plotted in the ordinate, and the ratio $\chi_{B,tot}/\chi_{A,tot}$ as indicated in each plot. *Middle Column*: Titration series of constant $\chi_{A,tot}$ as indicated; *Right Column*: Titration series of constant $\chi_{B,tot}$ as indicated. Concentration conditions where A is dominant and the faster-sedimenting component B forms the undisturbed boundary is shaded in gray. For comparison with the effective particle theory, the black/white dashed lines are $s_{A\cdots B}$-values predicted from Eq. (2.20).

the transition point, the boundary is very narrow at first. This is due to the fact that the relative change of free A across the asymptotic boundary is small, and most of it is part of the undisturbed boundary that provides a 'bath' keeping the fractional saturation of B fairly uniform. As we move up toward higher concentrations of B (arrow #2), we observe the now familiar decrease in velocity, but the velocity range becomes increasingly broad and stretches toward s_B as we approach the transition point. It is clear that closer to the transition point the relative changes in the free concentrations of A will be very large, causing the highest dispersion in the velocities. Close to the transition point, $d\hat{\chi}/dv$ develops a strong peak at the low-velocity end (arrow #3), which at the transition point develops in the undisturbed boundary. Once this has taken place and we move to still higher concentrations of B into the new branch (arrow #4), the latter quickly establishes a new main peak at the high-velocity end, increasingly narrower at higher concentrations above the transition. All concentration series exhibit this behaviour in a similar way. However, the region showing the broadened transition character is extended in the titrations where we increase A and B proportionally (*Left Column*), which can be understood again from a look at the phase diagram in Fig. 2.12 (p. 54): as mentioned above, the dilution series exhibit diagonal trajectories that remain closer to the phase transition line than the titration series.

Fig. 3.18 extracts the information on the relative width of the reaction boundaries, across the entire parameter space of loading concentrations at once. It may

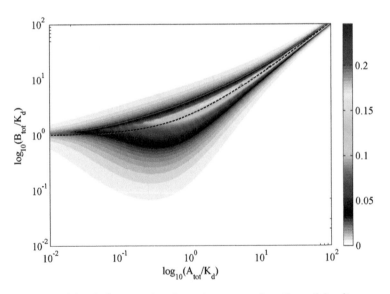

Figure 3.18 Relative width of the reaction boundary as a function of loading concentrations of A (abscissa) and B (ordinate). Plotted is the maximum boundary spread, as measured by $(v_{max} - v_{min})/\bar{v}$ with $\bar{v} = \int v(d\hat{\chi}/dv)dv / \int (d\hat{\chi}/dv)dv$, in the form of a color temperature plot. The data are based on the same system as shown in Fig. 3.17. The region with highest boundary spread is compressed into a very narrow band surrounding the phase transition line (dotted black line).

be discerned that the boundary spread overall is actually quite small, and that relative widths above 10% (cyan and warmer colors) occur only in a narrow region surrounding the phase transition, in a band that narrows at higher loading concentrations. This emphasizes the role played by the undisturbed boundary to suppress polydispersity in the reaction boundary by reducing the relative gradients in the non-dominant component across the reaction boundary.

To put the predicted velocity distributions in relation with experimental data, it is useful to define an average velocity s_{asy} as the signal-weighted integral over the asymptotic boundary [109]

$$
s_{asy} = \frac{\int_{s_B}^{s_{AB}} \left(\varepsilon_A \frac{\hat{\chi}_A}{dv} + \varepsilon_B \frac{d\hat{\chi}_B}{dv} + \varepsilon_{AB} \frac{d\hat{\chi}_{AB}}{dv} \right) v dv}{\int_{s_B}^{s_{AB}} \left(\varepsilon_A \frac{d\hat{\chi}_A}{dv} + \varepsilon_B \frac{d\hat{\chi}_B}{dv} + \varepsilon_{AB} \frac{d\hat{\chi}_{AB}}{dv} \right) dv}, \tag{3.52}
$$

in correspondence to $s_{A\cdots B}$ from effective particle theory (Eq. 2.20). It may be fitted to experimentally derived values s_{fast} determined by integration of $c(s)$ over the fast boundary component [109].

To contrast the isotherm of s_{asy} as a function of loading concentrations of A and B with the isotherm of the overall weighted-average s_w, Fig. 3.19 shows s_{fast} for the same system that was shown in Fig. 3.1 on p. 70. As we had already concluded from the effective particle model (e.g., Fig. 2.10 on p. 51), the reaction boundaries exhibit sedimentation velocities close to that of the complex whenever one component is in excess. This is different from the overall weighted-average s_w, which requires close to saturating concentrations in stoichiometric ratios for the s_w-value to approach that of the complex. Therefore, we can anticipate that the experimental determination and quantitative modeling of s_{fast}-isotherms will add significant information to the analysis of experimental systems (see below).

Numerical calculation of asymptotic boundaries was implemented in SEDPHAT in earlier versions, for example, for modeling of s_{fast} with s_{asy} [109]. This has been superseded by corresponding implementations of effective particle theory, which offers significant computational advantages for quantities that can be experimentally determined.

3.3.5 Relationship between Gilbert–Jenkins Theory and Effective Particle Theory

Gilbert–Jenkins theory is derived on the basis of differential boundary slices in the concentration gradient of the reaction boundary, leading to an asymptotic velocity distribution. In contrast, the effective particle model ignores the concentration gradients in the reaction boundary. The latter is not concerned with the reaction boundary shape but rather its overall sedimentation velocity, $s_{A\cdots B}$, as determined from the mass balance of material sedimenting within the reaction boundary. The foundation on the overall mass balance is shared by the weighted-average sedimentation coefficients s_w, although the latter accounts for all sedimenting material,

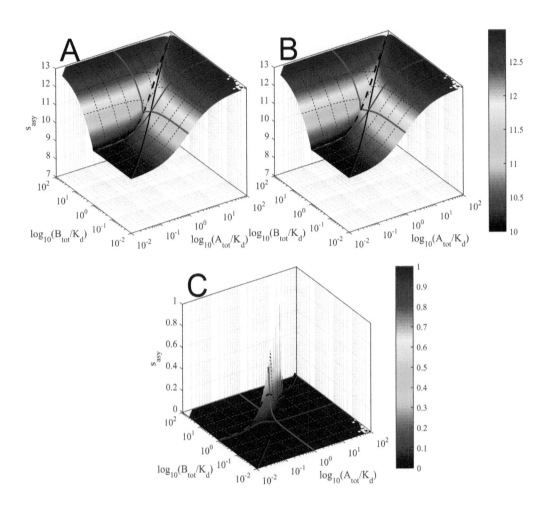

Figure 3.19 Signal-average s-values of the asymptotic boundaries, s_{asy}, for the same A+B↔AB system for which the overall signal-weighted s_w was presented in Fig. 3.1 on p. 70. Like in Fig. 3.1, to probe the dependence on ε_A and ε_B, *Panel A* is based on $\varepsilon_A = 0.1\varepsilon_B$, whereas *Panel B* is for $\varepsilon_A = 10\varepsilon_B$. *Panel C* shows the difference between the two. The red and pink lines indicate titration series of constant B with varying A and constant A with varying B, respectively, the black solid line highlights the isotherm for equimolar experiments, and the black dashed line indicates the s_{asy}-values at the phase transition line of the effective particle, $\ddot{c}_{\text{B,tot}}(c_{\text{A,tot}})$ according to Eq. (2.14).

whereas $s_{A\cdots B}$ is predicated on a division between reaction and undisturbed boundary. Because the effective particle model is simpler than Gilbert–Jenkins theory, it can yield straightforward and physically interpretable analytical solutions. Further, it can be extended to multi-site and multi-component mixtures for fast reactions. An important question is: How quantitatively consistent are these two theories?

A basic physical insight arising from the effective particle theory is that the time-average velocity of all components (or fractions thereof) co-sedimenting in the reaction boundary matches (Fig. 2.9). This imposes a molecular mechanism of transport of the reacting system that renders the velocity $s_{A\cdots B}$ independent of the signal increments. By contrast, the overall signal-weighted average s_w is strongly dependent on signal coefficients, as highlighted in Fig. 3.1. This is similar in Gilbert–Jenkins theory, as it predicts different velocity profiles for each species (Fig. 3.14), and therefore a shape of the asymptotic boundary signal that may depend on extinction coefficients (Eq. (3.51)). Actually, however, the signal-average s-value of the entire asymptotic boundary s_{asy} is virtually independent of signal coefficients, as illustrated in Fig. 3.19A and B, consistent with the predictions from effective particle theory. This validates the conclusions from effective particle theory and Gilbert–Jenkins theory about the *overall* propagation of the reaction boundary. It is satisfactory that, once we disregard the slight polydispersity of migration velocities, the migration mechanism governs detected s-values, independent of detected signals.

Where this consistency between the two theories becomes quantitatively poor is also an interesting point: As shown in Fig. 3.19C, deviations are found only very close to the phase transition line in the range of K_D, which is where the polydispersity of the asymptotic boundary causes large concentration differences in both components leading to large changes in fractional saturation of binding sites across the boundary, as indicated by the broad asymptotic boundaries in Figs. 3.17 and 3.18. This situation is the one most conflicting with the idea of step-function boundaries in the effective particle theory. Similarly, this is also the situation in which diffusional broadening will have the largest impact on the accuracy of the predictions of Gilbert–Jenkins theory.

A quantitative comparison of the predicted sedimentation boundary properties can be made with respect to several common aspects: (1) we can compare the values $s_{A\cdots B}$ from Eq. (2.20) with s_{asy} predicted by Eq. (3.52); (2) we compare the predicted total stoichiometry $S_{A\cdots B}$ from Eq. (2.21) with the integral over the asymptotic boundaries

$$S_{asy} = \frac{\int \left(\dfrac{d\hat{\chi}_A}{dv} + \dfrac{d\hat{\chi}_{AB}}{dv} \right) dv}{\int \left(\dfrac{d\hat{\chi}_B}{dv} + \dfrac{d\hat{\chi}_{AB}}{dv} \right) dv} ; \tag{3.53}$$

and (3) we can compare the amplitude of the undisturbed boundary, expressed as a fraction of the secondary component that does not participate in the reaction boundary, based on the predictions of Eq. (2.22) in effective particle theory with

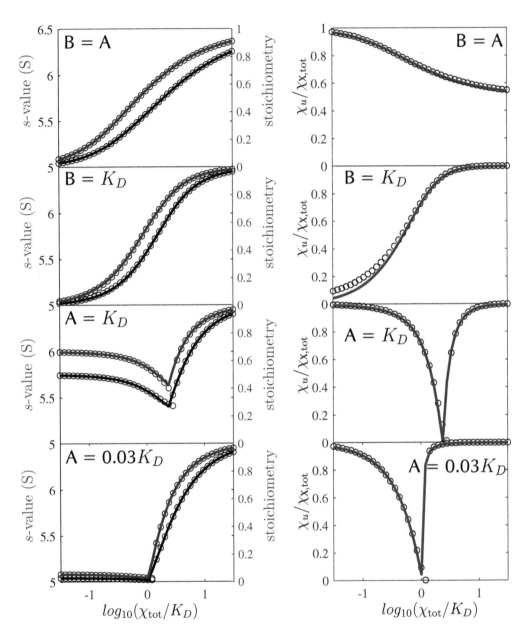

Figure 3.20 Comparison between the predictions from Gilbert–Jenkins theory (circles) and the effective particle model (lines) regarding parameters describing the overall boundary structure. Loading concentrations were chosen along lines in parameter space that correspond to equimolar dilution series (*Top Row*), or titration series at constant B (*Second Row*) or constant A (*Third and Lowest Row*). Calculations are based on a system with $s_A = 3.5$ S, $s_B = 5.0$ S, and $s_{AB} = 6.5$ S, calculating with equal signal increments for both components in Gilbert–Jenkins theory. Corresponding asymptotic boundary profiles can be found in Fig. 3.17. *Left Column*: The average s-value of the reaction boundary, s_{asy} predicted by Gilbert–Jenkins theory following Eq. (3.52) (blue circles) and $s_{A\cdots B}$ from effective particle theory Eq. (2.20) (red line), superimposed with the corresponding predictions for stoichiometry (S_{asy} from Eq. (3.53) as circles and $S_{A\cdots B}$ from Eq. (2.21) as magenta line) using the right ordinate. The root-mean-square deviations of the s-values and stoichiometry values are 0.013 S and 1.7%, respectively. *Right Column*: Fractional amplitude of the undisturbed boundary as predicted from Gilbert–Jenkins theory (circles) and the effective particle model based on Eq. (2.22) (red lines). The root-mean-square deviation is 7.6%. Reproduced from [41].

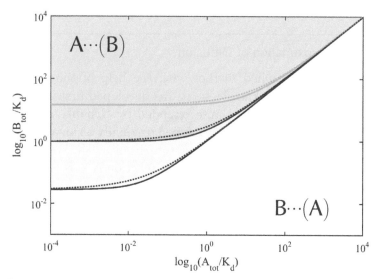

Figure 3.21 Phase transition in the concentration space for the system $A + B \leftrightarrow AB$ for different size ratios of A and B, comparing the predictions from Gilbert–Jenkins theory (solid lines) with those of the effective particle theory following Eq. (2.16) (dotted lines). Sedimentation parameters underlying the predictions are as in Fig. 2.12 on p. 54.

the equivalent quantity from Gilbert–Jenkins theory. Fig. 3.20 shows these comparisons for concentration series along a number of different representative trajectories, presenting the data from Gilbert–Jenkins theory as circles and effective particle theory as lines [41].[29] For orientation, corresponding contour plots from the asymptotic boundary profiles are shown in Fig. 3.17. Overall, the correspondence is remarkably good, with the r.m.s. differences close to the typical experimental precision. Finally, we can compare the phase transition line observed in Gilbert–Jenkins theory with the prediction of Eq. (2.16), which is shown in Fig. 3.21.

In summary, the additional information provided by the Gilbert–Jenkins theory over the effective particle model is essentially that of the dispersion of the boundary. This is at the cost of considerably higher numerical complications, and the departure from physically insightful predictions across the concentration space. For simple bimolecular reactions, the dispersion will often be very difficult to detect experimentally, in particular far from the phase transition point, where the dispersion will be small and masked in real experiments by the presence of diffusion. More extended reactions, however, such as two-site models, may create more significant polydispersity of the sedimentation velocity. This has not been systematically studied yet.

---

[29]Corresponding data from additional trajectories can be found in the supplemental material to [41].

3.4 CONSTANT BATH THEORY AND EXTENDED EFFECTIVE PARTICLE THEORY

3.4.1 Constant Bath Approximation of Diffusion

The role of diffusion can be studied further with the help of the constant bath approximation. To the best of our knowledge it was developed first in the Riessner laboratory [64] and later re-established and extended by Urbanke and colleagues [28,65,66]. The constant bath theory was developed originally as an approach to describe the binding of small ligands to larger macromolecules [64]. We will illustrate the principle with a bimolecular hetero-association reaction A + B ↔ AB in instantaneous equilibrium, as described by the Lamm equation for bimolecular reactions (Eq. 1.31 on p. 16) with large k_{off}.

The basic idea is that if A is small, it sediments more slowly and diffuses more rapidly than B or the complex AB. In a Gedankenexperiment, we ask "What would happen if this would cause the concentration gradients of free A to vanish"? Although strictly impossible, let us examine this question as a limiting case. Obviously, the properties of the sedimentation boundary would be determined by the evolution of B and the complex AB. In Section 1.3.4 we have worked out the exact Lamm equations in constituent concentrations for a rapidly reversible A + B ↔ AB system, using the notation $\bar{\chi}_B(r,t)$ for the total constituent concentration $\chi_B(r,t) + \chi_{AB}(r,t)$. After inserting the approximation of a constant concentration $\chi_{A,0}$ into the second line of Eq. (1.33) we have

$$\frac{\partial \bar{\chi}_B}{\partial t} + \frac{1}{r}\frac{\partial}{\partial r}\left(\bar{s}_B^* \omega^2 r^2 \bar{\chi}_B - \bar{D}_B^* \frac{\partial \bar{\chi}_B}{\partial r} r\right) = 0 , \qquad (3.54)$$

with the local coefficients from Eq. (1.34) likewise in the limit $\partial \chi_A/\partial r = 0$

$$\begin{aligned}
\bar{s}_B^*(\chi_{A,0}) &= \frac{s_B + \chi_{A,0} K s_{AB}}{1 + \chi_{A,0} K} \\
\bar{D}_B^*(\chi_{A,0}) &= \frac{D_B + \chi_{A,0} K D_{AB}}{1 + \chi_{A,0} K} .
\end{aligned} \qquad (3.55)$$

We recognize this to be a single-species Lamm equation. In contrast to the exact description in Eq. (1.33), the sedimentation and diffusion coefficients are constant, dependent only on the loading concentration $\chi_{A,0}$, which describes the constant value in the constant bath picture. This result has been shown to hold also for reactions with multiple binding sites for A on B [65,66]. Both the sedimentation and the diffusion coefficient of the reaction boundaries are population-averages between B and the complex AB, dependent on the fractional saturation of B.

Originally, the isotherm of the amplitudes and s-value of the reaction boundary was the primary purpose for the development of the constant bath theory. Interestingly, we find that $\bar{s}_B^*$ is identical with the prediction for the phase B⋯(A) of the effective particle (Eq. 2.4). The isotherm $\bar{s}_B^*(\chi_{A,0})$ of Eq. (3.55) is also identical to what we would expect for a slowly equilibrating system, if we were to integrate

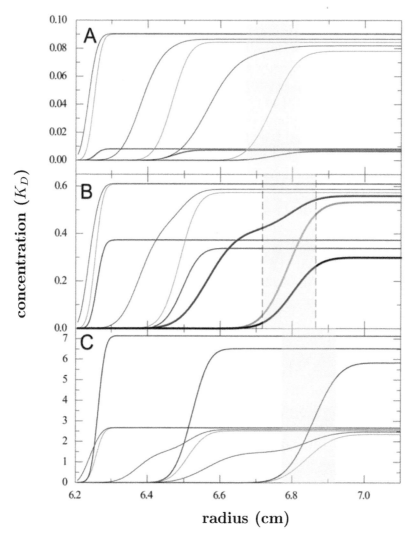

Figure 3.22 Concentration profiles of free A (red), free B (blue), and complex (black) species during sedimentation of the interacting system A + B ↔ AB in the limit of instantaneous reaction, for the conditions at equimolar concentrations shown in Fig. 2.16. Total concentrations of component A and B were equimolar at 0.1-fold K_D (*Panel A*), K_D (*Panel B*, as in Fig. 2.16), and 10-fold K_D (*Panel C*). For clarity, only the concentration profiles from time-points 300, 1500, and 3000 sec are superimposed. The vertical dashed lines in *Panel B* indicate the radial range that covers 10 – 90% of the boundary of free B, and at the same time 76 – 97% of free A at 3,000 sec (indicated by bold lines). Similar ranges are shaded red at all concentrations. At concentrations of 0.1-fold K_D, the relative gradients are smaller than at K_D, whereas they are of similar magnitude at 10-fold K_D. Reproduced from [28].

$c(s)$ over the peaks of free B and the complex AB. Besides fast systems in the constant bath approximation and slow systems, $\bar{s}_B^*(\chi_{A,0})$ applies equally to intermediate cases, as illustrated in Fig. 2.14 (p. 58).

An interesting qualitative result of Eqs. (3.54) and (3.55) is the prediction that in the constant bath picture the reaction boundary sediments with a single s-value and single D-value. The quality of this approximation was already illustrated above in Fig. 2.16 (p. 61), showing a single-species Lamm equation solution fit to a reaction boundary in combination with a second species describing the slower sedimenting undisturbed boundary of free A. It is noteworthy that the conditions of Fig. 2.16 were actually far from the assumptions of the constant bath approximation, with A being a 100 kDa, 7 S-particle, and B a 200 kDa, 10 S-particle. Although the value of $\bar{D}_B^*$ was not fixed to the predicted value of Eq. (3.55) and the fit is not perfect, it is of remarkable quality with a maximum deviation being on the order of 1% of the total signal.

Why does the constant bath theory work so well? After all, we know that the assumption about the absence of a gradient in A cannot be fulfilled in a real system due to the co-sedimentation of A in the reaction boundary. In order to examine this more closely, let us look at the real gradients of A for the example of Fig. 2.16. This is represented in Fig. 3.22B, where the concentration profiles of A, B, and AB are shown as red, blue, and black lines, respectively. Within the radial range that comprises from 10% to 90% of χ_B, shaded in red, the change in χ_A across the reaction boundary ranges from 76% to 97%. Following mass action law with $\chi_{AB}/\chi_B = K\chi_A$, this causes the fractional saturation of B to cover a range only from 30% to 35%. Clearly, the fact that there is a large constant background of free A from the plateau of the undisturbed boundary, on top of which there will be only some small relative change of A in the reaction boundary from the added concentration of co-sedimenting A, diminishes greatly the impact of $\partial\chi_A/\partial r$ on the reaction boundary. Ultimately, this is due to the nonlinear concentration dependence of the saturation following by mass action law [28].

3.4.2 Concentration Gradients and Average Diffusion Based on Effective Particle Theory

Since the effective particle theory predicts the relative concentration changes across the reaction boundary, it provides a tool to examine the concentration gradients further [42]. Despite the fact that the Lamm equation solution of the effective particle in Eq. (3.12) was based on step-functions (Eq. 3.13) resting on the explicit exclusion of diffusion, it is tempting to ask how effective particles would diffuse, particularly in view of the physical picture of the effective particles of the reacting system evoked earlier (Fig. 2.9).

To this end, let us consider the exact equation (1.12) for the constituent diffusion coefficients $\bar{D}_A(\chi_A, \chi_B)$ and $\bar{D}_B(\chi_A, \chi_B)$. Phrased in the terminology of the effective particle model these are $\bar{D}_X$ and $\bar{D}_Y$, where X and Y denote the dominant and secondary component, respectively. Because the components are instantaneously linked by the mass-action law at all times, it should be possible to approximate the diffusive broadening of the reaction boundary by a single diffusion coefficient of the system, $D_{A \cdots B}$. We expect that both components contribute to diffusion, and that the magnitude of $D_{A \cdots B}$ is weighted by the components' fluxes. Thus, it is hypothesized that the average diffusion coefficient should take the form of a gradient average

$$D_{A \cdots B} = \frac{\bar{D}_X \frac{\partial \chi_X}{\partial r} + \bar{D}_Y \frac{\partial \chi_Y}{\partial r}}{\frac{\partial \chi_X}{\partial r} + \frac{\partial \chi_Y}{\partial r}} . \tag{3.56}$$

At this point, we linearize the gradients $\partial \chi / \partial r$ as an average slope across the reaction boundary $\Delta \chi / \Delta r$. This allows us to exploit our knowledge of the boundary heights from the effective particle model, specifically the co-sedimenting fraction $\tilde{c}_Y$ in Eqs. (2.7) and (2.10), as depicted in Fig. 3.23. At the same time the hypothetical width assigned to these gradients is irrelevant, as Δr cancels out in Eq. (3.56).

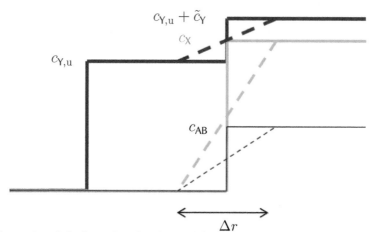

Figure 3.23 Schematics of the boundary heights and linear gradients across a boundary width Δr in the extended effective particle theory [42].

Likewise, the component average sedimentation coefficients are independent of the hypothetical boundary width in the approximation of linearized gradients: For example, we can rephrase $\bar{D}_A$ in Eq. (1.34) as

$$\bar{D}_A(\chi_A, \chi_B) = \frac{\frac{\partial \chi_A}{\partial r} D_A + K \left[\chi_B \frac{\partial \chi_A}{\partial r} + \chi_A \frac{\partial \chi_B}{\partial r} \right] D_{AB}}{\frac{\partial \chi_A}{\partial r} + K \left[\chi_B \frac{\partial \chi_A}{\partial r} + \chi_A \frac{\partial \chi_B}{\partial r} \right]} = \frac{\frac{\partial \chi_A}{\partial r} D_A + K \frac{\partial \chi_{AB}}{\partial r} D_{AB}}{\frac{\partial \chi_A}{\partial r} + K \frac{\partial \chi_{AB}}{\partial r}} . \tag{3.57}$$

In the present context we approximate the slopes $\partial c_X / \partial r \approx c_X / \Delta r$, $\partial c_Y / \partial r \approx \tilde{c}_Y / \Delta r$ and $\partial c_{XY} / \partial r \approx K \left[(c_X / \Delta r)(c_{Y,u} + \tilde{c}_Y / 2) + (\tilde{c}_Y / \Delta r)(c_X / 2) \right] = K c_X c_Y / \Delta r$, leading from Eq. (3.56) to [42]

$$D_{A \cdots B} = \frac{c_X \frac{D_X c_X + D_{AB} c_{AB}}{c_X + c_{AB}} + \tilde{c}_Y \frac{D_Y \tilde{c}_Y + D_{AB} c_{AB}}{\tilde{c}_Y + c_{AB}}}{c_X + \tilde{c}_Y} . \tag{3.58}$$

From here it is tempting to use the Svedberg equation (1.13) to define an 'apparent molar mass of the effective particle'

$$M_{\text{A}\cdots\text{B}} = \frac{s_{\text{A}\cdots\text{B}}RT}{D_{\text{A}\cdots\text{B}}(1 - \bar{v}\rho)}, \tag{3.59}$$

which should, of course, not be confused with a physical mass, but may be taken in the context of sedimentation velocity as a description of the diffusive property of the boundary in more familiar units [42]. This is useful considering that apparent molar mass values are naturally encountered in the boundary analysis with diffusion-deconvoluted sedimentation coefficient distributions such as $c(s)$ and $c(s, f_r)$.

Though rationally motivated, the extension of effective particle theory to estimate diffusion is not derived directly from the Lamm equation and is therefore an empirical estimate. In order to test the performance of the expressions Eqs. (3.56)–(3.59), exact Lamm equation solutions were calculated for our rapidly interacting A + B ↔ AB system at concentrations following an equimolar dilution series, as well as titration series at constant A or constant B, respectively. To extract the diffusion of the sedimentation boundaries, the resulting signal profiles were then fitted with a single-species non-interacting Lamm equation solution describing the undisturbed boundary, in combination with a continuous $c(s)$ distribution of non-interacting species modeling the reaction boundary. The distribution model for the reaction boundary is necessary to distinguish broadening due to the polydispersity of the s-value (predicted from Gilbert–Jenkins theory) from true diffusional broadening of the reaction boundary, which is captured in the refined f/f_0 of the $c(s)$ distribution analysis.[30] The signal-average values integrated across the $c(s)$ or $c(M)$ distribution peak can then be compared with the predictions of Eqs. (3.58) and (3.59). The results are presented in Fig. 3.24, with the Lamm equation based values shown as circles, and the predictions from extended effective particle theory displayed as solid lines. (Note that in these figures the solid lines are not fitted to the data points, but just show the independently made predictions in comparison with the observations from Lamm equation modeling.)

First, we examine simulations carried out for different selective or mass-weighted signal contributions, shown in Fig. 3.24 with red symbols. Their small scatter confirms that the extent of diffusion displayed in the reaction boundary does not much depend on the optical detection of the individual species. This is in line with the predictions of the effective particle model, and analogous to the reaction boundary s-value. Strictly, the concentration gradients in the reaction boundary predicted, for example, with Gilbert–Jenkins theory, are not proportional to each other, and accordingly one would observe slightly different average D-values across the reaction boundary when using a detection method that specifically follows A, B, or is more sensitive to the complex. However, the simulations show this influence is minor. Overall, the approximation of the diffusion and apparent molar mass of the

---

[30]The separation of the undisturbed boundary and the reaction boundary in this distribution model ensures that the focus is on the analysis of the diffusional property of the reaction boundary only, and not averaged with the diffusion properties of the undisturbed boundary.

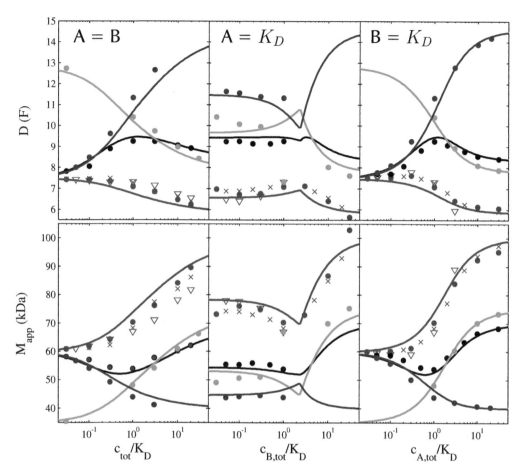

Figure 3.24 Diffusional properties of reaction boundaries (symbols) in comparison with the predictions of Eqs. (3.58) and (3.59) from extended effective particle theory (lines). Coupled Lamm equations were solved for A + B ↔ AB systems with different size and extinction parameters, and the calculated signal profiles were fitted with a discrete species for the undisturbed boundary in combination with a $c(s)$ distribution describing the reaction boundary. Where the $c(s)$ peak of the reaction boundary could be well-resolved, it was integrated to determine the average apparent diffusion coefficient (*Top Panels*) and apparent molar mass (*Bottom Panels*). Calculations were carried out in concentration series as equimolar dilution series (*Left Panels*), titration series of constant A (*Middle Panels*), and titration series with constant B (*Right Panels*). Simulated systems are: (*Red*) A component A with 40 kDa, 3.5 S binding a molecule B with 60 kDa, 5.0 S forming a 100 kDa, 6.5 S complex, using signal coefficients proportional to mass, as typically encountered in interference optical detection (*red circles*), with invisible component A (*red crosses*), or under the assumption of invisible complex as is possible with quenched fluorescence (*red triangles*); (*blue*) A with 10 kDa, 3.5 S binding B with 60 kDa, 5.0 S to form a 70 kDa, 6.5 S complex, using signal coefficients proportional to mass; (*green*) A with 40 kDa, 3.5 S binding B with 35 kDa, 5.0 S to form a 75 kDa, 6.5 S complex, using signal coefficients proportional to mass; and (*magenta*) A with 40 kDa, 3.5 S binding B with 40 kDa, 5.0 S to form a (unphysical) 40 kDa, 6.5 S complex, using signal coefficients proportional to mass. Although some parameter combinations are unphysical, they nonetheless are useful to probe the performance of the effective particle approximations under extreme parameters. Note that none of the predictions (solid lines) have adjustable parameters. Reproduced from [42].

effective particle is very reasonable. Next, we probe in more detail whether $D_{A\cdots B}$ in Eq. (3.58) is of the correct form by systematically assigning different values to D_A, D_B, and D_{AB}, even including unphysical parameter values. The close correspondence between different colored lines and symbols in Fig. 3.24 shows that $D_{A\cdots B}$ indeed follows closely the diffusive property of the reaction boundary, and its dependence on component concentrations and species' diffusion coefficients [42].

Further, if the approximation Eq. (3.58) holds for the average diffusion coefficient of the reaction boundary throughout parameter space, then additional conclusions can be drawn for the variation of $D_{A\cdots B}$ across an individual reaction boundary. Since the local diffusion coefficient is dependent on local concentrations and local concentration gradients, we may evaluate Eq. (3.58) for the range of concentrations expected across a single boundary. Consistent with our linearized simplification of the boundary in Eq. (3.58) we may parameterize the concentrations $c'_X = k c_X$, $\tilde{c}'_Y = k \tilde{c}_Y$, and $c'_Y = c_{Y,u} + k \tilde{c}_Y$ with $k \in [0, 1]$. Probing in this way across the concentration range encountered in the boundary, we estimate a maximum value $D_{A\cdots B,max}$ in the trailing edge of the reaction boundary where $k = 0$. For estimating a minimum value $D_{A\cdots B,min}$, we re-derive Eq. (3.58) with the reference point for $\partial c_{XY}/\partial r$ set at the leading edge where $c'_X = c_X$ and $c'_Y = c_Y$. The relative difference $(D_{A\cdots B,max} - D_{A\cdots B,min})/D_{A\cdots B}$ is plotted in Fig. 3.25 across the parameter space. It shows significant variation only in a very narrow range across the phase transition line. This range coincides with the greatest polydispersity of s in Gilbert–Jenkins theory (Fig. 3.18). In most of the parameter space, the estimated variation of the diffusion coefficient is far less than 10%. Thus, even though the description of $D_{A\cdots B}$ in the extended effective particle model is more complete than $\bar{D}_B^*$ in the constant

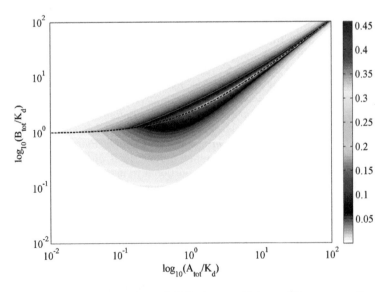

Figure 3.25 Estimated relative variation of diffusion coefficients $\left(D_{A\cdots B,max} - D_{A\cdots B,min}\right)/D_{A\cdots B}$ across the reaction boundary, calculated for the conditions of Fig. 3.26. Reproduced from [42].

bath approximation, we arrive at the same qualitative conclusion that the reaction boundary should evolve in good approximation with a single diffusion coefficient.

Having validated the quality of the estimate of $D_{A\cdots B}$, it is of interest to map the relationship between $M_{A\cdots B}$ from Eq. (3.59), the true mass of the complex M_{AB} and the weight-average mass M_w of the mixture. This is shown in Fig. 3.26. As can be discerned from *Panel A*, $M_{A\cdots B}$ only asymptotically approaches the molar mass of the complex, dependent on the saturation of the reaction boundary (compare with Fig. 2.10 showing corresponding $s_{A\cdots B}$-values). At the same time, it is always larger than the weight-average molar mass of the sedimenting system, but almost identical along the phase transition line, where the undisturbed boundary disappears (*Panel B*). This suggests that $M_{A\cdots B}$ may be similar to the weight-average molar mass of the material in the reaction boundary, $M_{w,rb}$. *Panel C* shows that this is accurate within 3.5% for conditions corresponding to the red circles in Fig. 3.24, though this will not generally hold true. In any case, the apparent molar mass values of the reaction boundary can certainly be useful to estimate the true complex mass in the limit of high saturation of the reaction boundary, i.e., with any or all components in excess far above K_D.

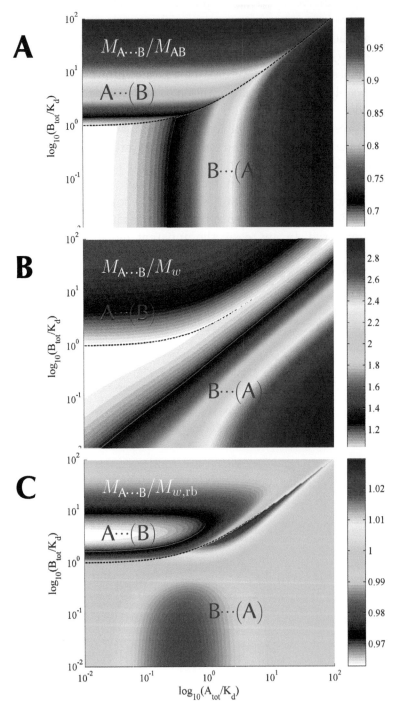

Figure 3.26 Apparent molar mass $M_{A\cdots B}$ of the effective particle in the reaction boundary as a function of component concentrations. *Panel A*: The ratio of $M_{A\cdots B}$ relative to the true molar mass of the complex. *Panel B*: The ratio of $M_{A\cdots B}$ relative to the overall weight-average molar mass of the entire sedimenting system. *Panel C*: The ratio of $M_{A\cdots B}$ relative to the weight-average mass of only the fraction of components co-sedimenting in the reaction boundary. Calculations are for a component A with 40 kDa, 3.5 S binding a molecule B with 60 kDa, 5.0 S to form a 110 kDa, 6.5 S complex, as shown in red circles in Fig. 3.24. Reproduced from [42].

CHAPTER 4

Sedimentation Coefficient Distributions

S EDIMENTATION boundaries of interacting systems show pauci- or polydispersity, as we have seen in the hydrodynamic separation of slowly exchanging populations of free and complex species, or in the separation of undisturbed and disperse reaction boundaries for rapidly interconverting reacting systems. As a consequence, sedimentation coefficient distributions are a natural approach for their analysis in either case. Building onto the definitions and properties of various types of sedimentation coefficient distributions discussed in Part II [2], we now recapitulate how the approximate solutions to the Lamm equations of interacting systems examined in the preceding chapter provide a solid framework to understand the potential of sedimentation coefficient distributions in their application to interacting systems.

The appeal of sedimentation coefficient distributions for data analysis of interacting systems is that they provide a tool to home in on the most robust aspects of sedimentation of interacting systems, which is the pattern of sedimentation boundaries with their characteristic s-values and amplitudes, while presenting the opportunity to exclude contaminating signals unrelated to the process of interest. In this way, sedimentation coefficient distributions turn out to be an indispensable practical approach to the analysis of interacting systems, alternative and complementary to the Lamm equation modeling of Chapter 1.

4.1 DIFFUSION-DECONVOLUTED DISTRIBUTIONS

Briefly, diffusion-deconvoluted sedimentation coefficient distributions $c(s)$ are defined by an integral equation to be least-squares fitted to the experimental data

$$a(r,t) \cong \int c(s)\chi_{1,ni}(s, D(s), r, t)ds , \qquad (4.1)$$

i.e., as a distribution of Lamm equation solutions of non-interacting species $\chi_{1,ni}(r,t)$ at a range of s-values and corresponding D-values related with a scaling law $D(s)$ [49]. Most commonly, the scaling law rests on the traditional assumption of a 2/3-power relationship between sedimentation coefficient and mass, which results in a single frictional coefficient as an additional adjustable parameter (see Part II [2]).

Slowly interconverting systems present no conceptual difference to mixtures of non-interacting populations, which can be analyzed naturally and appropriately using suitable models for sedimentation coefficient distributions. Considering the fact that there is only a narrow range of rate constants exhibiting intermediate sedimentation patterns (Section 2.6), the focus in the following is on systems with rapidly interconverting species compared to the time-scale of sedimentation. Further, we focus in the following largely on rapidly interconverting hetero-associating systems. This is because self-associations are similar to hetero-associations of equivalent molecules at equimolar concentrations (see p. 19), a special case that causes the phase transition line in effective particle theory to follow the equimolar line and the undisturbed boundary to disappear. The conclusions will hold for self- and hetero-associations alike.

Through the complementary approximations in the preceding chapter we have established that both the sedimentation and diffusion of the reaction boundary evolve largely normally — even though the interpretation of the sedimentation parameters requires careful consideration of the nature of the sedimentation process [109]. Specifically, the intrinsic polydispersity of asymptotic (diffusion-free) boundaries predicted from Gilbert theory and Gilbert–Jenkins theory are in perfect correspondence with continuous diffusion-deconvoluted sedimentation coefficient distributions fitted to reaction boundaries. The insight from constant bath theory and extended effective particle theory that in most cases the broadening of the reaction boundary is approximated well with an average diffusion coefficient ensures that diffusional deconvolution is likely to succeed. Similar is true, of course, for the undisturbed boundary. Therefore, we can profitably apply the $c(s)$ distribution of sedimentation coefficients, or the size-and-shape distribution $c(s, f_r)$ (using the same shorthand notation of f_r for the frictional ratio f/f_0 introduced in Part II [2]), to fit the experimental data and to deconvolute diffusion from sedimentation boundary data of interacting systems, just like in the application to sedimentation profiles from mixtures of non-interacting species.

We expect the sedimentation coefficient distribution then to reflect the boundary structure predicted by the effective particle model, showing a peak for the undisturbed boundary at the s-value of the secondary component, and a peak with the properties of the reaction boundary. Ideally, the peak of the reaction boundary should also reflect the underlying polydispersity of sedimentation coefficients as predicted by Gilbert theory or Gilbert–Jenkins theory, respectively (see below). However, considering the theoretical relationships in Section 3.4.2, it will not come as a surprise that the effective diffusion coefficient of the reaction boundary $D_{A\cdots B}$ does not follow the usual 2/3-power scale-relationship between s and M of globular

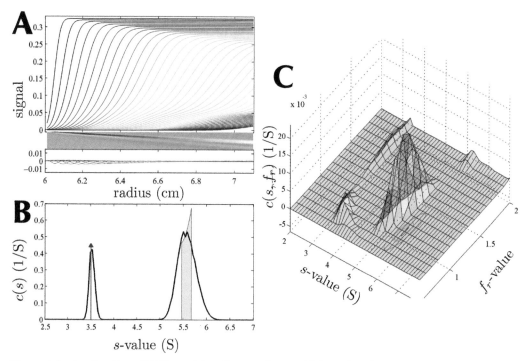

Figure 4.1 Application of differential sedimentation coefficient distributions to the signal profiles of an A + B ↔ AB system in the limit of instantaneous reaction. *Panel A*: Exact Lamm equation solutions (solid lines) were calculated for a molecule A of 40 kDa, 3.5 S rapidly interacting with a molecule B of 60 kDa, 5.0 S to form a 100 kDa, 6.5 S complex, at equimolar loading concentrations at K_D, sedimenting at 60,000 rpm in a 12 mm solution column. The $c(s)$ distribution was calculated with a single overall f/f_0 value refined for the best fit, which has an rmsd of 0.0017-fold the loading signal. The residuals of the fit are shown as residuals bitmap and overlay plot. *Panel B*: The best-fit $c(s)$ distribution is shown as a solid line. For comparison, the gray bar is the asymptotic boundary $d\hat{\chi}/dv$ from Gilbert–Jenkins theory, with the dropped line representing the undisturbed boundary. *Panel C*: Size-and-shape distribution $c(s, f_r)$ of the sedimentation process.

particles.[1] Nevertheless, this power law may be operationally applied in the $c(s)$ distribution with constant frictional coefficient in order to describe the monotonically decreasing D-value with increasing s-value. As described in Section 5.4.9 of Part II [2], over the limited range of polydispersity of s-values encountered in the reaction boundaries the 2/3-power scaling law implied with the constant frictional ratio model of $c(s)$ can be a good model, as long as the frictional ratio is adjusted to assume a best-fit value.

That this approach usually works well is illustrated, for example, in Fig. 4.1 for conditions in the B⋯(A) phase of a rapidly interacting A + B ↔ AB system at equimolar loading concentrations at K_D. It is important to note that the quality of fit of the exact coupled Lamm equation solutions (Eq. 1.33) with the $c(s)$ distribution is excellent, producing residuals with rmsd of only 0.0017-fold the loading

---

[1]The exact scale-relationship is given in Eq. 5.9 of Part II [2].

signal. The residuals bitmap shows the largest deviations in the reaction boundary, but with a maximum deviation far less than 1% of the loading signal — barely significant under typical signal/noise levels of experimental data (Part I, Section 4.3.5 [1]). The $c(s)$ distribution in *Panel B* shows the expected undisturbed and reaction boundaries, the latter being slightly broader than the former, which is consistent with the prediction from Gilbert–Jenkins theory. Fig. 4.2 shows analogous data from the same system at slightly higher concentrations where effective particle theory predicts a phase transition. Phase transitions are where the greatest range of complex saturation will be spanned in the reaction boundary. Nevertheless, $c(s)$ leads again to an excellent fit.

Additional applications of $c(s)$ to represent the sedimentation boundary structure were already shown above, prior to their theoretical justification, to efficiently present the sedimentation boundary structure (Figs. 2.1–2.5, 2.14, 2.15, and 2.17). Such applications of $c(s)$ to self- and hetero-associations are abundant

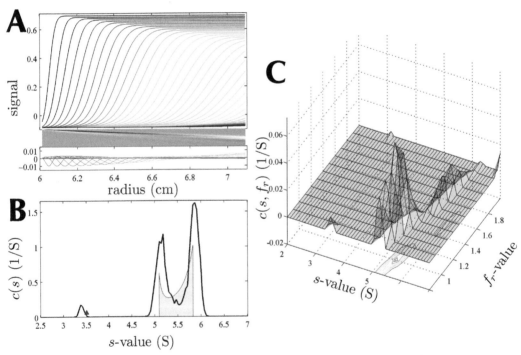

Figure 4.2 Application of differential sedimentation coefficient distributions to the same system as in Fig. 4.1, but at a loading concentration of A of 1.5-fold K_D and B of 3-fold K_D, which is at the predicted phase transition line in effective particle theory, and therefore in the region of strongest polydispersity of the reaction boundary. As in Fig. 4.1, the $c(s)$ analysis is shown in *Panels A* and *B*, with the $c(s, f_r)$ analysis in *Panel C*.

in the experimental literature. A common feature is their excellent fit to the raw experimental data within their typical data acquisition noise.[2]

The question arises to what extent the two-dimensional size-and-shape distributions provide additional advantages over $c(s)$. In principle, $c(s, f_r)$ might better capture the polydispersity in D deduced from the extended effective particle theory depicted in Fig. 3.25. In the present example of Fig. 4.1, when examining the $c(s, f_r)$ distribution in *Panel C*, it is interesting that the undisturbed boundary is well-resolved in s but does not lead to a well-formed peak in the f/f_0 dimension. As described in detail in Section 5.1 of Part II [2], such a behavior is an intrinsic feature of $c(s, f_r)$ when the signal/noise ratio is limiting, and reflects the fact that insufficient information is available on the polydispersity of the boundary spread. In the present case, the undisturbed boundary is indeed relatively small, but it is also possible that the remaining imperfections of the model in the present application to interacting systems — though small — additionally promote an apparent spread in the f_r dimension. For the more strongly populated reaction boundary in Fig. 4.2 the peak at the higher s-value also exhibits a higher frictional ratio, which would make sense on the basis of considerations of saturation gradients within the boundary. How the signal/noise ratio factors into the potential of resolving apparent diffusion parameters of interacting systems by fitting two-dimensional size-and-shape distribution is highlighted in a third example (Fig. 4.3). It shows a family of distributions obtained from simulations of an equimolar concentration series of smaller, more strongly diffusing components. Presented in the form of $c(s, M)$, the results highlight that modeling the diffusion of reaction boundaries will often not lead to easily quantitatively interpretable results, especially when exacerbated by broadening from unavoidable regularization at finite signal/noise ratios. This problem may be further complicated under conditions of intermediate reaction kinetics. Usually, therefore, barring other experimental factors favoring the analysis with independent degrees of freedom for diffusion and sedimentation information, it seems more commensurate with the information content of the data to apply the $c(s)$ analysis instead of $c(s, M)$ — especially if $c(s)$ provides a good fit of the experimental data.[3]

However, since the apparent molar mass $M_{A\cdots B}$ implied by s-value and apparent D-value of reaction boundary, as in Eq. (3.59), does not correspond to that of a

---

[2]Appropriate noise models for different data types may be required, as described in detail in Part I [1].

[3]However, there are certain situations where $c(s, f_r)$ can be a good choice. For example, Bekdemir and Stellacci have described the application of size-and-shape distribution plots mapped into $c(s, D)$ for the study of protein-nanoparticle interactions [131]. In this case, the sedimentation process is complicated by highly multi-valent binding and density differences of various complexes. Experiments were carried out under conditions with the slower sedimenting protein in large excess, satisfying the conditions of the constant bath theory, and frictional coefficients were determined from the average parameters measured with different degree of saturation [131].

Similarly, Chaton and Herr [132] have demonstrated how size-and-shape distributions $c(s, f_r)$ can be highly useful in the analysis of systems with complicated assembly mechanisms including species with widely varying shape sedimenting at similar sedimentation rates.

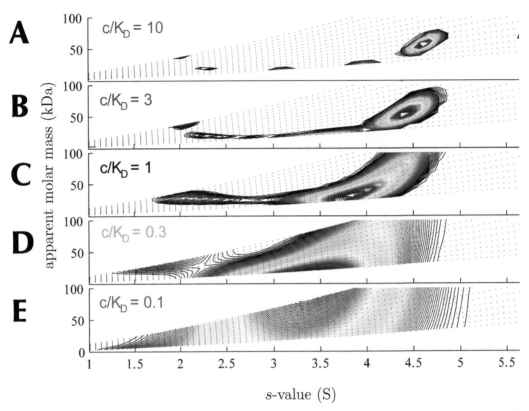

Figure 4.3 Application of the size-and-shape distribution $c(s, f_r)$ to a reaction mixture with rapid kinetics on the time-scale of sedimentation. Sedimentation velocity profiles were simulated for the interaction of a protein of 25 kDa, 2.5 S binding to a 40 kDa, 3.5S species forming a 5 S complex with an equilibrium dissociation constant $K_D = 3$ μM, and a dissociation rate constant $k_{off} = 0.01$/sec, studied at a rotor speed of 50,000 rpm. Interference optical detection was assumed, with conventional signal increments of 3.3 fringes/(mg/ml) at a noise level of 0.005 fringes. Concentrations were equimolar at 10-fold K_D (A), 3-fold K_D (B), K_D (C), 0.3-fold K_D (D), and 0.1-fold K_D (E). $c(s, f_r)$ distributions were calculated with s-values from 1 to 6 S and f_r-values from 0.8 to 2.0, and mapped into the $c(s, M)$ plane. Reproduced from [133].

physical particle, the corresponding frictional coefficient should not be expected to be hydrodynamically sensible, either. To illustrate this in our simulated model systems, the apparent frictional ratio implied by $D_{A\cdots B}$ and $s_{A\cdots B}$ for the conditions of the red line in Fig. 3.24 on p. 119 is shown in Fig. 4.4 (solid lines); interestingly, the prediction from effective particle theory is consistently higher than what is found from the best-fit average f/f_0-values in $c(s)$ fits to Lamm equation solutions (circles), although it is a better approximation than the signal-average frictional ratio of the species co-sedimenting in the reaction boundary (dotted lines in Fig. 4.4).[4]

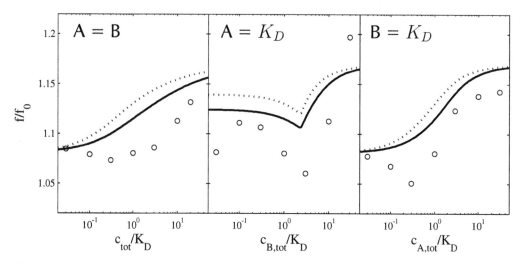

Figure 4.4 Apparent frictional ratio implied by the effective particle model values of $s_{A\cdots B}$, $M_{A\cdots B}$, and $D_{A\cdots B}$ (solid line) and observed values in the $c(s)$ analysis of exact Lamm equation solutions (circles) for the same standard system and titrations as shown in the red line of Fig. 3.24. For comparison, also shown is the frictional ratio calculated naïvely as the signal-average of frictional ratios of the species present in the reaction boundary (dotted line).

The lower apparent frictional ratio values observed from the exact Lamm equation model of the reaction boundary emphasizes the inherent limitations of the approximations of the effective particle model on one hand, and the $c(s)$ model on the other hand.[5] It occurs for self-associations and hetero-associations alike. The significance of this is that when refining f/f_0 in a fit of $c(s)$ to a reaction boundary, low values should be expected, and these should not be interpreted as that of a physical species. Conversely, unphysically low best-fit frictional ratios can be a hint for the possible presence of a reaction boundary when analyzing data from a system with unknown properties (see also Section 6.5.1.1).

The f/f_0 of the undisturbed boundary should reflect that of the physical species constituting this boundary, provided the diffusional spread of the undisturbed and reaction boundary are well separated. As described in Section 5.4 of Part II, within some tolerance any deviations from the best-fit frictional ratio in $c(s)$ can be absorbed into a broadening or sharpening of the s-distribution without causing a very significant increase in the rmsd of the fit (see Fig. 5.5 of Part II) [2]. However, if a

---

[4]This difference ultimately arises from the gradient-average diffusion coefficients determining the effective overall diffusion coefficients, as opposed to the signal-average.

[5]These deviations are sometimes compounded by familiar difficulties in the parameter estimation of f/f_0 when the undisturbed and reaction boundaries are close in s-value, and/or when the signal/noise ratio of the reaction boundary is small.

discrepancy between the apparent f/f_0 of the reaction boundary and that of the undisturbed boundary rises to a level that limits the quality of fit, segmented variations of $c(s)$ (Part II, Section 5.4.2 [2]) may allow sufficient flexibility in the model to capture both appropriately, without causing the over-parameterization inherent in the $c(s, f_r)$ approach.

In summary, when fitting diffusion-deconvoluted sedimentation coefficient distributions to reaction boundaries, the deconvolution of diffusion is a highly useful tool to clarify the sedimentation process, but not too much emphasis should be given to the precise numerical value of the diffusion parameter, or the implied frictional ratio or buoyant molar mass. However, this weakness at quantifying boundary broadening is acceptable in view of the goal to achieve a sedimentation analysis refraining from detailed analysis of boundary broadening, which we had set out to find in the conclusion of Chapter 1 as an alternative to the direct Lamm equation modeling.

> Different variations of the $c(s)$ distribution and the $c(s, f_r)$ distribution can be calculated in SEDFIT and SEDPHAT. For a comprehensive description see Part II [2].

4.1.1   Relationship to Asymptotic Boundaries

Since the asymptotic boundaries from Gilbert theory and Gilbert–Jenkins theory are ideal diffusion-free boundaries, there should be a close correspondence to the diffusion-deconvoluted $c(s)$ distributions, allowing for small discrepancies due to the rectangular sedimentation model underlying Gilbert and Gilbert–Jenkins theory [109]. In fact, for systems in the limit of small diffusion coefficients the shape of a distribution of non-diffusing particles ls-$g^*(s)$ fitted to the sedimentation profiles approaches the predicted asymptotic boundaries approach, as was already shown in Fig. 3.15 on p. 104. For systems showing ordinary diffusion, if diffusion deconvolution with $c(s)$ is successful, then one should likewise expect a close similarity between $c(s)$ and the asymptotic boundaries $d\hat{\chi}/dv$.

At first sight, this does not seem to be the case, due to the limited resolution of $c(s)$. For example, in Fig. 4.1 the narrow polydispersity of the reaction boundary cannot be resolved within the regularization envelope of the $c(s)$ distribution. This can also be observed in the previous examples, such as the $c(s)$ profiles of Fig. 2.4 (p. 32), none of which shows the characteristic sharp-edged features of $d\hat{\chi}/dv$.

However, where $d\hat{\chi}/dv$ spans a wide s-range, this is reflected in correspondingly broader reaction boundary peaks of $c(s)$, such as in Fig. 2.5 (p. 38). This is true also in the present example of Fig. 4.2 that additionally illustrates how the bimodal structure of $d\hat{\chi}/dv$ is mimicked in a double-peak in $c(s)$. It is important in practice to consider the possibility of multi-modal reaction boundary peaks when interpreting $c(s)$ distributions of unknown systems (Section 6.5.2.3).

It is possible to examine the relationship between $c(s)$ and $d\hat{\chi}/dv$ in more depth if we remember that the $c(s)$ distribution reports only one particular distribution from a large class of distributions that would fit the sedimentation boundaries equally well, selected by regularization. As described in Section 3.3 of Part II [2] and reviewed below in Section 4.1.4, both Tikhonov–Phillips and maximum entropy regularization select the most parsimonious distributions based on a criterion of smoothness [49]. They are excellent choices if we have no other knowledge about the measured distribution. In the present case, however, we can modify the regularization in a Bayesian approach [127] — termed $c^{(p)}(s)$ and described in Section 5.7 of Part II [2] — to probe specifically whether a given asymptotic boundary and $c(s)$ are consistent. Thus, we use $d\hat{\chi}/dv$ as a prior for the Bayesian regularization. An example is shown in Fig. 4.5. Except for a trivial scaling factor, and some small deviations at the highest concentration, the $c^{(p)}(s)$ distributions can take shapes quite similar to the prior. Thus, the sedimentation coefficient distributions $c(s)$ are

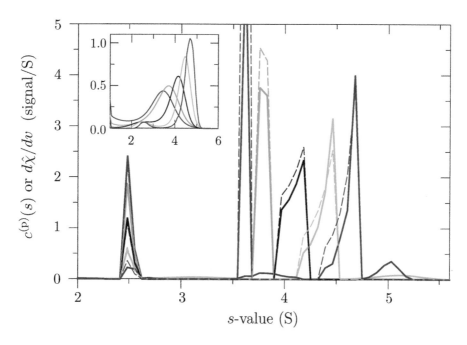

Figure 4.5 Sedimentation coefficient distributions $c^{(p)}(s)$ (solid lines) from the boundary analysis using asymptotic boundaries $d\hat{\chi}/dv$ (dashed thin lines) from Gilbert–Jenkins theory as Bayesian prior to shape the regularization (at invariant rmsd). Sedimentation data were simulated using Lamm equations incorporating reaction terms for the sedimentation of a 2.5 S, 25 kDa protein interacting with a 3.5 S, 40 kDa protein to form a 5 S complex with 1:1 stoichiometry with a K_D of 3 μM. Sedimentation was simulated to take place in a 10 mm column at 50,000 rpm, with signals that would be detected with the interference optical system from 12 mm centerpieces at a noise level of 0.01 fringes. Sedimentation profiles (not shown) were calculated for equimolar protein concentrations of 0.3 μM (red), 1 μM (green), 3 μM (dark blue), 10 μM (cyan), and 30 μM (magenta), leading to total signals of approximately 0.07, 0.22, 0.65, 2.2, and 6.5 fringes, respectively. The inset shows the conventional $c(s)$ analysis with maximum entropy regularization with P = 0.9. All distributions are scaled relative to the loading concentration. Reproduced from [127].

consistent with the asymptotic boundaries predicted by Gilbert–Jenkins theory, even though they are usually not selected in the standard regularization for their lack of parsimony.

4.1.2 Discrimination of Interacting Components

Based on our theoretical understanding of the sedimentation process gained in Chapter 3, and illustrated by our model systems, it is clear that no $c(s)$ peaks can arise from interacting systems at s-values that are smaller than that of the smallest interacting component in the free state, and at s-values that are larger than that of the largest complex species. If the diffusion-deconvoluted distributions achieve a good fit of experimental data and additional trace peaks occur outside this range, we can identify these as extraneous species and remove them from further consideration (see Section 6.5.1.1).

While the lower bound is usually easy to determine experimentally for hetero-associating systems by carrying out SV experiments of each component separately, the smallest non-dissociating unit in self-associating systems may not be trivial to determine. Likewise, the largest complex may be of unknown oligomerization state and feasible experimental conditions may remain far from saturation. However, the superposition of sedimentation coefficient distributions obtained at a range of loading concentrations, normalized by division with the loading signal, can contribute significantly to clarify these aspects by showing which peaks present a concentration-dependent fraction consistent with mass action law. Furthermore, the superposition will generally help to classify the kinetics of the interaction relative to the time-scale of sedimentation, and aid in building a model of the interacting system (see Chapter 6).

4.1.3 Quantifying s_w and Characteristic Boundary Features

In the overwhelming majority of cases it will be possible to obtain an excellent quality of fit of sedimentation profiles of reacting systems with diffusion-deconvoluted sedimentation coefficient distributions. Rooted in the transport method, as shown in detail in Sections 3.1 and 3.1.3, the fit ensures a faithful representation of the mass balance and thereby an accurate determination of the signal-weighted average s_w of the interacting system. This is accomplished through integration of the sedimentation coefficient distribution across the undisturbed and reaction boundary as in Eq. (3.2),

$$s_w = \frac{\int_{s_{min}}^{s_{max}} c(s)s\,ds}{\int_{s_{min}}^{s_{max}} c(s)\,ds}, \tag{4.2}$$

with s_{max} denoting the fastest sedimenting complex of the interacting system, and s_{min} comprising the slowest peak from the interacting system. For use in the analysis of the interaction, the integration should exclude signals from sample impurities or irreversible aggregates.

In addition, due to the removal of diffusion broadening from the sedimentation coefficient distributions, we can generally discern the undisturbed and the reaction boundaries of hetero-associating systems. This is greatly facilitated by the superimposed display of $c(s)$ distributions acquired at different concentrations, and obviously aided by theoretically familiarity with the expected phenomenology. If we restrict the integration of $c(s)$ to the reaction boundary,

$$
s_{\text{fast}} = \frac{\int_{s_{\text{min,react}}}^{s_{\text{max}}} c(s)s\,ds}{\int_{s_{\text{min,react}}}^{s_{\text{max}}} c(s)\,ds}, \tag{4.3}
$$

we obtain with s_{fast} an experimental counterpart of the average velocity s_{asy} of the asymptotic boundary in Eq. (3.52), and/or the sedimentation coefficient $s_{\text{A$\cdots$B}}$ predicted from effective particle theory in Eq. (2.20), respectively. This again rests on the intimate relationship between integration of distributions in the s-dimension and the mass transport of boundaries across the radial dimension.

Further, for hetero-associations it is of interest how much of the total signal appears in the reaction boundary and the undisturbed boundary. This can be extracted from the $c(s)$ distributions by carrying out the integrations

$$
a_{\text{slow}} = \int_{s_{\text{min}}}^{s_{\text{min,react}}} c(s)\,ds \qquad a_{\text{fast}} = \int_{s_{\text{min,react}}}^{s_{\text{max}}} c(s)\,ds\,. \tag{4.4}
$$

Since the s-value of the undisturbed boundary equals that of the free species, the four quantities s_w, s_{fast}, a_{slow}, and a_{fast} fully represent the boundary structure of rapidly reversible hetero-associating systems. The dependence of these quantities as a function of concentration establishes isotherms, which can be fitted to binding models as described in more detail in Chapter 6.

In this description, the precise shape of the sedimentation boundaries, as well as the precise shape of $c(s)$ peaks and the apparent frictional coefficient used for deconvolution of diffusion are ignored. This is useful since the latter features are very susceptible to microheterogeneity of the sample, and most sensitive to imperfections in the sedimentation process, ambiguities from limited signal/noise ratio, as well as errors arising in the approximation of the sedimentation process as a superposition of non-interacting species in $c(s)$.

Integration of the $c(s)$ distribution can be accomplished in SEDFIT and SEDPHAT by graphically specifying a region or by specification of the integration limits in a text file. This will report both the signal amplitude a and average s for the selected region. For a detailed description see Part II [2].

4.1.4 Embedding Prior Knowledge

As we have seen in Section 5.7 of Part II [2], regularization is unavoidable when calculating diffusion-deconvoluted sedimentation coefficient distributions from experimental data. This is due to the fact that for noisy data a large family of very different distributions can fit the same data statistically indistinguishably well. Adventitious noise amplification often governs the appearance of the overall best-fit distribution from this family. In order not to be misled by this, the application of an additional criterion that favors a parsimonious distribution is essential.

Usually parsimony is defined as high information entropy or low total curvature of the distribution. However, this can be detrimental for recognizing known features of the distribution, especially at low signal/noise ratio. Therefore, from the family of distributions that fit the data with the same rmsd, it can be advantageous to bias the selection of the distribution differently, to report the one that is most consistent with a particular known feature, rather than the overall smoothest one. This is accomplished using Bayesian regularization [127]. Importantly, since by design the quality of fit cannot change when implementing Bayesian prior, it cannot force a fit to a particular model. Rather, information contained in the data will be able to override any prior that is inconsistent with the data. But, to the extent that the prior is consistent with the data, the particular choice can facilitate further analysis.

> Bayesian priors can be implemented in SEDFIT in the menu Options ▷Size Distribution Options ▷Use Prior Probabilities, and in SEDPHAT in the model parameter box through the push-buttons labeled PP adjacent to each branch of the s-distribution. Gaussians can be specified by providing a peak location, width, and an associated amplitude factor. The latter scales the relative importance of different priors, and otherwise can be chosen arbitrarily, for example at a value of 1.0.

We have used this strategy already in Section 4.1.1 for examining the relationship between $c(s)$ and the asymptotic boundaries from Gilbert–Jenkins theory (Fig. 4.5). Beyond the theoretical interest, however, the precise peak shapes of the $c(s)$ distribution are not of further importance in the integrals in Eqs. (4.2)–(4.4), and therefore the adjustment of the regularization to favor $c(s)$ mimicking $d\hat{\chi}/dv$ is normally not further profitable.

What will be known in the study of hetero-associating systems are the s-values of the free components, since they can usually be studied separately. Based on the effective particle theory, as well as the initial superposition of $c(s)$ distributions as a function of loading concentration along the experimental trajectory in concentration space, it will usually be safe to predict which component provides the undisturbed boundary. For example, for equimolar loading concentrations it is always the component that sediments slower. Therefore, this knowledge can usually be incorporated safely into the $c(s)$ analysis in a second stage. This can be accomplished by offering a prior consisting of a narrow Gaussian centered at the s-value of the free component. This will be highly useful at low loading concentrations and poor signal/noise conditions to counteract the broad features of $c(s)$ driven by standard regularization. The resulting $c^{(p)}(s)$ distributions will usually exhibit a clear separation of undisturbed and reaction boundary, and thereby facilitate the integration *via* Eqs. (4.3)–(4.4) to determine s_{fast}, a_{slow}, and a_{fast} [106].

4.1.5 Limitations

The most obvious limitation encountered when applying the diffusion-deconvoluted sedimentation coefficient distributions to interacting systems, as opposed to non-interacting systems, is that the diffusion parameter cannot be easily interpreted under partially-saturating conditions, unless the reaction is slow on the time-scale of sedimentation. This was already discussed in detail above.

A second important caveat is that for the distribution to be meaningful, the quality of fit to the experimentally observed boundaries must be satisfactory. A number of different criteria for the quality of fit have been discussed in Part I, Section 4.3.5 [1]. Perhaps the most useful in practice is the residuals bitmap and the residuals histogram, as well as the error propagation toward the distribution integrals as in Eq. (3.3).

Even though the quality of fit to interacting systems is usually unproblematic within the given experimental data acquisition noise, conditions that sometimes lead to comparably poor performance of the $c(s)$ fit are rapidly reversible systems close to the phase transition, in particular with concentrations in the range of K_D. When the undisturbed boundary is smallest, the reaction boundary will exhibit largest gradients in saturation, and concomitant gradients in sedimentation and diffusion coefficients.

Similarly problematic are conditions where a small molecule is the dominant component in the reaction boundary, by virtue of an excess of a larger binding partner that does not contribute significantly to the signal. In this case, diffusion-driven concentration gradients will 'liberate' a fraction of the small molecule from the reaction boundary, causing boundary shapes that are not always modeled very well with $c(s)$.[6]

Additional deviations from boundaries that mimic superpositions of non-interacting species occur in systems where the signal is not conserved in the chemical reaction, such as in the presence of large quenching. Misfits of $c(s)$ models applied to such systems are also exacerbated at concentrations near K_D.

---

[6]This correlates with largest deviations from predictions of effective particle theory (see Fig. 3.4).

4.2 APPARENT SEDIMENTATION COEFFICIENT DISTRIBUTIONS

Various methods to determine apparent sedimentation coefficient distributions have been devised and can be found in the historic and current literature. The ls-$g^*(s)$ method [130], the radial-derivative and the time-derivative methods [134–136], and the van Holde–Weischet extrapolations [125] were already introduced and discussed in detail in Part II [2]. In the following we restrict the discussion solely to their utility for the study of interacting systems.[7]

4.2.1 Distributions without Consideration of Diffusion: ls-$g^*(s)$ and $g^*(s)$

An attractive feature of the apparent sedimentation coefficient distribution is their greater simplicity, afforded by the neglect of diffusion. The ls-$g^*(s)$ distribution was introduced in Part II, Chapter 4 as a distribution of non-diffusing particles [2]. It is based on the same principles of direct least-squares boundary modeling as $c(s)$ and $c(s, f_r)$, but with constant $D = 0$ (or equivalently $f/f_0 = \infty$) for all species [130]. This naturally eliminates any considerations of how to parameterize the model in this regard.

The ls-$g^*(s)$ distribution can be an excellent tool for the description of very large particles. Unfortunately, the evolution of experimental boundaries even of non-interacting diffusing macromolecules cannot be fully described, since the model is unable to accommodate the $\sqrt{t}$-dependent component of the migration, as discussed and illustrated extensively in Part II [2]. However, over a limited time-range, the boundary shapes can be described well as an apparent distribution of non-diffusing particles (Part II, Section 4.2). A straightforward objective criterion for acceptable time intervals is the quality of fit. As long as residuals are small and do not affect the mass balance considerations, the transport method to determine s_w by integration can be applied in the same way to ls-$g^*(s)$ as for $c(s)$ [47]. However, disadvantages of ls-$g^*(s)$ are the smaller number of scans it can fit and consequently the larger error in s_w. Another significant disadvantage is the lack of resolution in ls-$g^*(s)$ compared with $c(s)$, such that extraneous signal contributions may not be recognized and excluded. Furthermore, the division into undisturbed and reaction boundaries, can be less clearly observed, and complete separation of undisturbed and reaction boundaries is achieved only in favorable cases of large and slowly diffusing macromolecules. This hampers the classification of the kinetics of the interaction and the model building in various ways.

Departing from direct least-squares modeling of the sedimentation data, apparent sedimentation coefficient distributions can be derived from the radial-derivative dc/dr, via the Bridgman equation (Part II, Eq. 4.7 [2,134]) and the time-derivative dc/dt, via the temporal Bridgman equation (Part II, Eq. 4.11 [2,136]). They suffer

---

[7]Another sedimentation coefficient distribution occasionally found in the current literature is termed '2DSA' [137]. It is procedurally defined outside the realm of mathematical relationships [138], with key elements being in conflict with matrix algebra, and therefore ill-suited for discussion in the present framework.

from the same limitations in the ability to arrive at a consistent sedimentation coefficient distribution of diffusing particles from boundary data at different time points, and in the drawbacks from lack of resolution, as ls-$g^*(s)$. In addition, in practice, the differentiation in the radial dimension is invariant with regard to RI noise but sensitive to TI noise, whereas the temporal differentiation is sensitive to RI noise but invariant to TI noise, each requiring *ad hoc* adjustments potentially introducing bias and errors. By contrast, ls-$g^*(s)$ naturally allows for both TI and RI noise to be accounted for.[8]

Stafford has described a popular iterative approach to determine an apparent sedimentation coefficient distribution $g^*(s)$ via the time-derivative, referred to as the 'dc/dt'-method [135], which is described in detail in Part II, Section 6.2 [2]. It was well adjusted to analyze newly digitally acquired interference optical data in the early 1990s at a time predating abundant computational resources. In theory it is compatible with the transport method, and integration of $g^*(s)$ can be used to determine s_w. However, a significant problem arises from the approximation of the time differential dt with finite scan time differences Δt (Part II, Section 6.2.2). This leads to artificial, asymmetric broadening of $g^*(s)$ causing underestimates of s_w [47, 130, 136]. The problem is exacerbated at small s-values and large scan time ranges. Unfortunately, no residuals to the original boundary data are made available in the dc/dt-method to rationally assess the performance of the model.

4.2.2   The Integral Sedimentation Coefficient Distribution $G(s)$ by the Method of van Holde and Weischet

The van Holde–Weischet method is an extrapolation scheme to determine a sedimentation coefficient distribution allowing for a $\sqrt{t}$-dependent component of migration, based on a single-species diffusion, superimposed to sedimentation [125]. Originally it was thought of as a useful tool to demonstrate the homogeneity of the sedimentation boundary [125], but its applications have later been inflated to the determination of sedimentation coefficient distributions [139]. A fundamental drawback is a conflict of the basic underlying equations with polydispersity [140]. See Section 6.1 of Part II [2] for a detailed introduction and discussion.

With regard to interacting systems, even though the possibility for qualitative detection of interactions has been demonstrated [141], it cannot be applied to their quantitative study. At the root of this limitation is the initial boundary division scheme, where data are converted into relative boundary fractions and thereby scaled in a way that is inherently incompatible with the transport method and mass balance considerations. Further, the empirical nature of the temporal extrapolation clashes with the rigorous quantitative basis of the transport method. Therefore, no well-defined s_w can be derived from the van Holde–Weischet distributions [47].

---

[8]For a detailed description of the systematic signal contributions of RI and TI noise, see Part I, Sections 4.1.4 and 4.2.5 [1].

4.3 MULTI-COMPONENT DISTRIBUTIONS

Most hetero-associating systems involve components that are spectrally not identical. This includes a large fraction of protein-protein interactions — solely due to differences in aromatic amino acid contents — and virtually all interactions between proteins and other macromolecules or particles. In addition, a temporal signal dimension, as in photoswitchable molecules, can offer similar component-specific characteristic information [39]. Signal differences can be created and further enhanced, for example, through the introduction of extrinsic chromophores or protein fusion with fluorescent proteins.[9] Section 4.4.4 of Part I [1] describes in detail how data can be acquired with different optical systems and/or different absorbance wavelengths quasi-simultaneously. This additional data allows us to exploit the different optical signatures of the binding partners to generate additional information on the interaction, for example, the stoichiometry of hydrodynamically resolved complex species, or the composition of reaction boundaries, respectively.

In principle, all the strategies above for applying diffusion-deconvoluted sedimentation coefficient distributions — usually $c(s)$ — can be applied separately to data containing the different absorbance and interference optical signals, which would result in separate $c(s)$ distributions for each signal. However, relating different $c(s)$ distributions to each other to analyze the signal ratios in correspondence to component extinction ratios can be non-trivial or even impossible. It is complicated, for example, by different signal/noise ratios leading to different hydrodynamic resolution and different peak shapes. For example, this would render simple ratios $c_{\lambda_1}(s)/c_{\lambda_2}(s)$ ill-defined.

Therefore, the extension of the diffusion-deconvoluted $c(s)$ distribution to the spectral dimension in the multi-signal $c_k(s)$ (MSSV) [38, 142, 143], and analogously to the temporal dimension in monochromatic multi-component (MCMC) distribution $c_k(s)$ [39], both introduced in detail in Chapter 7 of Part II [2], are extremely useful tools that can greatly enhance both the component identification and hydrodynamic resolution. Briefly, in their simplest form, multi-component coefficient distributions are defined by a global fit of all signals at once

$$a_\lambda(r,t) \cong \sum_{k=1}^{K} d\varepsilon_{k,\lambda} \int c_k(s)\chi_{1,ni}(s, D(s), r, t)ds \quad \text{for all } \lambda , \qquad (4.5)$$

accounting for the signal contribution of component k to signal λ with the predetermined signal coefficients $\varepsilon_{k,\lambda}$, thereby resulting in molar component sedimentation coefficient distributions $c_k(s)$. Analogously, temporal modulation of signal coefficients leads to

$$a_\lambda(r,t) \cong \sum_{k=1}^{K} \int d\varepsilon_{k,\lambda}(s, r, t)c_k(s)\chi_{1,ni}(s, D(s), r, t)ds \quad \text{for all } \lambda , \qquad (4.6)$$

---

[9]Information on characteristic macromolecular signal increments for different detection systems can be found in Part I, Section 4.4.2 [1].

with an additional weak dependence of the observed temporal signal $\varepsilon_{k,\lambda}(s, r, t)$ on the sedimentation trajectory [39]. In either case, for this decomposition, the number of spectrally or temporally distinguishable signals must be equal or greater than number of components, and the extinction coefficients must not be proportional. Details on the requirements, analysis possibilities, practical implementation, and examples can be found in Part II, Chapter 7 [2].

Building onto the principles and capabilities of MSSV in non-interacting systems, in the present context it is important to recognize that, since the spatio-temporal evolution is based on superpositions of Lamm equation solutions of non-interacting species, $c_k(s)$ in MSSV inherits from the single-signal $c(s)$ all relationships to the Lamm equation solutions of reacting systems.

With regard to the transport method, since s_w-values are typically modeled as a signal-weighted population-average (Eqs. (3.7) and (3.22)), they do not generally benefit from any spectral decomposition at the stage of $c(s)$ prior to integration.[10] Similar is true for the s_{fast}-values of reaction boundaries, which in the effective particle approximation of $s_{\text{A}\cdots\text{B}}$ of Eq. (2.20) are completely independent of signal coefficients.

4.3.1 Determining the Complex Composition

The unique strength of multi-component analysis emerges clearly in situations when the identity and/or the stoichiometry of molecular complexes is in question. For slowly interacting systems on the time-scale of sedimentation, the peaks represent physical species, either free or in complex, and the potential applications and capabilities are identical to those discussed in Part II [2].

A simple example for the results of multi-signal analysis can be found in Fig. 4.6, which is the $c_k(s)$ analysis of the slowly interacting mixture of SLP-76 and PLC-γ peptides previously shown with a direct boundary model in Fig. 1.7 on p. 17. Since both proteins have very distinct signal contributions in UV absorption and refractive index detection, they can be easily distinguished. Their component sedimentation coefficient distribution shows a peak at a higher s-value than either free species. The equal magnitude of $c_k(s)$ in this complex species motivates a 1:1 model, which can be further tested by constraining the spectral space in the s-range of complexes to species with 1:1 and 2:1 composition (see Part II, Section 7.1.4 [2]). Clearly, the 1:1 complex is the predominant complex species. Thus, the data can be fitted with coupled Lamm equation solutions with explicit reaction terms accounting for a 1:1 complex, as shown in Fig. 1.7. In this way, MSSV analysis can precede the Lamm equation fitting or isotherm analysis and rationalize the interaction model to be chosen.

The identification of the binding mode is increasingly difficult with larger numbers of components, additional self-associations, and higher complex stoichiometry.

---

[10]They will, however, require extinction coefficients (or signal increments, respectively) which can be determined experimentally best by multi-signal SV (Part I, Section 4.3.2 [1]).

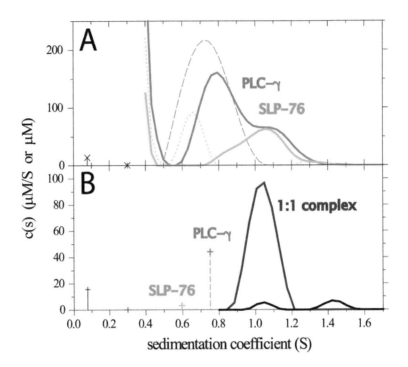

Figure 4.6 Multi-signal SV analysis of data from a slowly interacting mixture of peptides derived from SLP-76 (green) and PLC-γ (blue), acquired simultaneously by absorbance and interference optical detection as shown in Fig. 1.7. *Panel A*: The signal coefficients acquired from MSSV analysis of the individual proteins ($c_k(s)$ distributions shown as dashed lines) were used in the determination of the component $c_k(s)$ in the mixture (solid lines). The thin black crosses and dropped lines represent buffer contributions modeled as discrete species. In the mixture, a new peak emerges with apparent 1:1 stoichiometry. *Panel B*: In a further refinement, the *s*-range was subdivided into discrete species describing the free species (+ symbols with dropped dashed lines) and continuous branches of $c_k(s)$ with constrained 1:1 molar ratio (red) or 2:1 molar ratio (dark blue). For more details, see [38].

Here MSSV can be essential. As an example, Fig. 4.7 shows the multi-signal analysis of a mixture of HLA-A2, which is monomeric (3.5 S), and U21, which exists in a slow equilibrium between dimers and tetramers at 6 S and 8.5 S, respectively [144].[11] Controls with the individual proteins side-by-side in the same run allow signal coefficients $\varepsilon_{U21,\lambda}$ and $\varepsilon_{HLA,\lambda}$ to be determined, for use in the multi-signal analysis. In the $c_k(s)$ results from mixtures, a faster-sedimenting complex can be discerned at ~11 S with both components co-sedimenting in best-fit molar ratios, determined by integration of this $c_k(s)$ peak, of 1.87:1 to 2.33:1, within error consistent with a ratio of 2:1. The peak positions are concentration-independent, which shows that the complex lifetime is long on the time-scale of SV, and that the peak reflects a complex species (as opposed to a reaction boundary). Although

---

[11]For this example, raw scan files at the different signals and their fit were already shown in Fig. 4.37 of Part I [1]; here we discuss in more detail the interpretation of the component distributions.

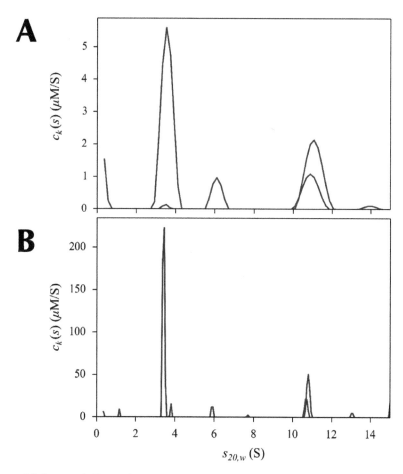

Figure 4.7 Multi-signal SV analysis of a protein HLA-A2 (blue) and U21 (red) based on a global fit of data sets recorded simultaneously by Rayleigh interference, absorbance at 280 nm and 250 nm, analyzed in terms of a spectrally and diffusionally deconvoluted component sedimentation coefficient distribution $c_k(s)$ [144]. In control experiments, HLA-A2 sediments as a single monomeric species at ∼3.5 S, and U21 is found to sediment as a mixture of dimers and tetramers at ∼6 S and ∼8.5 S, respectively. From the same control experiments, signal coefficients $\varepsilon_{U21,\lambda}$ and $\varepsilon_{HLA,\lambda}$ were determined. They exhibit significant spectral differences due to differences in aromatic amino acids and exacerbated by glycosylation of U21. Experiments with mixtures are shown with loading concentrations of 4.9 μM HLA-A2 and muM U21 (*Panel A*), and with a loading concentration of 39.4 μM HLA-A2 in the presence of 11.8 μM U21 (*Panel B*). Data reproduced from [144].

the component ratio in the complex is 2:1, the possibility of a 2:1 stoichiometry of the complex can be ruled out in view of the high s-value of the peak of 10.75 S. Rather, a 4:2 U21/HLA-A2 complex is hydrodynamically reasonable. Further, it is consistent with the measured apparent molar mass of the 10.75 S species. Thus, a reaction mechanism emerges where U21 tetramers are the preferred binding site for two copies of HLA-A2 each [144].

Similarly, information on the binding mode has been gained, and the identity of complexes has been determined by multi-signal SV and mono-chromatic multi-component SV for many other two-component and three-component systems studied in many laboratories [142, 143, 145–158]. *The importance of this information from MSSV and MCMC multi-component analysis on the binding scheme cannot be overstated, since the determination (or at least hypothesis) of the number and stoichiometries of complexes is a prerequisite for any direct Lamm equation or isotherm modeling.*

4.3.2   Multi-Component Analysis of Reaction Boundaries

For rapidly reversible interacting systems, the application of $c_k(s)$ requires additional considerations to account for the nature of the coupled sedimentation process. Most importantly, from our understanding of the reaction boundaries derived from Gilbert–Jenkins and effective particle theories, it is clear that the reaction boundary consists not only of complex but also free forms of all components in ratios that are dependent on concentrations, species s-values, and binding constants. Therefore, while the multi-component $c_k(s)$ analysis can easily provide the composition of the reaction boundary through simple peak integration, it does not necessarily equal the stoichiometry of the complex. In fact, the theoretical values span the entire range from zero to the complex stoichiometry (in terms of A/B the usual definition where free A sediments slower than free B).

For our standard model system, the 1:1 interaction with $s_A = 3.5$ S, $s_B = 5.0$ S, and $s_{AB} = 6.5$ S, we have already seen examples for the reaction boundary composition as a function of component concentrations, comprehensively represented in the color contour plot of Fig. 2.10C on p. 51. As highlighted in Fig. 2.11, the saturation curve takes a similar shape in s-value and composition, and spreads over the largest concentration range in the diagonal, i.e., for equimolar concentrations, whereas saturation is attained already at lower concentrations for titrations when a large excess of either component is loaded, as illustrated in Fig. 3.20 on p. 112 for various titration and dilution trajectories (magenta lines, *Left Column*).

There is also a strong dependence of the reaction boundary composition on the s-values of the components, as predicted by Eq. (2.21) and illustrated in Fig. 4.8 for systems with different size ratio of the binding partners. It is noteworthy that for any of the systems, when the larger component is at concentrations below K_d while the smaller one is in >5-fold molar excess, the reaction boundary composition is at >~0.8 converging closely to the complex stoichiometry. This, jointly with hydrodynamic predictions for differently sized complexes in comparison to the reaction boundary s-value, may be sufficient to correctly deduce the reaction stoichiometry. Furthermore, with more similar-sized components a greater concentration range exists where the reaction boundary composition exhibits values close to the stoichiometry (*Panel C*). Conversely, if a larger component is in molar excess over a much smaller component (*Panel B*), the reaction boundary composition is far below the complex stoichiometry at concentrations close to K_D. Thus, the experimental requirements necessary to produce reaction boundaries with composition close to the reaction stoichiometry depend on the macromolecules of interest, and whether

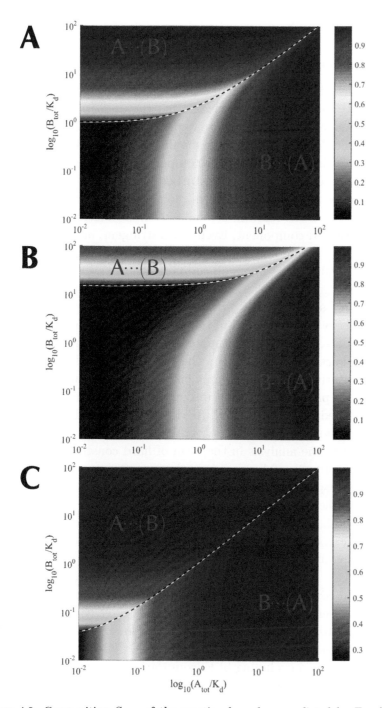

Figure 4.8 Composition $S_{A\cdots B}$ of the reaction boundary predicted by Eq. (2.21) in the effective theory for a system with A + B ↔ AB with values $s_A = 3.5$ S, $s_B = 5.0$ S, and $s_{AB} = 6.5$ S (*Panel A*), a system with a small ligand $s_A = 0.5$ S, $s_B = 5.0$ S, and $s_{AB} = 5.3$ S (*Panel B*), and more similar-sized molecules with $s_A = 4.9$ S, $s_B = 5.0$ S, and $s_{AB} = 8.5$ S (*Panel C*). The phase transition line according to Eq. (2.16) is shown in all panels as a black-and-white dashed line.

the composition is close to saturation will be best assessed in concentration series (see Chapter 6).

4.3.3 Exclusion of the Undisturbed Boundary

With regard to the undisturbed boundary, the multi-signal/multi-component $c_k(s)$ analysis offers the possibility to utilize our knowledge that the undisturbed boundary is composed of only a single component, and to exploit the prediction based on effective particle theory which of the components should provide the undisturbed boundary. For example, in equimolar mixtures the component with the slower sedimenting free form will always constitute the undisturbed boundary.

A straightforward strategy to embed this information is through the construction of a suitable $c_k(s)$ branch in the relevant sedimentation coefficient range that only provides for the undisturbed component. Even more stringent, only a discrete species at the pre-determined s-value may be allowed. This can be combined with a second branch at higher s-values that offers all components. A caveat of this approach may be imperfections in the predictions of effective particle theory in the A···(B) phase with moderately saturated A which may allow some A to escape (see above). It is advisable to verify that the rmsd of the fit does not increase from the application of these constraints, otherwise pointing to errors in the assumptions, for example, that the reaction is rapidly reversible on the time-scale of sedimentation.

The exclusion of the putative dominant component from the s-value region of the undisturbed boundary is particularly useful in conjunction with mass conservation prior knowledge, termed MC-MSSV [159]. As described in detail in Section 7.1.5 of Part II [2], expectations of the total integral of $c_k(s)$ over a certain s-interval can be embedded into the analysis in the form of hard constraints or as a soft Bayesian regularization criterion. Mass conservation constraints can improve, or even entirely substitute for spectral discrimination of components [159].[12] Essentially, MC-MSSV allows transcending the realm of multi-signal analysis by virtue of the known undisturbed boundary composition and known loading concentrations.

To illustrate this point, Fig. 4.9 shows the multi-signal data analysis of a mixture of *Treponema pallidum* protein Tp34 and bovine lactoferrin (bLF) [159]. Based on their extinction coefficients and refractive index signal coefficients, the MSSV analysis of SV data acquired by absorbance at 280 nm and Rayleigh interferometry alone does not sufficiently discriminate the two components. This leads to a false attribution of the peak at 2 S to bLF. However, this s-value is lower than that of the free species of bLF measured separately. In a revised analysis, MC-MSSV with the known loading concentrations, and excluding bLF from the range between $0 - 4$ S, leads to an estimate of the composition of the higher s-value peak of 0.7:1, typical for a reaction boundary of a 1:1 complex species under given conditions [159].

---

[12]See, for example, Appendix B3 of Part II for a mathematical derivation of this fact [2].

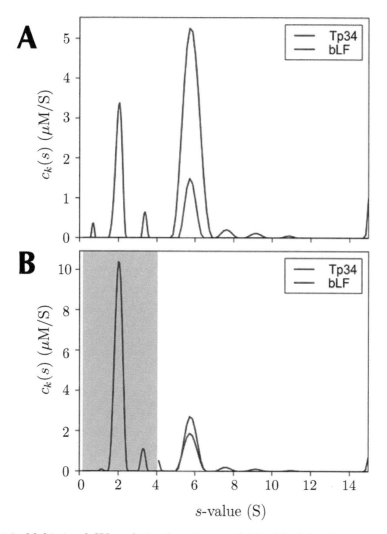

Figure 4.9 Multi-signal SV analysis of a mixture of 6.9 μM of the *Treponema pallidum* protein Tp34 and 2.3 μM bovine lactoferrin (bLF) recorded simultaneously by absorbance at 280 nm and by refractive index [159]. Since the D_{norm} value for this system is only 0.016, spectral decomposition alone is predicted to fail (see Part II, Section 7.1.2 [2]). *Panel A*: The standard MSSV analysis leads to $c_k(s)$ where the 2 S species is attributed to bLF. The misassignment can be recognized from the fact that free bLF sediments at 5.2 S, whereas free Tp34 sediments at 2.0 S. Furthermore, from integration of the $c_k(s)$ the total concentration in the fit is 5.7 μM Tp34 and 2.8 μM bLF, respectively, which is 22% in error. *Panel B*: Using the knowledge that Tp34 constitutes the undisturbed boundary, bLF is excluded from s-values smaller than 4 S. In addition, mass conservation constraints are applied such that total concentrations cannot differ by more than 5% from the pipetted loading concentrations. The resulting complex peak has a molar ratio of 0.7, consistent with the reaction boundary of a 1:1 complex. Reproduced from [159].

CHAPTER 5

Nonideal Sedimentation Velocity

U P to this point, we have considered systems where the forces acting on the sedimenting species — the gravitational field (centrifugal force), buoyancy, and hydrodynamic friction — can be considered to arise from a single particle. This was extended by the idea that chemical conversion of different species may occur, following the laws of chemical kinetics and equilibrium to slowly or rapidly redistribute locally the population of species, but in a way that does not change the basic sedimentation and diffusion of each species. This is an idealized picture valid in 'dilute systems.' At higher concentrations, additional forces can become significant and the movement of one particle is dependent on the presence and migration of other particles.

Additional forces may occur, for example, from obligate short-range repulsive interactions when molecules, which exclusively occupy a certain space, are in close encounter. Even the mere fact that volume-occupying molecules are sedimenting in a finite solution volume will cause a solvent counter-flow that is felt by all macromolecules. At short range, weak interactions that might be driven by electrostatic attraction or repulsion, or by depletion forces, can also become highly relevant [160–164]. In solutions of high macromolecular concentrations hydrodynamic interactions are significant, where the disturbance of the fluid velocity field from one particle affects the frictional forces acting on another particle. For polyelectrolytes, interactions may also couple macromolecular flows to those of sedimenting ions, exerting additional attractive or repulsive interactions.

Some of the phenomenology of sedimenting particles at high concentrations was already introduced in Section 2.2.2.2 of Part I [1]. In the present chapter we discuss in more detail the consequences for the analysis of sedimentation velocity. While some of the scenarios above can be approached theoretically, and approximate relationships can be established for simple cases of single macromolecular species at not too high concentrations, a general description of most of these effects for interacting multi-component systems would be very difficult, and many aspects are subject of current research [165]. Closed-form expressions incorporating all relevant detail may not be possible, though significant progress has been made recently on the numerical simulation of hydrodynamic interactions of suspensions of particles [166], and, independently, in the simulation of sedimentation of a large statistical ensemble of particles exhibiting Brownian motion [167]. Besides the hurdles in the theoretical description of systems at high concentrations, another concern is the pragmatic question whether the information content of an SV experiment would suffice to define all relevant parameters.[1] These difficulties are formidable. Therefore, for practical data analysis in the present context we will only attempt to capture the onset of hydrodynamic nonideality in semi-dilute conditions.

It is interesting to note that there is a vast amount of theoretical knowledge and practical work on nonideal sedimentation in the historic analytical ultracentrifugation literature, but nonideality is far less frequently encountered and considered in the modern experimental SV literature concerned with protein interactions. It appears that the combination of improved detection systems and more detailed data analysis approaches have improved opportunities to study systems with higher affinities, and to work under sufficiently dilute conditions where nonideal effects are negligible, or, when they do occur, can be captured in a linear, first-order approximation. Nevertheless, nonideality remains an important determinant of the sedimentation properties of many polymers and macromolecules that, due to their extended hydrodynamic volume and/or charge, are difficult or impossible to study under close-to-ideal conditions [169–174]. Furthermore, renewed interest in highly concentrated samples has arisen for the study of polymer conformation [170, 175], proteins in pharmaceutical formulations [176], as well as crowded biological fluids such as serum and even cell lysates [177–181]. Distinct from analytical ultracentrifugation, going back to the work of Jean Baptiste Perrin [182], sedimentation of particle suspensions is also traditionally of great theoretical interest in statistical physics and fluid dynamics and has been intensely studied due to its importance as an industrial separation tool [165, 183]. The latter literature traditionally discusses larger particles than typical biological macromolecules, and higher volume fractions, and is therefore highly relevant for analytical ultracentrifugation at nonideal conditions.

The aim of the following chapter is twofold. First, we will introduce qualitatively and semi-quantitatively the origin and phenomenology of the most common forms of nonideal sedimentation, such that they can be anticipated and recognized. Second, we discuss how the analysis methods developed so far for determining size-distributions and characterizing macromolecular interactions can be applied or need to be adapted, respectively.

---

[1]This situation is different in sedimentation equilibrium, even for mixtures with an arbitrary number of components with attractive and/or repulsive interactions [168].

5.1 PHENOMENOLOGY OF NONIDEAL SEDIMENTATION VELOCITY OF SINGLE SPECIES

5.1.1 Hydrodynamic Theory

With focus on the concentration-dependence of sedimentation behavior of single macromolecular component not subject to self-association, we may formally write the sedimentation and diffusion coefficients in a first-order, linear expansion as [19, 21]

$$s(w) = \frac{s^0}{1 + k_s w + \ldots} \quad \text{and} \quad D(w) = D^0 \left(1 + k_D w + \ldots\right) , \qquad (5.1)$$

w denoting the weight concentration and s^0 and D^0 the 'ideal' sedimentation and diffusion coefficients in infinite dilution, k_s the hydrodynamic nonideality coefficient for sedimentation, and k_D the nonideality coefficient for diffusion. The ellipses indicate sequentially higher-order terms (as in a Taylor series or virial expansion). Often, a different expression is used for the sedimentation coefficient

$$s(w) = s^0 \left(1 - k_s w\right) , \qquad (5.2)$$

which for small k_s and w is essentially equivalent to Eq. (5.1).

In order for these formally introduced flow coefficients to become more meaningful, they need to be put in the context of a molecular theory explaining their origin. To the extent that the concentration-dependence arises from inter-particle interactions, k_s and k_D are often related to the second virial coefficient A_2, as sketched out in the following [184, 185]: A_2 is a result of all inter-particle interactions, and for quasi-spherical particles can be written as [186, 187]

$$A_2 = \frac{2\pi}{M^2} \int \left(1 - g(r)\right) r^2 dr , \qquad (5.3)$$

where $g(r)$ is the pair correlation function that expresses the probability of finding particles with a separation r from each other, which, in turn, can be related to the mean particle-particle interaction potential $W(r)$ [187]

$$g(r) = e^{-\frac{W(r)}{kT}} . \qquad (5.4)$$

A_2 may be positive or negative, reflecting overall repulsive or attractive interactions, respectively. The osmotic susceptibility $(\partial \Pi / \partial c)_{T,\mu_3}$ (at constant temperature and chemical potential of the solvent) can be expanded in a power series [184, 188]

$$(\partial \Pi / \partial c)_{T,\mu_3} = RT \left(1 + A_2 M w + \ldots\right) . \qquad (5.5)$$

On the other hand, the osmotic susceptibility is also related to the diffusion coefficient in the generalized Stokes–Einstein relationship [189]

$$D = \frac{(\partial \Pi / \partial c)_{T,\mu_3}}{N_A f_c} , \qquad (5.6)$$

and the hydrodynamic nonideality of the frictional coefficient at finite concentration[2] f_c is related, in turn, to that of the sedimentation coefficient [184]

$$s/s^0 = f^0/f_c = 1/\left(1 + k_s w + \ldots\right) . \tag{5.7}$$

In the limit of infinite dilution, D approaches $D^0 = RT/N_A f^0$. With the assumption that the translational frictional coefficient and its concentration dependence is still the same for sedimentation and diffusion under nonideal conditions, we insert Eq. (5.7) into Eq. (5.6), and with Eq. (5.5) obtain

$$\left(1 + k_D w + \ldots\right)\left(1 + k_s w + \ldots\right) = \left(1 + A_2 M w + \ldots\right) , \tag{5.8}$$

or, at not too high concentrations [184, 185, 190],[3]

$$k_s + k_D \approx A_2 M . \tag{5.9}$$

Thus, it is possible to estimate the virial coefficient through the concentration-dependence of sedimentation and diffusion.[4] While the range of validity for flow conditions is limited,[5] due to the concentration-dependence of both the numerator and denominator in Eq. (5.6), at least k_D is generally expected to be smaller than k_s and $2A_2 M$ (i.e., the osmotically driven enhancement of diffusion will be partially compensated by the hydrodynamic retardation) [185, 188].

How large is k_s and what is its molecular origin? This question is not easy to answer for several reasons, and there are a number of contributing factors. Perhaps the best understood and in many practical situations most important contribution is the mutual hydrodynamic interaction of the particles, which can be approached rigorously with the tools of statistical mechanics and theoretical fluid mechanics.

A seminal work is that by Batchelor [194] on the sedimentation velocity of a suspension of hard spheres as a function of concentration. Assumptions are that the

---

[2]This ratio f_c/f^0 is the nonideality of the molecular parameter, not to be confused with the parameter value obtained when using the apparent frictional ratio f_r as an empirical fitting parameter describing the measured boundary spread in SV under nonideal conditions.

[3]The corresponding expression by Harding and Johnson features $\bar{v}$ as an additional term. This is due to an alternate definition of the nonideality coefficients based on solution density and viscosity corrected parameter values to the reference conditions of water at $20°C$.

[4]This relationship between k_s and k_D was confirmed experimentally by Kops-Werkhoven and Fijnaut for silica particles in cyclohexane as a model for hard spheres [191]. Later, it was demonstrated also for protein solution by Solovyova et al. using halophilic malate dehydrogenase in different solvents as a model system [185].

[5]A basis for this derivation is the assumption that the friction coefficient for sedimentation and diffusion is the same. In addition to the usual requirements of the absence of flow-induced shape changes in the particles, this also requires the probability distribution of inter-particle distances to be the same for diffusion and sedimentation. However, in the presence of hydrodynamic nonideality, the pair-distribution function of interacting particles will not be a Boltzmann distribution of the interparticle potential, but will be affected by additional hydrodynamic interparticle forces, such that for flow conditions the thermodynamic equilibrium relationships may not apply [192, 193].

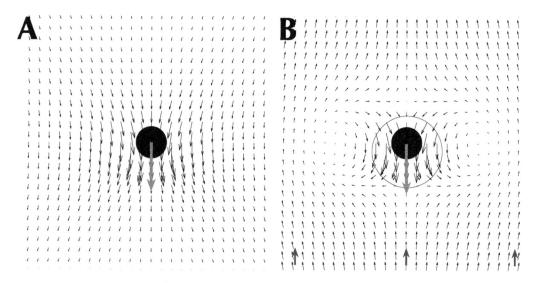

Figure 5.1 Laminar flow of a viscous fluid past a sedimenting spherical particle (circle) at low Reynolds number. The local fluid velocity is indicated by blue arrows, following the approximate solutions of the Navier–Stokes equation for creeping flows by Stokes [197]. *Panel A*: The ideal sedimentation of a particle through resting solution is depicted (gray arrow), which is correct for dilute conditions. Noteworthy is the long-range ($\sim 1/r$) effect of the sedimenting sphere dragging the surrounding fluid even far from the particle itself. *Panel B*: The same situation superimposed by a back-flow caused by the condition that the overall velocity of the solution remain at rest. The sedimentation of particles must then be balanced by an opposite flow of the fluid (red arrows). This is more appropriate for finite particle concentrations, since the counter-flow velocity is proportional to the number of sedimenting particles and their sedimenting volume fraction. The counter-flow acts to reduce the absolute velocity of the particle, since it generates increased frictional forces. Highlighted by the red circle is fluid dragged along due to the stick boundary condition, increasing the effectively sedimenting volume fraction.

spheres are equal in size, randomly uniformly distributed in the fluid, and that they exhibit sticky boundary conditions for the fluid flow.[6] The reason for the important contributions of hydrodynamic interactions is that the velocity disturbance of a single sphere decays very slowly with $1/r$, as indicated in Fig. 5.1 for a single particle in an infinite solution volume, with the fluid at infinite distance from the particle being at rest. The figure shows that even several diameters away, the fluid is dragged by the particle and moving with a significant fraction of the particle velocity. As a consequence, neighboring particles that are an appreciable distance away will feel the disturbance in the flow and experience altered frictional forces.

Fig. 5.1 highlights a second key observation: The reference frame for measuring sedimentation is not that of the fluid resting at an infinite distance from the particle, but that of the entire solution being at rest. In the latter, particles going one

---

[6]This may not always apply, for example for small molecules in organic solvents. The literature on this topic is largely concerned with rotational diffusion, as this is a more sensitive indicator of the boundary conditions [195, 196].

way must generate a counter-flow of fluid in the opposite direction. The back-flow velocity will depend on the total volume of fluid relative to the total number of particles, their velocity, and the volume they occupy. This velocity offset of the fluid leads to larger drag forces. For example, the flux from a single sedimenting particle will generate a negligible back-flow velocity in a solution contained in a volume — a situation depicted in Fig. 5.1A. In Fig. 5.1B the same velocity field is depicted as in 5.1A, only with an offset back-flow velocity of the fluid. Importantly, because of the long-range effect of the hydrodynamic drag on the fluid, the sedimenting volume relevant for the back-flow is not only that of the particle itself, but increased by a factor of $\sim$4.5 due to the 'co-migrating' fluid (encircled in red in Fig. 5.1B). Batchelor calculated the overall effect of the mutual hydrodynamic interaction of a statistical ensemble of mono-disperse hard spheres, randomly distributed through the solution (but two spheres are not allowed to overlap in the same space), and occupying a total volume fraction Φ. Accounting for two-particle interactions, valid for not too high volume fractions, the resulting decrease in sedimentation velocity is [194]

$$s = s^0 \left(1 - 6.55\Phi\right) . \tag{5.10}$$

Besides the total back-flow contribution of -5.5Φ, other contributions are those from repulsive near-field two-particle hydrodynamic interactions of -1.55Φ, and a small positive effect of $+0.5\Phi$ from the hydrostatic pressure gradient that slightly promotes sedimentation [187]. Eq. (5.10) is applicable in the dilute limit, for volume fractions below $\sim$5%. Several authors have calculated expressions for hindered sedimentation at higher volume fractions, making different assumptions and taking into account higher-order interactions [198–201]. Cichocki and colleagues have extended Batchelor's result to account for three-particle interactions [201] leading to $s = s^0 \left(1 - 6.55\Phi + 21.9\Phi^2\right)$. Brady and Durlofsky [200] describe an effective medium approach to account for many-body interactions leading to $s = s^0 \left(1 - 5\Phi\right)$ in the limit of small Φ, later extended by Hayakawa and Ichiki to include hydrodynamic lubrication terms, arriving at $s = s^0 \left(1 - 6.49\Phi\right)$ for small Φ [202].

It is important to appreciate the crucial effect of the inter-particle distance distribution on sedimentation velocity, because, for example, the magnitude of the overall contributions to sedimentation from near-field hydrodynamic interactions is dominated by the probability of close pairs [203, 204].[7] A difficulty is that higher-order hydrodynamic interactions, in turn, influence the distance distribution, i.e., contribute additional velocity-dependent terms to the pair-correlation function Eq. (5.4) [161,194]. Randomness of the distribution, as assumed in the derivation of Eq. (5.10), seems reasonable for rapidly diffusing small particles; however, Brownian motion of very large particles is much reduced relative to sedimentation. Thus, Cichocki and Sadlej [203] have calculated two-particle distance distributions of sedimenting non-Brownian suspensions of mono-disperse spherical particles, and found that,

---

[7]Different inter-particle distance distributions will be discussed in more detail in Section 5.3.1 when considering interacting systems (see Fig. 5.8 below).

even without attractive or repulsive inter-particle potential when at rest, hydrodynamic interactions alone will modulate the inter-particle distance-distribution. The
sedimentation process results in non-random arrangement of particles, with higher
probability of particles to be in close vicinity, in particular lateral to sedimentation [204], such that they experience reduced hydrodynamic nonideality [203]

$$s = s^0 \left(1 - 3.87\Phi\right) . \tag{5.11}$$

As mentioned above, the result Eq. (5.10) holds only for moderately high concentrations. Above volume fractions of >6%, the solution becomes so crowded that
the average gap between particles is on the order of a few particle diameters or less,[8]
and therefore more complex many-body interaction effects set in. For example, Xu
and colleagues have pointed out that the slower fluid flow between two particles
creates Bernoulli forces, bringing particles closer together, and creating a driving
force for particle aggregation [205]. This effect will counteract the concentration-
dependent retardation at high volume fraction.

With regard to the nonideality coefficient of diffusion, k_D, significant work
was done on the concentration-dependence of the hydrodynamic interactions
[191, 193, 195, 206, 207]. As indicated above, this topic is closely related to that
of sedimentation, since the particle flux due to gradient diffusion is analogous to
that resulting from a steadily applied force. However, the treatment of diffusion in
the presence of sedimentation is more complex. For large particles, in addition to
the ordinary diffusion, a 'hydrodynamically induced diffusivity' occurs [208], arising from the fluctuations of sedimentation velocity of individual particles. These
velocity fluctuations are driven by fluctuations in the local particle concentration.
For example, particles that by chance are transiently at a small distance from each
other will experience a higher sedimentation velocity (due to reduced friction in
their mutual flow field), whereas a particle that has transiently been depleted of
near neighbors will experience higher hydrodynamic friction. This variance will lead
to excess diffusive migration. Another, equivalent explanation for the excess diffusivity is the advection of volume elements of the solution with transient density
fluctuation driven by random concentration fluctuations. The correlated regions
have been observed to span ~20-fold the mean interparticle spacing [209]. The velocity fluctuations arising from correlated regions are thought to be dependent on
container size [210], but suppressed in the presence of concentration gradients (i.e.,
in the boundary region). The magnitude of this diffusivity is still the subject of
theoretical studies and computer simulations [183, 204, 211]. In any event, it appears that such hydrodynamic diffusivity effects can become a dominant source

---

[8]In a simple back-of-the-envelope calculation we assume a simple cubic grid of spheres with
uniform spacing. A volume fraction of 5% (for example, ~70 mg/ml of a globular protein with
$\bar{v}$ of 0.73 ml/g) corresponds to a gap between neighboring particles of ~2.4 particle radii. For
comparison, 10 mg/ml of a globular protein with $\bar{v}$ of 0.73 ml/g would leave average gaps of 6–7
radii, whereas 1 mg/ml of such a protein would leave more than twice as much space for fluid to
flow in between, in the latter case leading to almost negligible hydrodynamic interactions relative
to experimental accuracy.

of boundary spreading for very large, non-Brownian particles that have very low diffusion coefficients, but for smaller particles, where ordinary diffusion fluxes are large or of comparable order of magnitude as sedimentation, we expect that the hydrodynamically induced diffusivity will be much smaller or negligible, due to the short life-time of the concentration fluctuations. In this case, the framework for k_D as derived in the absence of macroscopic sedimentation should be applicable.

As is apparent from above, the complexity of the phenomena involved seems usually to exceed the information content that can be easily extracted from experimental data in analytical ultracentrifugation, as well as the knowledge we usually have about the properties of the sample in practical experiments. Keeping the potential complexities in mind, we will generally resign in the following to use what seems to be the most advanced, yet at the same time experimentally tractable, case as a reference point. This is the case of purely hydrodynamic interactions of randomly distributed, mono-disperse, small hard spheres at volume fractions less than ∼5%. Using this as the reference case, we will interpret discrepancies with this theory as an indicator for additional attractive or repulsive forces (or other effects as the case may be).

5.1.2 Comparison with Experimental Data

Let us turn our attention to the comparison of the theoretical predictions of hydrodynamic interactions with experimental results. To this end, it is useful to express the coefficient in front of the volume fraction Φ in Eq. (5.10) with the nonideality coefficient k_s from the phenomenological description of Eqs. (5.1) and (5.2) that can be experimentally measured. For spheres, it should ideally be

$$\Phi = \bar{v}w \ , \tag{5.12}$$

which would lead to a value of k_s of $6.55\bar{v}$ or ≈ 0.005 ml/mg for spherical particles with partial-specific volume of 0.73 ml/g, such as proteins.

As a practical matter, there is a potential source of confusion about k_s-values arising from the correction of the sedimentation coefficients s_{xp} to $s_{20,w}$, as described in Chapter 1 (Eq. 1.5) of Part I [1], if carried out prior to the linear regression $s_{20,w}(w)$ that measures the slope $-k_s$. While the experimental viscosity value for this correction must always be taken as that of the solvent, there are two different schools of thought whether the experimental density should be that of the pure solvent, or that of the solution including macromolecules at concentration w (see Part I, Chapter 2 [1]). If the solvent density is used for the correction to standard conditions, then the correction factor is concentration-independent, and the linear regression of $s_{20,w}(w) = s_{20,w}^0(1 - k_s w)$ will lead to the same k_s-value as the uncorrected s-values under experimental conditions, $s_{xp}(w) = s_{xp}^0(1 - k_s w)$. However, if the solution density is chosen then a distinct concentration-dependence appears in the correction factor, which will alter the observed slope. We will denote the values obtained from using the solution density as k_s^*, to distinguish it from that using the solvent density, k_s, which we have assumed so far and will continue to refer to in the following unless noted otherwise. As shown by Rowe [212] and Harding [184], the resulting slopes can be converted by $k_s^* = k_s - \bar{v}$ in the limit of close to dilute conditions.

When comparing the theory with experiment, we have to account for the problem that most particles are not exactly spherical. In fact, it is well established experimentally that elongated macromolecules exhibit significantly larger hydro-dynamic nonideality. This motivates the following *ad hoc* correction for roughly globular, but not perfectly spherical particles. Since ∼80% of the back-flow arises not from the actual volume of the macromolecule (with or without hydration), but from the effective volume of the dragged fluid, we may accordingly calculate the macromolecular volume fraction as the volume of the hydrodynamically equivalent sphere [212], which leads to the estimate

$$k_s \approx 6.55\,\bar{v}\,(f/f_0)^3\;. \tag{5.13}$$

Certainly, this is a simplistic approximation of the hydrodynamics of non-spherical particles,[9] but we believe a useful account for a shape-dependence for not extremely extended, globular macromolecules in the regime close to dilute solution. The value of (f/f_0) may be determined experimentally by SV in dilute conditions.[10]

This approximation leads to a value of 0.0093 ml/mg predicted for a protein with $\bar{v} = 0.73$ ml/g and $(f/f_0) = 1.25$, which is nearly twice the value of ∼0.005 ml/mg for a perfect sphere according to Eq. (5.12), but in good agreement with conventional wisdom and in the range of experimental data for many proteins [216, 217]. The formula of Eq. (5.13) highlights an experimentally well-established fact that elongated molecules will exhibit nonideality at much lower mg/ml concentrations than globular ones. Thus, where typical globular proteins as a rule of thumb would show a decrease of s by ∼1% at a concentration of ∼1 mg/ml, extended proteins with $f/f_0 = 2$ would show the same already at 0.25 mg/ml.

Unfortunately, besides the problem of non-spherical hydrodynamic shape, another limitation in the comparison of experimental k_s-values for proteins with theoretical predictions is the question of whether weak self-association may be superimposed on the concentration-dependence of $s(w)$ (see below).

---

[9]Dogic and co-workers have derived the expression

$$s = s^0 \left(1 - \frac{6.4 + (2/9)(L/D)}{2\ln(L/D)}\frac{L}{D} \times \Phi\right)$$

for rods with ratio of length (L) to diameter (D) greater than 20 [213]. Claeys and Brady have analyzed the hydrodynamic interactions and sedimentation of prolate spheroids [214].

[10]Different descriptions with regard to the consideration of hydration are possible, as discussed in Section 2.1.4 (Eq. 2.11) of Part I [1]. For example, the partial-specific volume may be taken as that of the hydrated particle [215] at a correspondingly reduced apparent frictional ratio. We believe which picture is adopted is irrelevant in the present context, as long as it is applied consistently to both $\bar{v}$ and (f/f_0) (as well as the weight concentration), such that the correct experimental buoyant molar mass and the correct Stokes radius is maintained.

Perhaps one of the most carefully conducted and most frequently cited early experiments is that by Cheng and Schachman [218], who studied the concentration-dependent sedimentation of polystyrene beads. These beads are thought to be close to ideally spherical (with a diameter of 0.26 μm) and sufficiently large for surface hydration to have a minimal influence. The effective partial-specific volume ϕ' was determined as 0.9506 ml/g by the neutral buoyancy method. NaCl at 0.098 M was added to the suspensions in order to suppress charge effects. At 25°C, the s-value at infinite dilution was 2410 S, and in experiments with volume fractions up to 7.5% a k_s-value of 0.0051 ml/mg was found. By contrast, with a frictional ratio of 1.0, the expected value from Eq. (5.13) for purely hydrodynamic interactions of Brownian particles following Batchelor's theory is 0.0069 ml/mg.

There are several factors probably contributing to this discrepancy. The first is rather trivial: The value reported by Cheng and Schachman was taken as the first term in a quadratic fit, whereas a linear fit would seem equally adequate within the error of the data points, leading to a $\sim$10% higher value for the average slope in the dilute regime for which Batchelor's predictions were later made. Second, not known at the time, it is likely that residual electrostatic repulsion may have led to altered distance distributions [187] (see below). This is quantitatively supported by the experimental and theoretical work of Goldstein and Zimm, who later determined the Hamaker constant for polystyrene particles in water [160] and obtained results that would provide a plausible explanation of Cheng and Schachman's data in view of hydrodynamic theory extended to interaction potentials [160, 161, 187]. Third, based on later theoretical work by Batchelor, polydispersity of the polystyrene beads should lead, at the high ratio of sedimentation to diffusion for the large particles, to slightly lower predictions for the coefficient than the ideal case of mono-disperse particles [161]. A fourth potential factor arises from most recent theoretical work: It may be that the polystyrene beads are of sufficient size to exhibit some degree of non-Brownian hydrodynamic behavior, where the hydrodynamic interactions themselves modulate the pair distribution function and thereby reduce the theoretically expected nonideality coefficient to the smaller values of Eq. (5.11) [201].

Other experiments were carried out to test Batchelor's prediction. Kops-Werkhoven and Fijnaut carried out comprehensive sets of experiments with up to 150 mg/ml of stabilized silica particles in cyclohexane, and found values consistent with theoretical expectations at $k_s/\bar{v} = 6 \pm 1$ [191] and $k_s/\bar{v} = 6.6 \pm 0.6$ [219]. Interestingly, at a radius 16.5 nm, these show significant diffusion, such that they should exhibit more random particle distance distributions.

In summary, despite thorny experimental details and theoretical improvements, Batchelor's theory presents an excellent starting point for expectations of the magnitude of nonideality in sedimentation.

5.1.3 Characteristic Boundary Features

The above considerations explain the overall sedimentation velocity of particles at given concentration, which could be measured, for example, from the propagation of the boundary mid-point, or better from mass balance considerations of the transport method. In addition, repulsive hydrodynamic interactions will have a profound impact on the shape of the sedimentation boundary.

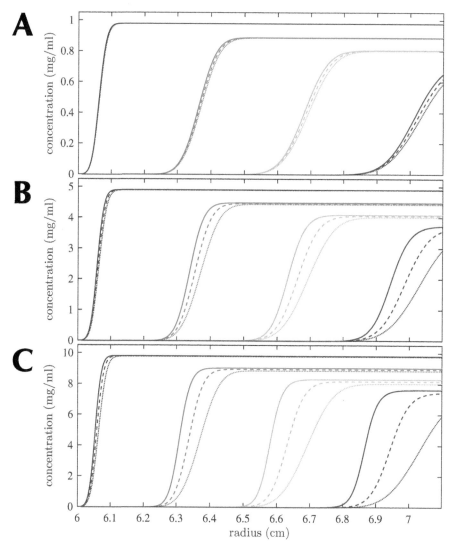

Figure 5.2 Effect of hydrodynamic nonideality on the boundary shapes of a 200 kDa, 10 S particle, sedimenting at 50,000 rpm. The nonideality coefficient k_s was assumed to be 0.02 ml/mg (solid lines), 0.01 ml/mg (dashed lines), or zero for ideal comparison (dotted lines). k_D was zero for all cases (see Fig. 5.3). Boundary shapes were calculated for loading concentrations of 1 mg/ml (*Panel A*), 5 mg/ml (*Panel B*), and 10 mg/ml (*Panel C*), and are shown at the same points in time for each color.

It is clear that the hydrodynamic, electrostatic and other interactions will act instantaneously as a function of the local occupied volume fraction. In a first approximation for small, globular particles at not too high volume concentrations, they therefore can be described by the Lamm equation with locally concentration-dependent sedimentation and diffusion coefficients [21]. Accordingly, a modified Lamm equation for nonideal sedimentation is

$$\frac{\partial \chi}{\partial t} = -\frac{1}{r}\frac{\partial}{\partial r}\left(\chi s(\chi)\omega^2 r^2 - D(\chi)\frac{\partial \chi}{\partial r} r\right) , \qquad (5.14)$$

analogous to Eq. (1.16), with the concentration dependence of the sedimentation and diffusion coefficient following the expansions of Eq. (5.1).

The theoretical concentration profiles of such nonideal sedimentation were examined in some detail by Fujita using approximate analytical Lamm equation solutions [220], and later by Yphantis and co-workers using numerical solutions [221]. It is straightforward to adapt modern finite element solutions described in Appendix A of Part II to this case [2], and representative shapes are shown in Fig. 5.2 for moderate and relatively strong nonideality for a protein at 1 mg/ml, 5 mg/ml, and 10 mg/ml.

> The nonideal Lamm equation (5.14) can be solved in `SEDFIT` for a single or multiple discrete species. k_s and k_D are specified as base-10 logarithm of the value in ml/mg units (e.g., -2 stands for 0.01 ml/mg). Since nonideality scales with the weight concentration but fitted data are in signal units, a conversion factor `signal/(mg/ml)` is required. This value will depend on the data acquisition system, optical pathlength, and molecular extinction coefficient.

The overall retardation of the boundary at higher concentrations following Eq. (5.2) can be readily discerned (compare in Fig. 5.2 the midpoints of equivalent boundaries in the order of increasing concentrations from A to C). Superimposed to that, for given loading concentrations, a more subtle effect of nonideality on the boundary movement is a slight acceleration of the boundary, increasing the s-value with time due to radial dilution. Cheng and Schachman [218] present an approximate expression for the magnitude of increase Δs relative to the initial s-value s_0 associated with boundary migration from meniscus m to a midpoint $r(t)$

$$\frac{\Delta s}{s_0} = \frac{r(t)^2 - m^2}{r(t)^2} \times \frac{k\,\Phi}{1 + k\,\Phi\,m^2/r(t)^2} . \qquad (5.15)$$

From the second term it follows that the acceleration is stronger at higher concentrations.[11]

Repulsive nonideality has a major effect on the boundary shape due to self-sharpening: The leading edge of the boundary will experience more retardation

---

[11] An explicit expression was later presented by Fujita [220].

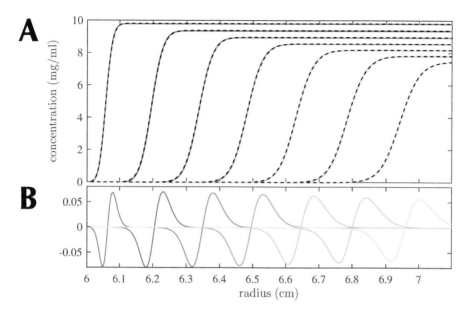

Figure 5.3 For the same conditions as in Fig. 5.2C, the impact of k_D is examined. *Panel A*: The simulation at $k_s = 0.01$ ml/mg is reproduced with $k_D = 0$ (colored solid lines) in comparison with $k_D = 0.01$ ml/mg (black dashed lines). *Panel B*: Residuals between the two simulations.

from the higher local concentration, as opposed to the trailing edge where the lower concentration allows sedimentation to be closer to the ideal velocity, causing the trailing particles to catch up with the rest of the boundary more than usual. In fact, in the limit of very strong repulsive nonideality, the boundary shapes become constant with time [222]. In this steady-state, the diffusional broadening is completely compensated by the differential sedimentation within the boundary. For the case of negligible concentration-dependence of D, Creeth [222] has found the limiting steady-state boundary shape

$$\frac{c_p - c(r)}{c(r)} = \exp\left[-\frac{c_p \omega^2 s^0 k_s}{2D}\left(r^2 - r_{1/2}^2\right)\right],\qquad(5.16)$$

where $r_{1/2}$ is the radius at the boundary half-height.[12,13]

Thus, the concentration-dependence of the s-value is a major determinant for the boundary spread, which in the ideal case is determined mostly by diffusion. By contrast, the effect of concentration-dependent nonideality of the diffusion

---

[12]This form was confirmed numerically first by Dishon and colleagues [221] and later used as a criterion for the correctness of a modern implementation of numerical Lamm equation solutions by Solovyova and colleagues [185].

[13]Creeth has also discussed possible practical applications of this, for example, in terms of the determination of s/D and sample homogeneity [222]. More recently, the steady-state was revisited as a tool to determine D by Scott, Harding, and Winzor [223], in particular, for unstructured proteins where conditions are favorable due to their high k_s, small D, and small s.

coefficient is much more subtle: As shown in Fig. 5.3, it does not affect the boundary midpoints, and positive values of k_D lead to only a slight broadening of the profiles.

If we were to fit the nonideal sedimentation profiles with a standard $c(s)$ model, besides the impact on s, we would obtain at higher concentrations increasingly worse fits. For example, for the data with $k_s = 0.010$ ml/mg in Fig. 5.2 (dashed lines) the rmsd is 0.0007 at 1 mg/ml, 0.015 at 5 mg/ml, and 0.056 at 10 mg/ml. Relative to the total loading signal, these deviations may still be considered subtle. Therefore, mass balance considerations for determining s_w through the integration of $c(s)$ may still be acceptable at not too high concentrations (especially in conjunction with Eq. (3.3) as error control). Still, one can discern the nonideal sedimentation pattern in comparison with the best-fit ideal $c(s)$ sedimentation model from the tell-tale signs of residuals: As illustrated in Fig. 5.4, ideal fits (dashed lines) systematically under-estimate the boundary width at early scans in combination with a systematic over-estimate of the boundary width at later scans. These residuals are the result of the nonideal boundary self-sharpening and spreading less, in conflict with the $\sqrt{t}$ behavior for ideal diffusion, and even stronger spread with $e^{\omega^2 st}$ for heterogeneous systems (see Part II, Chapter 3 [2]).

By contrast, $c(s)$ fitting parameters relating to the boundary spread are very severely impacted. Since the boundary shape is governed chiefly by the differential s-value in the concentration gradient, any *ad hoc* interpretation of boundary spread in terms of diffusion coefficient or frictional coefficient as for ideal systems would be misguided [19,222,224]. Rather, these values need to be considered 'apparent' values empirically describing the boundary shape, instead of assigning physical molecular meaning: The concentration-dependence of the observed boundary width should not be confused with the concentration-dependence of the diffusion coefficient referred to in Eq. (5.1). Similarly, the best-fit apparent frictional coefficients from $c(s)$ analysis will be strongly concentration-dependent, but in a different way than the molecular parameter f^0/f_c described in Eq. (5.7). For example, for the data in Fig. 5.2 the values of best-fit apparent frictional ratios of a $c(s)$ fit are 1.238 in infinite dilution, compared with 1.298 at 1 mg/ml, 1.61 at 5 mg/ml, and 2.11 at 10 mg/ml. Correspondingly, if this apparent frictional coefficient is combined with the boundary s-value and erroneously converted into apparent molar mass, values significantly higher than the true mass are obtained, with overestimates by 8% at 1 mg/ml, 42% at 5 mg/ml, and 100% at 10 mg/ml.

However, it should be noted that the concentration-dependence of the boundary spread parameters can still be extrapolated back to zero concentration, to arrive at well-defined physically meaningful extrapolations in infinite dilution, for example, of the frictional ratio [225]. These may be used, for example, to describe molecular shapes, or, in conjunction with the extrapolated sedimentation coefficient at infinite dilution, to calculate the macromolecular molar mass [170, 225–227]. More information on the application of these approaches to different types of polymers can be found in the review by Pavlov [225].

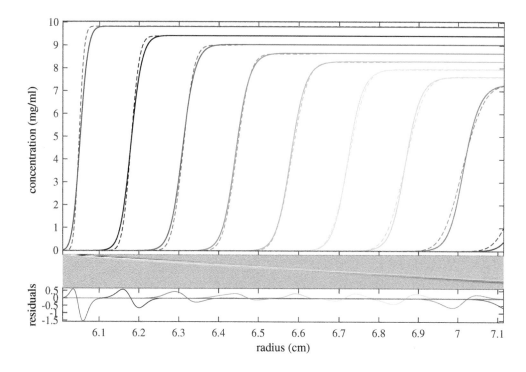

Figure 5.4 Mismatch between sedimentation data exhibiting hydrodynamic nonideality with an impostor ideal sedimentation model. The data from Fig. 5.2C with $k_s = 0.020$ ml/mg and $k_D = 0$ (solid lines) are fitted with a standard $c(s)$ model (dashed lines). Characteristically, the ideal sedimentation model cannot adapt to the relatively unchanged boundary shape, and the best-fit values results in an underestimate of the boundary width in the early scans and an overestimate of the boundary width for the late scans.

In the `Single Nonideal Species` model of `SEDPHAT` it is possible to fit experimental data from a nonideal sedimentation process of sufficiently homogeneous samples, based on the numerical solution of Eq. (5.14) [185]. With data from a sufficiently large range of concentrations, this allows one to obtain estimates for s^0, D^0, k_s, and k_D (if it is larger than 0.003 ml/mg). This works best in a global fit in `SEDPHAT` comprising data at a range of concentrations.

The hierarchical strategy for the non-linear regression is to set first k_s and k_D to zero, in order to obtain initial estimates for apparent s- and D-values, as well as the loading concentrations. These will be used as starting guesses of the second stage that additionally allows k_s to adjust, using a starting guess for this parameter of 0.01 ml/mg. The resulting estimates are then used as starting guesses for the final stage, together with a starting value for k_D of 0.01 ml/mg allowing now all parameters s^0, D^0, k_s, and k_D, in addition to the loading concentration, to refine and converge to the best-fit values.

A model for multiple discrete nonideal species is available in `SEDFIT` for analysis of a single data set at a time.

5.2 SEDIMENTATION OF HETEROGENEOUS NONIDEAL SYSTEMS

5.2.1 Predictions from Statistical Fluid Mechanics

The extension of the considerations so far to multi-component systems of different size and different density introduces significantly more complex behavior [192]. For example, in suspensions of large particles at above 15% volume occupancy the faster sedimenting particles may generate 'streaming columns' while sedimenting through lower density, slower sedimenting particles, which allow an increase in sedimentation rate as compared with a structure-less random mixture [228].

In the limit of low volume fractions, the nonideal sedimentation behavior of multiple species requires multiple mutual nonideality coefficients $k_{s,ji}$ describing the hydrodynamic interaction between species j and i in the form

$$s_i = s_i^0 \left(1 - \sum_j k_{s,ji} w_j\right). \tag{5.17}$$

Extending Batchelor's seminal work on the sedimentation of uniform spheres, Batchelor and colleagues have considered sedimenting polydisperse suspensions in detail [161, 192]. Values of the mutual nonideality coefficients can be quite different from the value $6.55\bar{v}$ of a single spherical component. The behavior depends very much on a mutual Péclet number

$$P_{ij} = \frac{1}{2}\left[(R_i + R_j)\left(s_j^0 - s_i^0\right)\omega^2 r\right]/\left[D_j^0 + D_i^0\right] \tag{5.18}$$

(with R denoting the particle radii) that in this case measures the relative separation due to different velocities in comparison with the average species diffusion. This number is very large for large macromolecules of very different size, where in ideal sedimentation we would see very strong separation of sharp boundaries. On the other extreme, the Péclet number is small for small macromolecules of similar size, where in dilute conditions the heterogeneous mixture remains within one diffusion-broadened boundary. Furthermore, Batchelor and Wen present numerical results indicating that samples containing particles of different density may generally behave very differently from those of homogeneous density [161, 192].

Fortunately, two simpler cases of practical relevance may be recognized for mixtures of particles with identical $\bar{v}$. First, for the case of large macromolecules (small diffusion coefficients) with high P_{ij}, provided they have not too dissimilar radius, the nonideality coefficient is lower with $k_{s,ji} = 2.52\bar{v}$ [161]. Second, in the case of smaller macromolecules (large diffusion coefficient) with $P_{ij} \approx 0$, the coefficient follows

$$k_{s,ji} = \left[(R_j/R_i)^2 + 3(R_j/R_i) + 1 - \alpha\right]\bar{v}, \tag{5.19}$$

with the number $\alpha \approx -2.5/(1+0.6(R_j/R_i))$, assuming values in the range between -2.17 and -0.67 for R_j/R_i within $0.25 - 4$ [161]. For example, spheres of equal size have $\alpha = -1.55$ leading to $k_{s,ii} = 6.55\bar{v}$, whereas at twofold different radii $R_j/R_i = 2$ a value of $k_{s,ji} = -11.09\bar{v}$ is predicted, and at $R_j/R_i = 0.5$ a value of

$k_{s,ji} = 3.94\bar{v}$. This illustrates that large, strongly size-dependent cross-coefficients may occur.

The difficulties involved in discriminating several nonideality coefficients seem to exceed the practical possibility for data analysis. However, the results above suggest that for polydisperse mixtures of proteins with molar mass differences not more than two-fold, it seems reasonable — in a first approximation — to reduce the nonideality coefficients to a single one, such that Eq. (5.17) simplifies to[14]

$$s_i \approx s_i^0 \Big(1 - k_s \sum w_i\Big) . \tag{5.20}$$

However, we also need to keep in mind that the relevant quantity underlying this approximation is the ratio of species' Stokes radii, i.e., not only mass but also differences in frictional ratio could potentially impact the dispersion of the $k_{s,ji}$-values.

5.2.2 The Johnston–Ogston Effect — Stratification of Multi-Component Solutions with Repulsive Nonideality

Even in the simplified form of Eq. (5.20), heterogeneous nonideal sedimentation can produce very interesting effects, and pose very challenging analysis problems. Let us just consider a system of two kinds of particles sedimenting at different velocities. Their migration is coupled by the mutual hydrodynamic repulsive nonideality. Let us also assume that the nonideality constant is the same for hydrodynamic self-interactions as for hydrodynamic hetero-interactions as suggested in Eq. (5.20). We may then write the resulting system of Lamm equations in weight concentration units as

$$\frac{\partial w_1}{\partial t} + \frac{1}{r}\frac{\partial}{\partial r}\left[w_1 s_1^0\Big(1 - k_s(w_1 + w_2)\Big)\omega^2 r^2 - D_1^0\Big(1 + k_D(w_1 + w_2)\Big)\frac{\partial w_1}{\partial r}r\right] = 0$$

$$\frac{\partial w_2}{\partial t} + \frac{1}{r}\frac{\partial}{\partial r}\left[w_2 s_2^0\Big(1 - k_s(w_1 + w_2)\Big)\omega^2 r^2 - D_2^0\Big(1 + k_D(w_1 + w_2)\Big)\frac{\partial w_2}{\partial r}r\right] = 0 . \tag{5.21}$$

An example for the solution is shown in Fig. 5.5 for the mixture at high concentrations of two proteins with dissimilar s-values. The most striking aspect is the formation of a plateau of the smaller species in excess of the loading concentration at the region in the solution that has been cleared from the larger species. Such behavior has been experimentally observed and first correctly explained by Johnston and Ogston [229].

The origin for this boundary anomaly lies in the lower s-values of the small species in the presence $(s_2^0 [1 - k_S(w_1 + w_2)])$ then in the absence $(s_2^0 [1 - k_S w_2])$ of the large species. To quantify this effect, let us look at the sketch of Fig. 5.6 where we

---

[14]For example, this form is considered by Frigon and Timasheff in their comprehensive analysis of tubulin self-association [57].

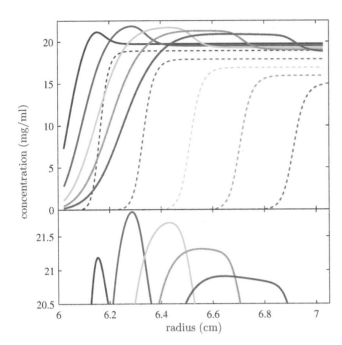

Figure 5.5 Lamm equation solutions for the nonideal sedimentation of two proteins at 20 mg/ml loading concentration each, with $s_1^0 = 1.5$ S, $D_1^0 = 7$ F, and $s_2^0 = 7$ S, $D_2^0 = 4$ F, respectively, corresponding to extended 20 kDa and 160 kDa proteins. The rotor speed is 50,000 rpm. Eq. (5.21) was solved with nonideality coefficients of $k_s = 0.014$ ml/mg and $k_D = 0.01$ ml/mg. The concentration profiles for the smaller species are shown in bold solid lines, and those for the larger species are shown in dashed lines, with the same colors corresponding to the same sedimentation time. The lower plot zooms in on the 'plateau' region of the small component in the zone free of the larger one.

have labeled the region containing no solute as α, that containing the small solute (2) only as β, and that containing the mixture of small solute (2) and large solute (1) as γ, respectively. We adopt the simplifying model of a constant force acting on molecules in a rectangular solution column with cross-section A. In this model, the $\beta\gamma$ boundary migrates with the velocity of the faster component, $v_{1\gamma}$, and is located at time t at the position $r_{\beta\gamma} = v_{1\gamma}t$. During the time-interval dt, it migrates by $dr_{\beta\gamma} = v_{1\gamma}dt$ and thereby uncovers $w_{2\gamma}Adr_{\beta\gamma}$ of the slower component, except for the part $w_{2\gamma}v_{2\gamma}Adt$ that will have co-sedimented to stay within γ. That leaves the accretion of the slower component into the β phase: $dm_{2+} = w_{2\gamma}(v_{1\gamma} - v_{2\gamma})Adt$. The $\alpha\beta$ boundary is at $r_{\alpha\beta} = v_{2\beta}t$, and the volume of the β phase is $(v_{1\gamma} - v_{2\beta})At$, increasing in the same time-interval by $(v_{1\gamma} - v_{2\beta})Adt$. In the linear geometry, since both the volume of the β phase and its mass of the slower component grow linearly with time, we find

$$w_{2\beta} = w_{2\gamma}(v_{1\gamma} - v_{2\gamma})/(v_{1\gamma} - v_{2\beta}) \tag{5.22}$$

Thus, since due to the repulsive interactions the velocity of the slow component is always higher in the absence of the fast component than in its presence, $(v_{1\gamma} - v_{2\gamma}) > (v_{1\gamma} - v_{2\beta})$, and therefore we have $w_{2,\beta} > w_{2\gamma}$. This means that the smaller component seems to be expelled from the solution region occupied by both components, by virtue of its retardation of the sedimentation rate in the co-occupied region. This result is valid in the same form also for sector-shaped geometry [230], and has been generalized to an arbitrary number of components by Smith [231,232].

To evaluate the magnitude of $w_{2\beta}$ *via* Eq. (5.22), a difficulty remains in the fact that $v_{2\beta}$ will depend on $w_{2\beta}$. Unless the slow boundary is very dilute such that $v_{2\beta} \approx s_2^0$, the concentration-dependent retardation of the $\alpha\beta$ boundary counteracts

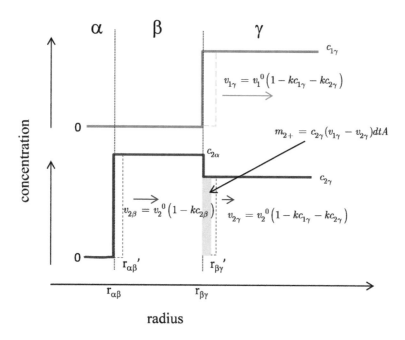

Figure 5.6 Sketch of conceptualized sedimentation boundaries, at a particular point in time t, of a two-component system with repulsive nonideality. We can distinguish three regions of the solution: α is the region of pure solvent; β is the region where only the slower component 2 is present; and γ is the mixed solution. The velocity of the faster sedimenting component 1 (blue) determines the migration of the $\beta\gamma$ boundary, and the velocity of the slower sedimenting component 2 (red) determines the migration of the $\alpha\beta$ boundary. The dashed blue and dotted red lines sketch the boundary positions at a later time $t + dt$. Highlighted as shaded red area is the material of component 2 that is transferred from the γ into the β region during this time interval.

the Johnston–Ogston effect. Expressing all velocities as those at infinite dilution retarded with a uniform nonideality coefficient leads to a quadratic equation for $w_{2\beta}$ with the solution

$$
w_{2\beta} = \frac{1}{2v_2^0 k_s} \left[-\left(v_1^0 \left(1 - k_s w_{tot}\right) - v_2^0\right) \right.
$$
$$
\left. + \sqrt{\left(v_1^0 \left(1 - k_s w_{tot}\right) - v_2^0\right)^2 + 4v_2^0 k_s w_{2\gamma}\left(v_1^0 - v_2^0\right)\left(1 - k_s w_{tot}\right)} \right],
$$
$$(5.23)$$

where $w_{tot} = w_{1\gamma} + w_{2\gamma}$.

This model is quantitatively a reasonable first approximation, but it does not

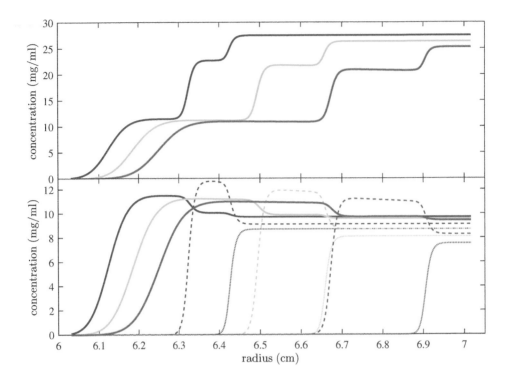

Figure 5.7 Lamm equation solutions for the nonideal sedimentation of an equal weight concentration mixture of three extended polymers with $s_1^0 = 2$ S and $D_1^0 = 4$ F, $s_2^0 = 10$ S and $D_2^0 = 2$ F, and $s_3^0 = 15$ S and $D_3^0 = 1.0$ F, with $k_S = 0.05$ ml/mg and $k_D = 0.05$ ml/mg. The loading concentration of each species is 10 mg/ml. *Panel A*: Total concentration distributions at three points in time. *Panel B*: At the same time points, the distributions of the individual components (smallest in solid lines, middle in dashed lines, largest in dotted lines).

describe more subtle points.[15] For example, it relies on the assumption that there are plateaus in the β phase. However, a closer inspection of the concentration profiles in the lower plot of Fig. 5.5 shows these to exhibit slight negative gradients in the β phase. These gradients can be rationalized by considering that radial dilution will steadily lower all concentrations, and thereby leads to a reduction of the amounts dm_{2+} transferring to the β phase with time. For most macromolecules, diffusion will not be able to diminish these gradients significantly. This is an intrinsically unstable situation because density gradients invariably accompany the concentration gradients. This density inversion could lead to convection [233], unless stabilized by positive density gradients from sedimenting buffer salts and solvent compressibility.

Based on the same principles, the extension of the mechanism of the Johnston–Ogston effect for multi-component mixtures can easily be envisioned to produce multiple steps for components sedimenting behind multiple faster species. An

---

[15] A more detailed review of results in radial geometry and their discussion can be found in Fujita's second book [21].

example for the sedimentation of a three-component mixture at equal 10 mg/ml loading concentration in the presence of moderate nonideality is shown in Fig. 5.7. Although the total signal that would be measured with a non-discriminating signal gives little indication for the presence of the Johnston–Ogston effect, the individual concentration distributions of all but the fastest sedimenting component show a clear increase of their concentration above the loading concentration in the partially depleted regions. Thus, a naïve analysis of the partial concentrations of each species would significantly overestimate the loading concentrations of the smaller components, and significantly underestimate that of the largest component. Clearly, the same happens in the continuous case, which leads to a potentially strong distortion of the sedimentation coefficient distribution if analyzed without accounting for the stratification from hydrodynamic nonideality.

5.3   CHEMICAL REACTIONS IN NONIDEAL SEDIMENTATION VELOCITY

The extension of the strategies for studying chemically reacting systems by SV into the regime where nonideal hydrodynamic interactions take place is highly desirable for many reasons. Chiefly, this is necessary for studying very weak interactions, which can be observed only at high concentrations, yet may play a very important role in the physical behavior of concentrated solutions. Weak interactions between biological macromolecules may be critical, for example, for protein function in the cytosolic environment. Even under more dilute conditions they can become highly relevant, as they may be amplified in multi-valent binding, for example, for recognition of ligands on the cell surface by pattern recognition receptors. Further, highly elongated molecules will exhibit hydrodynamic nonideality even at low concentrations. Finally, since the analysis of reacting systems often requires probing a large concentration range, even the study of moderate interactions may require experimental conditions where corrections for hydrodynamic nonideality at the highest concentrations would be in order.[16]

We will discuss two different scenarios. First, for interactions resulting from weakly attractive potentials, it may be more useful to express them in the framework of inter-particle potential and virial coefficients. Second, for chemical reactions that (at least transiently) produce discrete oligomers, we explore how the tools for the analysis of interacting systems introduced previously are affected by additional repulsive hydrodynamic interactions arising from high concentrations of the interacting macromolecules. Because of the seemingly intractable complexities of hydrodynamic interactions at higher volume fractions for quantitative analysis,

---

[16] An important practical virtue of SV analytical ultracentrifugation that makes it an attractive experimental choice for studying weak interactions is its ability to separate adventitious irreversible aggregates from the species of interest: The formation of such aggregates is a practical problem for many preparations of biological macromolecules, and their formation is usually promoted at high concentration. Despite the overall nonideal retardation of sedimentation, their typically substantially higher mass will allow the aggregates to rapidly sediment and clear the sample during centrifugation.

we restrict our considerations again to the regime of $\Phi < 0.05$, where comparatively simple linear approximations should hold.

5.3.1 Statistical Fluid Mechanics Picture of Chemical Interaction Potentials in Hydrodynamic Nonideality

As we have seen above, a key determinant for the magnitude of hydrodynamic interactions is the particle distance distribution. Batchelor's result $s = s^0 \left(1 - 6.55\Phi\right)$ was derived under the assumption of random positions of spheres, except for volume exclusion, as would be expected for rapidly diffusing hard spheres without other particle interaction potentials [194]. Let us look at this in more detail in order to see how particle interactions alter sedimentation.

The strongest near-field hydrodynamic effect stems from the contributions from individual pairs of spheres that happen to be in close vicinity. This situation will occur by virtue of their statistically random position in solution. Such a pair of spheres will not be retarded from the back-flow as much, since the pair will be more strongly influenced by the mutual flow-field of the dragged fluid in the vicinity of the pair. Actually, we can estimate from elementary hydrodynamic bead modeling that the hydrodynamic friction in the limit of touching spheres is only 0.7-fold that of two separate spheres. Thus, the number of 'close encounters' in the distribution will be very significant in modulating the overall hydrodynamic interactions.

Under conditions of moderate concentrations where Batchelor's results in Eq. (5.10) hold, it may be surprising that a completely random distribution of spheres, at reasonable concentration, will already exhibit a remarkable number of pairs of spheres in close vicinity as well as higher clusters just by chance. This is depicted in the computational experiment of Fig. 5.8B, which shows a random placement of particles on a two-dimensional grid with a 10% coverage. As can be expected, the number of pairs decreases in the presence of long-range repulsive pair-potentials, and it increases for attractive potentials, as shown in Fig. 5.8A and Fig. 5.8C, respectively. This picture is suggestive for the role of the interparticle potential on the hydrodynamics of the suspension.

Batchelor's statistical fluid dynamics analysis of the sedimentation of interacting spheres [192] leads to an expression for the impact of an excess number of (nearly) touching spheres caused by interaction potentials, relative to the random distribution. With the excess number of nearly touching spheres expressed in units of Φ as a factor $\alpha\Phi$, he found the overall sedimentation velocity will follow

$$s = s^0 \left(1 - \left[6.55 - 0.44\,\alpha\right]\Phi\right). \tag{5.24}$$

Thus, a 15-fold enhancement of the population of sphere pairs would create a situation where the hydrodynamic repulsive interactions are canceled out by the increase in sedimentation from the dimerization. On the other hand, repulsive forces reducing the random occurrence of close pairs will exacerbate the concentration-dependent decrease of s. For a given interparticle potential $W(r)$, α can be evaluated

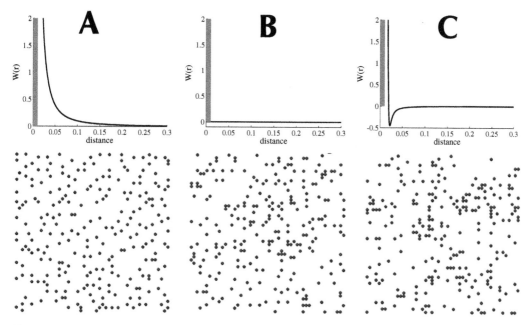

Figure 5.8 Effect of inter-particle interaction on the abundance of particles in close vicinity. This is the result of three computer experiments placing 250 particles onto a rectangular mesh of 50×50 positions evenly subdividing the interval $[0, 1] \times [0, 1]$. Different inter-particle potential energies were assumed in the different panels, as shown in the upper plots. In all panels, the chance occurrence of particles occupying neighboring grid points is highlighted by coloring these particles red. *Panel A*: The probability of particle position is calculated on the basis of Eq. (5.4) with a hard-sphere potential (random but not occupying the same place), with a superimposed long-range repulsive term $10^{-3}r^{-2}$. *Panel B*: Pure hard-sphere potential, without any other interactions. *Panel C*: Hard-sphere potential with an added term for weak long-range attraction $-10^{-4}r^{-2}$ plus a Lennard–Jones potential in close vicinity potential energy of magnitude $(0.02/r)^{12} - (0.02/r)^{6}$.

as

$$\alpha = \frac{24}{a^3} \int_a^\infty \left(e^{-W(\rho)/kT} - 1 \right) \rho^2 d\rho , \qquad (5.25)$$

for particles of radius a [192, 234]. α relates to the second virial coefficient in Eq. (5.3) such that

$$s = s^0 \left(1 - \left[6.55 - 3.52 \left(1 - B_2/B^{\mathrm{HS}} \right) \right] \Phi \right) , \qquad (5.26)$$

where B^{HS} is the virial coefficient for hard spheres [163, 187, 192, 235].

There are many potential sources of weak attractive and repulsive interactions that can modulate the particle distance distribution. A particularly interesting case is the presence of long-range electrostatic forces. In general, electrostatic repulsive interactions will tend to increase the overall magnitude of the hydrodynamic repulsive interactions, and lead to lower sedimentation velocities. Strong long-range repulsive interactions can force particles to be maximally apart, causing highly ordered, 'crystalline' states which can be observed even at very low volume fractions [206]. This state will minimize the occurrence of nearby particles that benefit from each other's hydrodynamic drag. For highly charged spheres under conditions of low ionic strength, the reduction of the sedimentation velocity approximately follows [162]

$$s = s^0 \left(1 - 1.8\,\Phi^{1/3} \right). \tag{5.27}$$

Unfortunately, the case of weakly charged spheres is computationally more complex. The addition of neutral salts, though diminishing long-range electrostatic interactions, may also introduce additional phenomena because now we have a multi-component system with coupled migration of species of different mobilities. Unless high concentrations of neutral salt are added, effects from the primary charge effect will be encountered (see Part I, Section 2.1.5 [1]), where the difference in the migration velocity of the small counter-ions will be slowing down the macromolecular sedimentation [236–238]. This is intrinsically not dependent on macromolecular concentration, but it does change the local fluid velocity (hence concentration-dependent hydrodynamic interactions). In the limit of very small volume fractions and very small charges, an expression for the reduction in sedimentation velocity has been presented by Petsev and Denkov [239]

$$s = s^0 \left(1 - \left[6.55 + \frac{3Z^2 e^2}{\epsilon kT \kappa a^2 (1 + \kappa a)^2} \right] \Phi \right), \tag{5.28}$$

where a is the radius of a sphere with effective charge Z, the dielectric constant ϵ, and the screening parameter $\kappa^2 = (4\pi e^2/\epsilon kT) \sum n_i z_i$, with n_i and z_i being the number density and charge of ions in solution. However, Watzlawek and Nägele have noted that this expression is likely correct only at extremely low volume fractions $\Phi < 0.0001$ [162].

5.3.2 Discrete Chemical Reactions Not Influenced by Hydrodynamic Interactions

We proceed with the consideration of stronger interactions that result in the formation of discrete (homo- or hetero-) oligomeric structures, governed by a chemical reaction that is unrelated to hydrodynamic interactions of sedimentation. In other words, the rate of formation of encounter complexes may well be affected also by mutual (local) hydrodynamic interactions during the approach of the reactants, but these are assumed to be identical to those of a dilute system at rest. We assume that there is a distance over which the concentration is sufficiently constant such that chemical reactions will take place locally as in the absence of concentration gradients. In this picture, the distribution of inter-particle distances is influenced by the chemical interactions and diffusion, but not by hydrodynamic effects. We do not expect these assumptions to hold at high volume fractions or where fluxes from sedimentation greatly exceed those of diffusion.

5.3.2.1 Nonideal Lamm Equation of Interacting Systems

In this approximation, in principle, it is possible to incorporate hydrodynamic nonideal interactions into the Lamm equation of reacting systems (Eq. 1.2), by accounting for the mutual hydrodynamic concentration-dependence in the sedimentation and diffusion fluxes of all species

$$\frac{\partial \chi_i}{\partial t} + \frac{1}{r}\frac{\partial}{\partial r}\left(\chi_i s_i^0\left(1 - \sum_j k_{s,ij}M_j\chi_j\right)\omega^2 r^2 - D_i^0\left(1 - \sum_j k_{D,ij}M_j\chi_j\right)\frac{\partial \chi_i}{\partial r}r\right) = q_i.$$

(5.29)

We have seen already for the ideal case how the Lamm equation of reacting systems requires an exquisite interpretation of the boundary spread, as it is governed by diffusion, differential sedimentation of the reaction boundary, and reaction kinetics, which makes it particularly susceptible to practical imperfections such as microheterogeneity of the sedimenting species. The hydrodynamic nonideality terms add yet another very dominant factor into the boundary shape analysis. These may be reduced to two unknown parameters if using the simplifying assumption that the nonideality coefficients are the same for all species, but as we have seen above, this pragmatic assumption can be questioned, but not easily validated. The boundary sharpening effect has the potential to mask the broadening effects from finite reaction kinetics, diffusion, and heterogeneity. Therefore, while the practical data analysis with Eq. (5.29) seems computationally possible and is theoretically perhaps attractive at first sight, the number of parameters and assumptions governing the boundary shape seem too many for meaningful analysis.[17]

An added difficulty in the study of interactions under nonideal conditions is that all species need to be accounted for in Eq. (5.29), whether or not they contribute to the signals, participate in chemical interactions, or even sediment. Importantly, chemically 'inert' species that are present in significant concentrations will contribute to the hydrodynamic interactions, and thereby can control the sedimentation behavior of the interacting system. This case will apply to sedimentation in 'crowded solutions.'

Therefore, to circumvent these formidable problems, we will again explore alternative approaches to the direct fit with Lamm equation solutions. In particular, we will examine how the approximate solutions to the ideal Lamm equation for reactive systems discussed above can be productively applied under nonideal conditions.

5.3.2.2   The Transport Method and the Weighted-Average s_w-Value

Of all the isotherm approaches for analyzing reacting systems, the weighted average sedimentation coefficient s_w is the only approach where hydrodynamic interactions can be relatively easily accounted for. This is due to the fact that its definition rests on the transport flux of materials across an imaginary plane in the solution plateau (see Section 3.1 of the present book and Part II, Section 2.3 [2]), which renders s_w independent of the distortions of the boundary shape from concentration-dependent hydrodynamic interactions, and independent of distortions in the heights of the sedimentation boundaries from the Johnston–Ogston effect. Therefore, in the still

---

[17]This does not mean that it would not be possible to achieve a good fit with Eq. (5.29).

relatively dilute regime considered of $\Phi < 0.05$, the weighted-average s_w-value can be used in the same fashion as in the absence of hydrodynamic interactions, as long as the concentration-dependent retardation $s_i = s_i^0 \left(1 - k_{s,i} w_i\right)$ is accounted for.

For example, the self-associating system in Eq. (3.8), can be written as

$$s_w(c_{\text{tot}}) = \frac{1}{c_{\text{tot}}} \sum_i i K_{1,i} c_1^i s_i^0 \left(1 - \sum_j k_{s,ij} j M K_{1j} c_1^j\right), \qquad (5.30)$$

M being the monomer molar mass, and using molar protomer concentration units for c_{tot} and c_1. The term $j M K_{1j} c_1{}^j = j M c_j$ describes the weight concentration of the oligomer, w_j. Again, since the different nonideality coefficients $k_{s,ij}$ cannot be independently measured, a possible approximation — for not too high concentrations of species with not too dissimilar shapes — is that of a single (average) nonideality coefficient $\bar{k}_s$, which leads to

$$s_w(c_{\text{tot}}) = \frac{1 - \bar{k}_s M c_{\text{tot}}}{c_{\text{tot}}} \sum_i i K_{1,i} c_1^i s_i^0, \qquad (5.31)$$

i.e., a concentration-dependent multiplicative correction to s_w from Eq. (3.8).

Hydrodynamic nonideality can be considered in SEDPHAT using the Options ▷Corrections for Hydrodnamic Interactions menu. This allows $\bar{k}_s$ to be entered as base-10 logarithm of the value in ml/mg units. Importantly, the components' molar mass values will be used to calculate c_{tot} from the customary molar concentration basis used in SEDPHAT. Some of the global parameter boxes will also offer an alternate way to specify hydrodynamic nonideality: In this case, a check-box labeled log10[(ks/(mg/ml)] nonid sed toggles hydrodynamic nonideality on/off, and the associated field allows entering the numeric value. A second check-box next to this field determines whether to fix or refine this value.

Fig. 5.9 illustrates the shape of these isotherms for the case of a monomer-dimer self-association with different equilibrium constants. When the dimerization constants K_D are in the range of between 1–10 mg/ml (magenta and blue line), it may in favorable cases be possible to estimate $\bar{k}_s$ from experimental isotherm data jointly with K_D. For weaker interactions, however, within the concentration range that is experimentally accessible — and considered in our theoretical approximation — the two parameters k_s and K_D will usually become strongly correlated. Notably, dimerization with a dissociation constant K_D in the range of 100 mg/ml may completely offset the hydrodynamic interaction in that the resulting s-values may be virtually concentration independent (red line). In this regime it is often necessary to make a theoretical prediction for $\bar{k}_s$, e.g., via Eq. (5.13), such that K_D can be determined.[18] Even though uncertainties in K_D correlate with those of $\bar{k}_s$,

---

[18]For example, such an approach was taken in the study of weak carbohydrate interactions in [240], using the prediction of nonideality by Rowe [212].

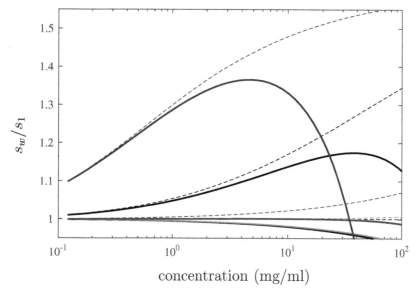

concentration (mg/ml)

Figure 5.9 Weighted average sedimentation coefficient s_w predicted for a monomer-dimer interaction at different binding constants (different colors) under ideal conditions (dotted) and with hydrodynamic nonideality with $\bar{k}_s = 0.01$ (solid lines). Shown are the s_w values in units of the monomer s-value (assuming $s_2^0/s_1^0 = 1.6$), as a function of mg/ml loading concentration. Lines are for $K_D = 1$ mg/ml (magenta), 10 mg/ml (blue), 100 mg/ml (red), 1000 mg/ml (green), and 10,000 mg/ml (black).

interactions with dissociation equilibrium in this K_D range can still be confidently recognized. At K_D-values of 1000 mg/ml and above, however, hydrodynamic nonideality will completely dominate. For example, this limit corresponds to 100 mM for a 10 kDa protein.

Hetero-associations may, in principle, afford the opportunity to determine independently the hydrodynamic interaction coefficients of the two interacting components, but the coefficient for the complex that forms in the mixture is still an unknown, as are it's cross-coefficients. Therefore, analogously to the self-association case above, we will have to resort to the simplest approximation of using a single, average $\bar{k}_s$ value for the hydrodynamic nonideality, to describe the heterogeneous interaction of similar-sized particles with similar $\bar{v}$ in the dilute region of low hydrodynamic interactions. The isotherm corresponding to Eq. (3.9) for a bi-molecular hetero-association would then be

$$s_w(c_{A,\text{tot}}, c_{B,\text{tot}}) = \left(1 - \bar{k}_s \left(M_A c_{A,\text{tot}} + M_B c_{B,\text{tot}}\right)\right)$$
$$\times \frac{\varepsilon_A c_A s_A + \varepsilon_B c_B s_B + \varepsilon_{AB} K c_A c_B s_{AB}}{\varepsilon_A c_A + \varepsilon_B c_B + \varepsilon_{AB} K c_A c_B}. \tag{5.32}$$

Examples are plotted in Fig. 5.10 for small interacting proteins with K_D-values between 1 and 30 μM. They display the expected reduction at higher total concentrations. The onset of hydrodynamic nonideality effects will be at higher molar concentrations for proteins of smaller molecular weight.

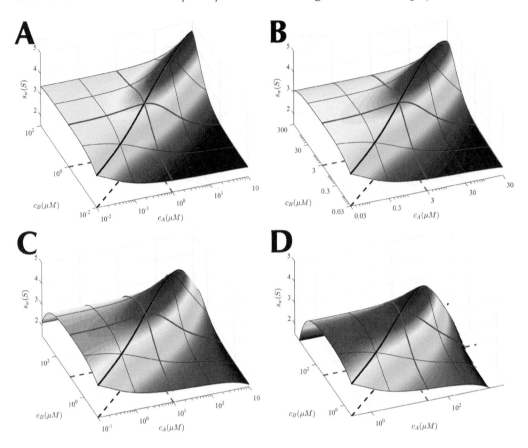

Figure 5.10 s_w isotherms for a bimolecular hetero-association reaction A + B ↔ AB, with $s_A = 2$ S, $s_B = 3.5$ S, and $s_{AB} = 5$ S. The isotherms are similar to those of Fig. 3.1 on p. 70, but now we consider absolute concentrations and hydrodynamic interactions with $\bar{k}_s = 0.01$ ml/mg. Equilibrium dissociation constants are $K_D = 1$ μM (*Panel A*), 3 μM (*Panel B*), 10 μM (*Panel C*), and 30 μM (*Panel D*). The conversion from molar to weight concentrations assumes molecular weights of $M_A = 20$ kDa and $M_B = 40$ kDa, and the extinction coefficients of all species are weight-based signals. The red and pink lines indicate titration series of constant B at K_D with varying A and constant A at K_D with varying B, respectively. The black line highlights the isotherm for equimolar experiments.

5.3.2.3 Slow Reactions and the Population Isotherms

In the absence of nonideality, slow reactions on the time-scale of sedimentation allow free and complex species to sediment virtually independently. One could regard the sedimentation process merely as one of several conceivable tools for the empirical separation of species, which may be followed by the quantitation of species' populations and the determination of the equilibrium constant *via* mass action law. This approach to slow interactions is trivial in the absence of hydrodynamic interactions. In the presence of hydrodynamic interactions, however, the sedimentation process does not provide simple resolution of species even with clearly resolved sedimentation boundaries. This is due to the fact that the species will not be physically separated during sedimentation, because all species are present in the zone of the fast boundaries. As shown in Figs. 5.5 and 5.7, the velocity difference of the slow material in the absence and presence of the faster material causes an increase of the slower species' concentration in the slow boundary, known as the Johnston–Ogston effect (see above).

The topic of how this modulates apparent amplitudes of slowly interacting species has not been completely explored in the literature. In simple cases, it may be possible to use the predictions from Eqs. (5.22) and (5.23) to estimate the expected magnitude of this effect. For example, with Eq. (5.22) for a slow monomer-dimer system that exhibits only two boundaries, the increase in the concentration of the slower boundary can be phrased as

$$c_{m,\text{corr}} = c_{m,\text{obs}} \frac{s_{d,\text{obs}} - s_{m,\text{obs}}}{s_d^0 - s_m^0} \left(1 - \bar{k}_s w_{\text{tot}}\right)^{-1}, \tag{5.33}$$

where $c_{m,\text{obs}}$ and $c_{m,\text{corr}}$ are the concentration of the slow boundary (here the monomer) measured or back-corrected for the Johnston–Ogston effect to reflect concentrations in the loading mixture, respectively, w_{tot} is the total loading concentration in mg/ml units, $s_{d,\text{obs}}$ and $s_{m,\text{obs}}$ denote the measured s-values of the fast and slow boundary, respectively, and s_m^0 and s_d^0 denote the s-values of the monomer and dimer species in infinite dilution. This makes the assumption that nonideality of the monomer and dimer is the same. Even though the latter s_d^0 may not be measurable, it may be back-calculated from $s_{d,\text{obs}}/\left(1 - \bar{k}_s w_{\text{tot}}\right)$. The dimer boundary would appear correspondingly:

$$c_{d,\text{corr}} = c_{d,\text{obs}} + c_{m,\text{corr}} - c_{m,\text{obs}}. \tag{5.34}$$

For example, a mixture of monomer and dimer at a total concentration of 10 mg/ml with $\bar{k}_s = 0.01$ ml/mg, assuming a ratio of dimer/monomer s^0-values of 1.6, would show an apparent increase of the monomer concentration by 9.2% for a 1:1 weight ratio of monomer to dimer, and 19.5% for a 1:9 weight ratio of monomer to dimer.

Analogous corrections based on extensions of Eq. (5.22) might be possible for more than two species.

5.3.2.4 Fast Reactions: Effective Particle Theory and Gilbert–Jenkins Theory

In the analysis of the impact of hydrodynamic nonideality on the framework of effective particle theory, in a first approximation, similar considerations can be applied. Again, we assume that the hydrodynamic interactions among all species can be sufficiently well approximated by a single, average coefficient $\bar{k}_s$ in the dilute regime. In this case, hydrodynamic nonideality will lead to a uniform retardation $(1 - \bar{k}_s \sum_\kappa w_{\kappa,\text{tot}})$ dependent only on the total weight concentrations of all components in any zone of the solution. In the 'chemical approximation' considered here, i.e., that the chemical reactions take place in a locally homogeneous solution, this means that the mechanism of co-transport of the effective particles will be unaffected.

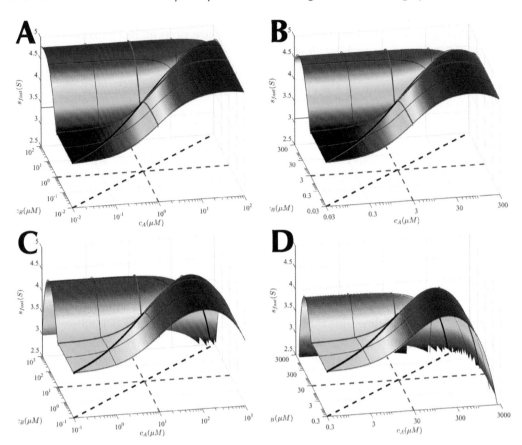

Figure 5.11 $s_{A\cdots B}$ isotherm for the same system as in Fig. 5.10, again considering the same hydrodynamic interactions with $\bar{k}_s = 0.01$ ml/mg. Equilibrium dissociation constants are $K_d = 1$ μM (*Panel A*), 3 μM (*Panel B*), 10 μM (*Panel C*), and 30 μM (*Panel D*).

In the zone delimited by the highest (fastest) reaction boundary, total concentration is at the loading value. In the expression for $s_{A\cdots B}$ of Eq. (2.20) on p. 52, multiplication of the species velocities with $(1 - \bar{k}_s \sum_\kappa w_{\kappa,\text{tot}})$ leads to a reduction of $s_{A\cdots B}$ by the same factor. Fig. 5.11 illustrates the resulting shape of $s_{A\cdots B}$ in concentration space for a two-component system. With regard to the reaction boundary stoichiometry, inspection of Eq. (2.21) shows that the stoichiometry $S_{A\cdots B}$ will be unaffected by a uniform retardation factor $(1 - \bar{k}_s \sum_\kappa w_{\kappa,\text{tot}})$. As a consequence, in this approximation the nonideality will not influence the phase transition between the highest and next lower boundary.

Unfortunately, this does not mean that the strategy of using the composition of the reaction boundary to probe the complex stoichiometry will hold in the same way as in the absence of nonideality. This is due to the fact that the measured quantities in the differential sedimentation coefficient distributions correspond to the measurable boundary amplitudes. The amplitude of the fastest reaction boundary will depend also on the concentration — and concentration increase relative to the ideal case — of the lower zones and the slower boundaries. In the zones associated with the slower boundaries, the lower total concentration will result in less retardation, and the Johnston–Ogston effect will cause an increase of the component concentrations.[19]

For a two-component system where the undisturbed boundary is a single free species only, we can estimate the magnitude of the effect with the help of Eq. (5.22). The assumptions of a single average coefficient $\bar{k}_s$ leads from Eq. (5.22) again to an implicit equation for the increased concentration, $c_{u,obs}$, building up due to the lower retardation in the undisturbed zone:

$$c_{u,obs} = c_u^0 \frac{s_{A\cdots B}^0 - s_Y^0}{s_{A\cdots B}^0 - s_Y^0 \dfrac{(1 - \bar{k}_s c_{u,obs} M_Y)}{(1 - \bar{k}_s (c_{A,tot} M_A + c_{B,tot} M_B))}} . \tag{5.35}$$

Here the superscript '0' refers to values in the absence of hydrodynamic nonideality, and the component Y is the one supplying the undisturbed boundary. Specifically, c_u^0 is the concentration of the undisturbed boundary in the zone of the reaction boundary, i.e., the concentration fraction that co-exists in the highest zone but cannot be dragged along with the reaction boundary.

A consequence of the concentration increase of the slower boundary is, as in the case of non-interacting species, that the faster boundary will appear at correspondingly lower amplitude. The magnitude of this effect additionally depends on signal coefficients, and thereby modulates the measured apparent reaction boundary composition.

These distortions of the boundary amplitudes and apparent reaction boundary composition will be minimal when $s_{A\cdots B}^0 \gg s_Y^0$, i.e., under conditions of large excess of either component where $s_{A\cdots B}^0 \approx s_{AB}$, and for systems with species of dissimilar size where the smaller component is in excess. Also, the effect will be minimal for cases where most of the total weight concentration arises from the secondary component (such that the fraction in the denominator of Eq. (5.35) is close to unity). This will be the case when one of the components is in very large excess. For our model system of Figs. 5.10 and 5.11, assuming A and B could be measured with independent signals, the resulting shift in the measured reaction boundary stoichiometry $S_{A\cdots B,obs} - S_{A\cdots B}^0$ is shown in Fig. 5.12. We conclude that for concentrations below 100 μM the magnitude of the effect is generally negligible.

---

[19]In the case of two-component systems where the slower boundary is a single species, an added complication is that the increase in total concentration will invariably cause a readjustment of chemical equilibria in this zone, relative to the ideal case, which will additionally impact the velocity of the slower boundary.

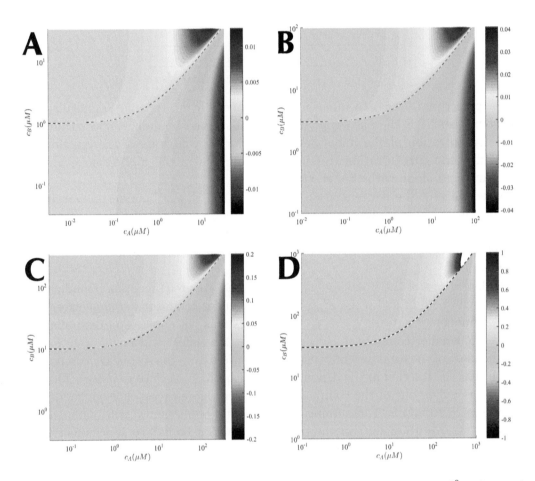

Figure 5.12 Shift in the measured reaction boundary stoichiometry $S_{A\cdots B,obs} - S^0_{A\cdots B}$ due to the Johnston–Ogston effect for the same system as in Fig. 5.10, again considering the same hydrodynamic interactions with $\bar{k}_s = 0.01$ ml/mg. Equilibrium dissociation constants are $K_D = 1$ μM (*Panel A*), 3 μM (*Panel B*), 10 μM (*Panel C*), and 30 μM (*Panel D*).

CHAPTER 6

Practical Analysis of Interacting Systems

THE goal of the following sections is to weave together the different theoretical approaches of the preceding chapters, in order to present specific practical approaches for the study of interacting systems, from designing the experiments to analyzing the data.

In practice, the analysis of interacting systems by SV requires the consideration of the experimental capabilities described in Part I [1], as well as intimate knowledge of the practical analysis of non-interacting systems that was described in Part II, Chapter 8 [2]. Building onto this, additional aspects arise from the characteristics of interacting systems, such as concentration dependence, complex solution composition, and binding kinetics coupled with the dynamics of physical movement.

SV probes particle transport under strong centrifugal fields, such that heterogeneity in macromolecular composition and assembly states will be revealed, due to their distinct hydrodynamic features. As we have seen, interactions impact the sedimentation process on different levels: the boundary velocities, boundary patterning, and boundary spreading. With regard to the first, for interacting systems, the signal-weighted average sedimentation coefficient s_w will depend on the composition of the sample, i.e., the concentration of the different components as well as the binding affinity and stoichiometry. Next, the formation of complexes complicates the boundary patterning because association/dissociation reactions take place during sedimentation, and their dynamic coupling determines the collective migration. Finally, where the interpretation of boundary spread for non-interacting systems was governed by the difficult decomposition into its contributions from diffusion *vs.* differential sedimentation and microheterogeneity, in the study of interacting systems, this problem is exacerbated by the often significant or even dominant contributions from conversion of molecules between free and bound forms. In addition, we must be attentive to the effects of repulsive hydrodynamic interactions and back-flow, which lead to concentration-dependent sedimentation and diffusion, and result in overall retardation and distortions of boundary velocities, amplitudes, and shapes.

If all these factors were present at the same time, the sedimentation data would currently be impossible to analyze from scratch without significant prior knowledge. Fortunately, in practice we usually have a slightly different scenario to handle. For one, we will assume in the following sections that we are primarily studying the interactions of sufficiently purified components, rather than characterization of a potentially polydisperse sample of the interacting macromolecules. This means, for example, that little or no contamination or degradation products are sedimenting in the s-range of the interacting molecules and their complexes.[1] To this end, prior characterization using AUC or other biophysical methods is very helpful. In heterogenous interactions the characterization by SV of each component separately is a prerequisite of a well-designed experiment. Furthermore, unless explicitly mentioned we will assume that hydrodynamic interactions are negligible, i.e., that the experiments can be conducted under sufficiently dilute conditions. This will usually be applicable as long as the interaction is not too weak, and the solvent composition ensures minimal long-range ionic interactions.

Even so, it is not possible to give a single recipe to follow that will universally apply.[2] Therefore, the following strategies need to be adapted to the system at hand, dependent on the binding schemes, affinity ranges, size ratios of the components, available signals and concentration ranges. The analysis may also have different primary goals, for example, to test whether there is an interaction at all, to measure the affinity, number of sites, cooperativity, or hydrodynamic shape of the complexes. Together, these factors often lead to a variety of different experimental and data analysis approaches. In order to sketch at least some common ideas, unless otherwise mentioned, the following comments refer to interacting systems of moderately-sized proteins, with the goal to determine the complex stoichiometry, binding constants, and maybe hydrodynamic friction of the complex. Special considerations apply, for example, to systems with extremely dissimilar sizes, such as small-molecule binding or protein fibrillation; obvious deviations from the described routes in special situations are indicated where appropriate.

---

[1]One thing to note is that it is indeed possible to study macromolecular interactions in a polydisperse environment if the optical signal is selective for a mono-disperse subset of molecules of interest, which may be achieved with absorbance and fluorescence detection systems.

[2]Some general recommendations and analysis steps can be found in [241]. Detailed sedimentation velocity step-by-step protocols have been developed exemplifying specific strategies for focused topics, including the interactions of protein-protein interactions of single-stranded DNA-binding proteins [242], high-affinity protein interactions using the fluorescence optics [243], the study of the self-association state of membrane proteins [244, 245] and intrinsically unfolded proteins [246], the analysis of polyglutamine aggregates by fluorescence-detected SV [247] and the homogeneity of monoclonal antibodies [248].

6.1 EXPERIMENTAL DESIGN

The discussion of the experimental design parameters in Part I, for example, with regard to rotor speeds, detection limits, as well as the need for calibration of the ultracentrifuge, applies similarly to the study of interacting systems [1]. Generally, high rotor speeds and long solution columns are favorable to achieve optimal hydrodynamic resolution. In this section, we will highlight the additional aspects specifically relevant to interacting systems.

The first sets of experiments with a new sample are usually SV runs to establish purity and verify the concentrations of all components, sometimes accompanied with pilot experiments to establish binding properties. These can be helpful to identify appropriate concentration series for the comprehensive study of the interaction. For planning of this stage, we discuss the advantages of different experimental designs, as well as how to optimize the experimental plan with simulations.

6.1.1 Sample Concentrations

For heterogeneous interacting systems, the goal of a well-designed binding experiment series is to vary the solution composition by mixing the individual components in different ways to result in variations of the populations of their complexes. Therefore, assessing a concentration range where the complexes can be populated is a key design factor. Similar applies to self-associating systems, where we also want to aim for loading concentrations that produce different oligomeric populations in different samples.[3]

Accurate knowledge of sample concentrations during sample preparation is crucial. It is generally recommended to measure stock concentrations by dual-beam spectrophotometry in cuvette-based benchtop instruments. In addition, since SV does often involve direct measurement of sample concentration through absorbance or refractive index detection, wherever possible we should reassess the actual sample concentrations of the observed sedimenting material after the SV analysis through peak integration after $c(s)$ analysis. Significant differences between the actual and the nominal design concentrations can reveal inadvertent loading mistakes, or errors in the benchtop measurement from contributions of small non-sedimenting molecules or scattering of large particles.

Prior to characterizing the binding of heterogeneous interactions between different components, it is useful to acquire data on the concentration-dependence of the sedimentation behavior of each component separately, particularly covering the possible concentration range envisioned for later study of their heterogeneous

---

[3]As we have seen in Section 2.2, rapidly equilibrating self-associating systems will exhibit a concentration gradient in the diffusion broadened boundary in which the species populations locally adjust, such that the species populations range from maximal population of oligomers at loading concentration to a virtually infinitely dilute, completely monomeric sample in the trailing edge near the solvent plateau. Nevertheless, it is always advantageous to carry out experiments at a range of loading concentrations, such as to eliminate the need to rely on detailed boundary modeling for the analysis of binding constants.

interactions. This may not only reveal the onset of significant hydrodynamic interactions and/or weak self-association, but also report on purity, microheterogeneity, aggregation properties, and sometimes solubility limits. Further, it can help to reassess the veracity of stock concentrations and signal increments, e.g., through combined analysis of absorbance and interferometry data of the sedimenting material (see Part II, Section 7.1.2 [2]).

Considering the range of concentrations for the mixtures in the binding study, as a rule of thumb, at least one of the components needs to bracket — with as wide a range as possible — the equilibrium dissociation constant. Dependent on the purpose of the experiment, the existing knowledge about the system, and the relative size of the molecules, concentrations below K_D for one or two of the reactants may be sufficient. When working with previously uncharacterized samples, it is often useful to perform a preliminary analysis with a few samples exploring as wide a concentration range as possible to obtain a thorough understanding of the potential interactions, before characterizing the interactions in more detail with more informed, optimized design.

A special scenario arises if the interest is mostly in the hydrodynamic properties of the final complex. Instead of studying a full range of concentration, one can focus on conditions where the complex formation is almost saturated in the reaction boundary and s_{fast} becomes virtually concentration-independent. As illustrated in Fig. 2.10C on p. 51, this may be achieved, for example, by applying a large molar excess of the small component over the large component, at concentrations $>$ 100-fold K_D, but still below significant contributions from obligate hydrodynamic nonideality (Fig. 5.11). The focus on the highest concentrations is usually not as essential for the measurement of binding constants in the isotherm analysis (see below).

In addition to the requirements imposed by mass action law, there are other practical constraints that need to be considered in the choice of concentrations. For strong interactions, because low concentrations are needed to maintain significant fractions of free species in equilibrium with complexes, a major consideration in practice is a sufficiently high sensitivity of optical detection for the molecule(s) of interest. For weak interactions, on the other hand, high concentrations are needed to sufficiently populate the complexes. Since optical signals may be reduced by suitably short pathlength centerpieces, in this case specific emphasis needs to be put on the hydrodynamic nonideality interactions. Obviously, the available stock concentrations and volumes may need consideration, as well.

For self-associations, the choice of concentrations is relatively easy[4] — it needs to span an adequate range, sampled preferably in a roughly logarithmic scale. For hetero-associations, the series of concentrations is best considered in the space of component concentrations relative to the binding isotherm, represented, for

---

[4]Even self-associations become more complicated if a mixture of labeled and unlabeled species are studied. This should be considered a competitive two-component system. If both have the same association properties, tracer experiments can be carried out.

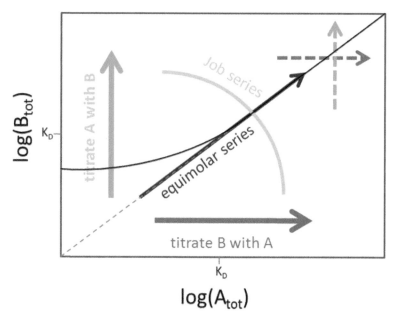

Figure 6.1 Illustration of titration (blue and green arrows) and dilution (red arrow) schemes, as well as a Job series (orange) in the parameter space of loading concentrations. The equimolar line is indicated as a thin blue dotted line, and the phase transition as a solid black line. Titrations high above K_D (bold dashed blue and green arrows) lead to stoichiometric binding, whereas all schemes with variations in the range of K_D generate equilibrium information.

example, by s_w or complex formation as in Figs. 3.1 and 2.10, respectively. Any concentration series that sufficiently samples the binding isotherm even in a non-systematic fashion will work. However, it is usually more effective to follow a rational trajectory in concentration space (Fig. 6.1; see also Fig. 6.3 below). Some trajectories can have advantages in minimizing pipetting errors and establishing constraints that aid data analysis. Popular choices are titration series, dilution series, and Job plot series.

Titration series can be pipetted such that experimental concentration differences of the constant component(s) are minimal. Their constancy can be used as a constraint in the data analysis even when concentrations otherwise are treated as unknowns. Titration series highlight particularly nicely whether or not there is an interaction in the case that the species at constant concentration is the only one contributing to the signal, a situation often presented when using fluorescence data acquisition. In this case, it has the additional advantage of revealing changes in the signal increments (such as fluorescence quantum yield) associated with complex formation, which will cause a systematic change in the total signal [102, 243]. Reaction boundaries of titration series also can define complex s-values particularly well, as well as the complex stoichiometries, often leading to a better determination of binding constants.

Dilution series are often attractive as they can be prepared in fewer pipetting steps than other strategies. In addition, they offer benefits for the data analysis in providing the constraint that the molar ratio of component concentrations is identical across all samples, if the samples are created by a dilution of the mixture. This is ensured even if the stock concentrations have ordinary concentration errors, and the molar ratio is different from the expected value. As will be described in more detail below, dilution series can be particularly powerful for deciding whether or not there is an interaction between components that have no exclusive signal. On the other hand, dilution series have slight drawbacks over titration series, in that their path keeps the sedimenting system closer to the phase transition line, where reaction boundaries are broader, undisturbed boundaries smaller, and reaction boundary s-values closer to those of the undisturbed boundaries (Fig. 2.10). This can make it more difficult to discern the boundary structure. Preparatively, diluting samples from a mixture at high concentration can also introduce potential problems for slowly equilibrating interactions, since the sample is initially in a maximally assembled state in the stock, and dissociation of complexes required to re-establish the equilibrium at the more dilute conditions may take a long time. Therefore, unless the interaction is known to be rapidly equilibrating, there is a danger that the sample is not in chemical equilibrium at the start of the experiment. This drawback is avoided when each sample is mixed from its components.

Another alternative rational concentration design is a Job plot design (also known as method of continuous variations) where the total molar concentration is held constant [249, 250]. Under some conditions, if a signal proportional to complex formation is plotted against molar ratio, this offers an opportunity to graphically interpret the stoichiometry from the maximum measured complex formation (e.g., Fig. 6.4D, and Section 6.4.4 below). In the present context, the multiple pipetting steps at different volumes required for the sample preparation in a Job series eliminates the opportunity of precise quantitative constraints reducing the number of concentration parameters.

Clearly, if neither the number of available rotor positions nor sample amounts are limiting, a combination of equimolar mixture series, titration series, and a Job trajectory will be the optimal sampling of the parameter space of loading concentrations. While the opportunity for such exhaustive experiments is rare, with a reasonable experimental design it should not be necessary in most cases.

For a desired titration or dilution scheme for two-component mixtures, the SEDPHAT function Pipetting Planner from the Options ▷Interactions Calculator menu can be used to produce a table of volumes of components at given stock concentrations to be mixed with buffer in order to produce mixtures along a certain trajectory.

6.1.2 Optical Detection and Signal Increments for the Analysis

Considering the signal increments of the macromolecules of interest and the dynamic range of the different detection systems (see Part I, Section 4.3.2 [1]), even a ballpark estimate of K_D is usually sufficient to determine which optical systems will be able to produce data with partial assembly of complexes $(0.1 - 10\ K_D)$.

6.1.2.1   Absorbance and Interference Optical Detection

Samples with imperfect purity may be easier to analyze on the basis of absorbance data, where some classes of contaminating species do not show. Likewise, unanticipated buffer mismatches can complicate the analysis of interference optical data but not absorbance data, even though the former is often providing better quality data in the ideal case. Therefore, it is advantageous to acquire data with both systems simultaneously if possible. Furthermore, for heterogeneous associations, more often than not, some combination of absorbance wavelengths and interference signals can be found where the different components have different relative contributions, thus creating information for the spectral discrimination of each component in the mixture. This can help greatly to unravel the stoichiometry of complexes and improve the determination of K_D.[5]

For the data analysis, we will need molar signal increments for each of the acquired signals (Part I, Section 4.4.4 [1]). When using a single signal, any error in the extinction coefficients will propagate into a proportional relative error in the binding constant. When using multiple signals, the signal increments should be predetermined, if possible, from data acquired in the same AUC instrument as used for the mixture studies, and ideally side-by-side in the same SV run. In addition, in the context of multi-signal analysis we need to have very precise relative extinction coefficients (Part II, Section 7.1.2 [2]). This is particularly important for direct boundary fitting of multi-signal data sets with Lamm equation models with embedded interactions. Here, even small errors in the relative extinction coefficients that would be inconsequential as errors in K_D, will cause large misfits in the boundary amplitudes and therefore significantly bias the analysis. Similarly sensitive to errors in relative extinction coefficients is the multi-signal $c_k(s)$ analysis of heterogeneous

---

[5] As discussed in Section 4.4.4 of Part I, the time requirements for radial absorbance scanning in Optima XLA/I analytical ultracentrifuges may be limiting for the number of scans and wavelengths that can be acquired during the sedimentation process [1]. However, the availability of fewer scans at any given wavelength is usually more than compensated by the fact that the information from the different signals can be integrated into a global analysis. For example, when scanning at two different absorbance wavelengths, the time between sequential scans of the same cell at the same wavelength is twice that required when scanning only at a single wavelength. However, the data at the second wavelength are reporting from the same solution at times in between those at the first wavelength, and in fact, the time intervals between successive observations of the solution column are about the same as when using only a single wavelength. By contrast, the interference optical detection does not add significant time-delays to absorbance scans. For each signal, as few as 15–20 scans representing evenly the sedimentation across the solution column are usually sufficient. Therefore, the acquisition of data at three different wavelengths for 7 samples in an 8-hole rotor is usually possible.

As described in Part I, wavelength reproducibility in multi-wavelength scans in the Optima XLA/I is not a limiting factor if wavelengths are chosen near minima or maxima of the extinction spectra [1]. This is a prudent choice also for ensuring linearity of the absorbance signal [1].

mixtures. Especially, it should be noted that if the ratio of extinction coefficients at the different signals is similar among components, the extinction coefficient matrix is close to singular, and spectral deconvolution can amplify errors strongly. This can potentially cause large errors in the estimated complex stoichiometries.

As discussed in Section 4.3.2.1 of Part I [1] and Section 7.1.2 of Part II [2], obtaining sufficiently precise extinction coefficients for each component at each signal can be accomplished best via a global multi-signal analysis $c_k(s)$ of multi-signal SV data sets, fixing one extinction coefficient to a known value and letting all others float.[6] It is critical that these control experiments are carried out with concentrations that provide good signal/noise ratio. Considering the often imperfect wavelength calibration, if a wavelength in a narrow maximum of the macromolecular extinction is selected, e.g., an absorbance peak of a chromophore in the VIS, the exact peak wavelength should be determined in a preliminary wavelength scan in the particular AUC instrument(s) to be used for the interaction study.

6.1.2.2 Fluorescence Detection

Different from the conventional interference and absorbance optical detection systems, the fluorescence detection system [251] has many specific features that impact the data and analysis (see Part I, Sections 1.2.3, 4.1.2.3, and 4.3.2.3 [1]). In particular, geometric factors and temporal drifts in signal magnification impose a special structure on the fluorescence data that we may need to computationally model to enable a detailed quantitative analysis [252]. Also, photophysical effects causing changes in the fluorescence quantum yield [102] need to be kept in mind;[7] these can be accounted for in special models described in Section 4.2.6 of Part I [1] and Section 1.1.3.1 of Part II [2], and in Section 3.2.2 of the present volume, respectively.

When using fluorescence detection for studying interacting systems, one can take advantage of the virtually unlimited dynamic range in component concentration when using a tracer strategy, where only a fraction of the molecules carry a fluorescent label. In the other extreme, the capability of detecting extremely low concentrations of the fluorescent molecules offers a unique opportunity for studying high-affinity interactions [14, 243, 254–256]. However, for performing experiments with very low concentrations, the addition of a carrier macromolecule to the buffer is required to minimize surface adsorption of the molecules of interest.[8] This necessitates corresponding control experiments to ensure the carrier is inert. This may be

---

[6]It is not possible to use *a priori* estimates of extinction coefficients and signal increments for *all* signals, since the experimentally determined signal ratio — with its high intrinsic signal/noise ratio and large statistics of data points — is usually much more precisely determined from the SV experiment than independent estimates could provide.

[7]Irreversible photobleaching, however, is rare and was not observed with current instrumentation for FITC and GFP in standard phosphate buffered saline pH 7.4 [253].

[8]Bovine serum albumin is a common choice [14, 257], but has the disadvantage of providing a fluorescence signal with 488 nm excitation [14, 255]. Chromatographically purified lysozyme and κ-casein have been found equally effective [14, 257].

probed in experiments that test whether the s-value of all interacting components are identical with and without the carrier. Similarly, it is necessary to establish experimentally the signal contributions of the carrier in suitable control experiments of buffer with/without carrier. On the other hand, for high concentrations due to the relatively low degree of adsorbed fraction of molecules, it is possible to leave out the carrier. However, it is a good practice to use the same strategy across the entire concentration range in the experiment to ensure consistency.

Also, the work at low concentrations requires particular attention to the reaction kinetics, to ensure that sufficient time is provided for the sample mixtures to reach binding equilibrium [14]. This is particularly important for high-affinity interactions since these are prone to involve slow complex dissociation. Experiments with additional incubation time, and/or comparisons of stock dilution *vs.* additive mixing of the same final concentration can be useful approaches [14].

The signal increment of the molecules in fluorescence detection is dependent not only on the concentration, labeling ratio and the photophysical properties of the fluorophore, but also the geometry factors of the optics, laser power, photomultiplier, and amplifier settings. Therefore, it is usually not straightforward to directly link the signal and concentration with one signal increment. Rather, different settings will require different signal increments; and signal ratios can be experimentally determined by scanning the same solution columns at multiple settings.[9] It can be particularly useful to apply a titration series of a constant concentration of fluorescent binding partner, which can be scanned with uniform settings. In this case, any quenching or enhancement can be directly deduced from changes of the signal with loading composition [243].

The selective fluorescence signal has great potential for studies in crowded environments such as serum and cell lysates [177, 180, 181, 247, 254, 257–259]. With regard to the quantitative analysis of binding equilibria, potential complications include a dependence of the fluorescence signal on the environment, as well as strong hydrodynamic nonideality.

Finally, even though the non-fluorescent binding partners do not contribute to the signal, their thorough characterization with interference and absorbance systems is indispensable for any quantitative analysis. Any errors in concentration and association state will directly impact the interaction analysis.

6.1.3 Sample Purity

Questions of sample purity have been discussed in general terms in Part I, Section 3.1.1 [1]. The often cited value of 95% purity requirement does give a good indication for the experimental goal. However, the tolerable amount of contaminating species depends on the nature of the impurities, their signal contributions, and their s-value. For biomacromolecules, purification by size exclusion chromatography

---

[9]Adjusting the gain can be considered a linear factor (at the same photomultiplier voltage), which allows to link concentration signals across a larger range.

immediately prior to the SV experiment is often very effective, and best combined with SDS-PAGE and mass spectrometry to assess the purity of the resulting sample. Obviously, the study of heterogeneous interactions requires that the purity of each component be characterized independently by SV.

Impurities can result from imperfect purification, degradation due to insufficient stability or susceptibility to trace amounts of proteases, as well as aggregation processes. Any such prior knowledge is important to take into consideration regarding the strategy of sample preparation, design and duration of experiments, and data analysis. Impurities that are of very large size, such as large aggregates that may be detected by dynamic light scattering, can be tolerable since they often sediment much faster than any species of interest, and may be even depleted already in the first scan after acceleration of the rotor and therefore go undetected. Nevertheless, strong aggregation may be obvious if the total signal of the sample from the first scan is significantly smaller than expected.

Impurities of very low molecular weight can be similarly inconsequential, as long as they do not produce significant signal offsets that cause the total signal to exceed the linear range. Below this threshold, any signal contributions will still need to be accounted for in the data analysis as a separately sedimenting species. However, high concentrations even of inert, invisible, non-sedimentable species must be accounted for with regard to their volume occupancy, which will impact the observed sedimentation process *via* hydrodynamic interactions and nonideality of the molecules of interest.

There are also different purity requirements dependent on the type of analysis conducted. The most stringent requirements apply to the direct modeling with Lamm equation systems for reacting systems. In heterogeneous interactions, SV data of each component, when studied by itself, must conform well to a model for a single discrete species and give the correct species molar mass (within the uncertainty arising from the partial-specific volume, usually ~5%).[10] A useful criterion is that the quality of fit with the discrete model is comparable to that of a $c(s)$ model. Otherwise, the direct Lamm equation modeling will not be applicable (or at least not give reliable answers). Common reasons for the failure of the purity test by this criterion are — besides imperfect purifications — the presence of micro-heterogeneity in size and/or shape of proteins due to, for example, non-uniform glycosylation or other post-translational modifications.[11]

Some allowance can be made for a discrete, non-participating species sedimenting distinctly slower or faster than the molecules of interest and their complexes, which may be accounted for separately in the Lamm equation modeling (see below). Similarly, the technique of partial boundary modeling (Part II, Section 8.1.4.1 [2])

---

[10]An exception, of course, is a component that exhibits self-associations. In this case, the SV data of this component by itself must conform to a Lamm equation solution of the self-associating systems.

[11]With regard to such micro-heterogeneity, the presence of a single band by gel electrophoresis will not be a sufficient criterion for purity.

can be helpful to exclude the signals from obvious large aggregates, if they are separated from the main boundaries of interest.

If the samples do not reach this level of purity, the isotherm based analysis models are the method of choice. Their requirements for purity are significantly less stringent because they consist of a two-stage analysis, where a first $c(s)$ analysis allows one to discern which peaks reflect sample imperfections, and to exclude these from consideration during integration. Owing to diffusional deconvolution, a relatively high resolution can be achieved in $c(s)$, such that impurities that sediment outside the s-range of the species of interest and their complexes are easily excluded in the second stage. Impurities of known signal contribution f_{imp} and s-value $s_{w,imp}$ within the s-range of interest can be accounted for in the s_w isotherm analysis, as described in Eq. (3.1) of Part I, with the correction

$$s_{w,\text{adj}} = \frac{s_{w,\text{tot}}c_{\text{tot}} - s_{w,\text{imp}}c_{\text{imp}}}{c_{\text{tot}} - c_{\text{imp}}} \qquad (6.1)$$

where $s_{w,\text{adj}}$ is the adjusted s_w-value, $s_{w,\text{tot}}$ and $s_{w,\text{imp}}$ are the $c(s)$ integrated s_w-values for the total system and impurities, respectively, and c_{tot} and c_{imp} are the respective integrated signals [1, 255].[12]

The least demanding experiments with regard to sample purity can be those to establish whether there is an interaction or not, by assessing the concentration-dependence of the weighted average s_w-value in dilution series (see below). In principle, assuming that any impurities constitute a constant fraction of the material and are non-interacting, this requires only that the molecules of interest are the majority species of those generating a signal.

For all purified samples, contaminations particularly difficult to detect are binding incompetent variants of the macromolecule of interest, which may not be hydrodynamically distinguished by SV. For example, such incompetent species may be proteins that did not undergo the same post-translational modification, are incorrectly folded, or were subject to some degradation processes. For both Lamm equation modeling and isotherm analysis, it is possible to introduce a parameter for the incompetent fraction of each component, which may be optimized in the fitting process.[13] The adjustment of incompetent fractions in the fit increases the uncertainties for other parameters, but is often required for a realistic model. In some cases, the conditions of complete saturation of the complex (by virtue of a large excess of one of the binding partners) can reveal that a fraction of binding-incompetent species

---

[12]Specifically, this applies to the signal contributions of the carrier protein BSA in fluorescence-detected SV [255], which need to be measured for this purpose in separate control experiments alongside the samples of interest.

[13]A similar consideration is usually made in isothermal titration calorimetry, although in SV we have the benefit of measuring at least the total component concentrations as part of the experimental data, whereas sample concentration errors will contribute to the estimate of 'incompetent fractions' in isothermal titration calorimetry [260].

must exist.[14] Unfortunately, for practical reasons such experiments are sometimes difficult to implement.[15]

Binding incompetent fractions can be specified in the `Global Parameters` boxes of SEDPHAT, in the fields labeled `incfA`, `incfB`, etc., in the hetero-association models, or the field `incompetent fraction` in self-association models. They exclude a certain fraction (with values between 0 and 1) of the total concentration of a component from participating in binding equilibria. In hetero-association models, the state of the incompetent fraction is assigned to be monomeric, whereas in self-associations they can be a stable monomer or oligomer, dependent on the input of the value n (which must be an integral number). The incompetent fraction can be refined in the fit, but generally only one incompetent fraction should be refined in order to avoid parameter correlation.

6.1.4 Experimental Planning with Simulation Tools

The computational simulation of experiments with hypothesized molecular parameters, binding scheme, and affinity is a very important tool for testing an experimental design. Data can be assigned a realistic noise level, such that re-analyzing the simulated data as if they were experimental data will reveal the information content of the given experimental design with regard to the parameters of interest. Simulated data can also be used to test how well different binding schemes can be distinguished.

A simulation requires specification of an interaction scheme, signal increments of all components, hypothesized binding constants, and prior estimates for the sedimentation parameters of all free species and complexes. It is advantageous if these values can be extracted from preliminary experiments, if available. Alternatively, complex s-values may be estimated from hydrodynamic scaling laws, or be derived from hydrodynamic modeling if structural data is available.

---

[14]Related, if incompetent species must be considered in s_w isotherm analyses, the constraint of free species and complex s-values to known or estimated values can be an effective strategy to prevent complete parameter correlations.

[15]A related problem is that of assigning correct component concentrations. In particular, if concentration data for the isotherm are taken from the nominal sample concentrations, rather than those actually observed in sedimentation experiments (if they can be observed at all), errors in the binding constants will arise, or the deduced incompetent fractions will be erroneous.

The SEDFIT menu Options ▷Calculator offers different variations for estimating s-values of complexes. For example, Calculate s(M) for spherical particle produces the s-value of a compact sphere of a given mass and partial-specific volume. This result may be divided by the frictional ratio (e.g., at a value of 1.3 for globular proteins in the absence of other knowledge) to arrive at realistic s-values for hydrated and asymmetric particles.

6.1.4.1 Lamm Equation Modeling

SV boundary profiles of interacting systems can be challenging to anticipate. This is true, in particular, if reaction kinetics leads to chemical conversion of species on the time-scale of sedimentation. Therefore, it can be very useful to examine the boundary features through simulation.

To simulate sedimentation boundaries in SEDPHAT, the interaction model must first be set in the Model menu. Then, the function Generate ▷Sedimentation Velocity Boundary xp will prompt the user to specify global model parameters, the total centrifugation time, as well as detailed experimental parameters, such as signal coefficients and noise level. This leads to a display of s_w in concentration space (which may be cropped to the feasible range, as shown in Fig. 6.3 below). A mouse right-click in the graph defines the concentration for the simulation, which is invoked with the exit and simulate button. Data are automatically saved and loaded for analysis.

The simulated data will default to 50 scans; data for other sedimentation/scan conditions can be obtained through use of the Generate function in SEDFIT as a first step to create a template data set, which may then be loaded in SEDPHAT for simulation with the Run and Save Fit Data functions.

For rapidly interconverting systems, a complementary tool to study the coupled sedimentation/reaction process is the simulation of the effective particle migration modus in an animation.

Maps of effective particle properties in concentration space can be generated in SEDPHAT, in the Effective Particle Explorer, which can be started from the Options ▷Interactions Calculator menu. As shown in the screenshots of Fig. 6.2, a first display will provide an overview of the concentration space, highlighting different features of the boundary pattern, from which the selection of a particular concentration mixture will lead to a schematic animation of particles migrating at different relative speeds while combining to form complexes. Pressing the ESC button will exit the effective particle explorer.

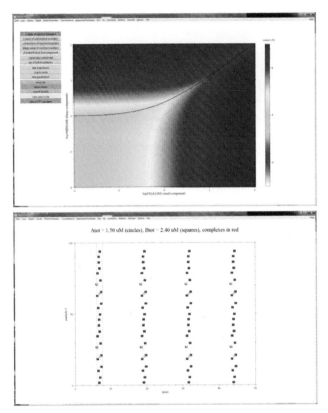

Figure 6.2 Screenshot of the `Effective Particle Explorer` in SEDPHAT. *Top Panel:* Color contour map of s_{fast} as a function of $c_{A,tot}$ and $c_{B,tot}$ of a rapidly interacting system. The plot can be switched to show other quantities including S_{fast}, s_w, and relative boundary amplitudes. The phase transition line is drawn in black. *Lower Panel:* Marking a particular point will produce an animation of the concerted propagation of the effective particle (only one frame of the movie shown here) for the particular conditions. Effective particles are depicted in columns of its co-sedimenting constituting species (A green, B blue, AB red), moving from left to right to represent sedimentation, and arbitrarily moving vertically to facilitate the display of their exchange. The process is faithful to the relative concentrations, K_D, fractional times spent in different states, and relative species s-values.

In particular, this can illustrate the fractional time each molecule in the reaction boundary spends in free and complex states, and how this relates to their migration, preserving identity of time-average velocities for all components (Fig. 6.2).

6.1.4.2 *Species Populations along a Trajectory*

Usually data from experiments at a range of loading concentrations are globally analyzed to determine the binding parameters. It can be very useful to verify that along the chosen trajectory of concentrations different species are actually populated, and to examine for which conditions their abundance is highest. For example, it is of interest to study whether intermediate species are significant and whether their population may even dominate under certain conditions.

For given loading concentrations and binding scheme, species concentrations can be calculated using the `Mass Action Law` function from the `Options▷Interactions Calculator` menu of SEDPHAT. More efficiently, the `Species Population Plot` function in the same menu will display, for each component, the fraction of protomer present in a certain complex state. To aid recording and presenting the results, the plotted data are also stored in a file in the `\temp` subfolder of the SEDPHAT home directory.

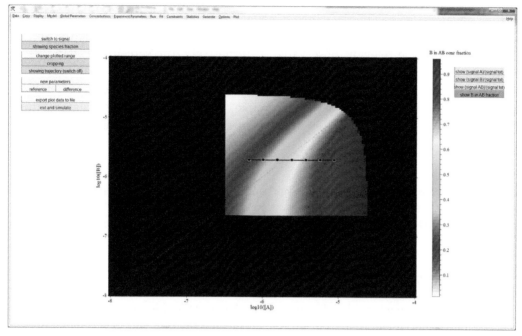

Figure 6.3 Screenshot of the SEDPHAT simulation tool for generating experimental concentration series along a certain trajectory in concentration space. For a given interaction scheme, binding constants, and s-values, different quantities of an interacting system can be mapped, dependent on the data to be generated (here showing the fraction of B that is in the complex state). A black mask can be generated to crop the display to feasible mixture concentrations, given stock concentrations and minimum acceptable concentrations for each component, and within constraints for acceptable optical signals for the mixture. (Three-component systems are plotted as two-dimensional slices parallel or diagonal to the concentration axes, controlled by a slider that enables scrolling along a line perpendicular to the slice; not shown). Using the mouse, a trajectory can be graphically specified, for example, to sample a change in the displayed feature. SEDPHAT will provide a pipetting plan such as to achieve a desired number of samples along the trajectory. Finally, a simulated data set with the expected signals for these mixtures will be generated and loaded into SEDPHAT for analysis.

6.1.4.3 Selecting a Trajectory in the Two-Dimensional or Three-Dimensional Concentration Space

Perhaps the most powerful design tool is the representation of characteristic boundary features of interest in concentration space. For two-component systems, a color contour plot can be generated with concentration of A and B as abscissa and ordinate, respectively, and a color temperature representation of s_w or fractional species populations (Fig. 6.3). For three-component systems, two-dimensional slices through the three-dimensional concentration space can be represented in the same way, with slices for different concentrations of C, or diagonally for constant concentration ratios A/C, etc. Much like the analogous simulation tool introduced for experimental planning in isothermal titration calorimetry [261], the concentration space can be cropped to feasible mixtures given stock concentrations and detection limits.

Favorable concentration series can be recognized as paths that lead across significant changes in the relevant observable. Alternatively, the screen can map differences in the observable relative to a reference map. The reference map can be defined by different models and/or different parameter values. This allows to determine the concentration regions that, for example, are most sensitive to changes in certain parameters of interest, or report best on the differences between alternate binding schemes [261].

The `Generate` function in SEDPHAT invokes representations of characteristic observables in two- or three-dimensional concentration space, as shown in Fig. 6.3, as an aid to simulate different kinds of SV data and to conveniently obtain pipetting plans.

6.2   ASSESSING WHETHER OR NOT THERE IS BINDING

Interactions between macromolecules in solution are inescapable, if only from hydrodynamic interactions, excluded volume, and back-flow even at low concentrations. Measurable interactions by SV continuously cover repulsive and attractive interactions. The question whether there is binding between macromolecules is understood in the following as the question whether there are attractive interactions. Those leading to structurally well-defined long-lived complexes are the easiest to identify, but for weak interactions transient formation of complexes cannot be distinguished from attractive potentials in the particle distance distribution.

6.2.1   Self-Associations

As mentioned above, for self-associations, there is not much choice regarding the setup of experiments. As large a concentration range as possible should be covered, which usually results in a serial dilution of the stock concentrations in 2–3 steps per decade down to (or below) the expected detection limit of the optical detection method. An obvious indicator of interactions is the emergence of new $c(s)$ peaks at higher s-values, or the increasing amplitude of existing peaks, as the concentration increases.

While this can be very clear at times, in many cases — in particular weak associations — the determination whether there are new peaks can be difficult to make because of limitations from signal/noise and the presence of low levels of irreversible aggregates. Further confounding can be the theoretically expected substructure of $c(s)$ peaks, and/or possible dependence of the details of $c(s)$ peak shape on the details of the fitting. Another difficulty is that rapid interactions on the time-scale of sedimentation will not necessarily lead to the formation of distinguishable new peaks at all, but may often show as shifts in peak positions. Finally, at high concentrations the shift in peak positions may be compensated for by increasing retardation from nonideality at higher volume occupancy. Therefore, unless it is very obvious, the method for assessing the presence of interactions by emerging new peaks may not be optimal and may be unsuccessful.

A much better approach is the analysis of the weighted-average s-values for the concentration series (Section 3.1; for practical determination of s_w see Section 6.5.1.1). The simple nature of s_w, which is directly rooted in the transport method and independent of boundary shapes, makes it a distinct and precise reporter on the average assembly state in macromolecular solutions. Isotherms of s_w can be determined easily and rigorously from the $c(s)$ distributions, as shown in Chapter 4, and further described below. Often the isotherm will look similar to Fig. 3.3B on p. 75, although at high concentrations emerging repulsive nonideality will lead to an isotherm with negative slopes, such as those in Fig. 5.9 on p. 173. It is obvious that if a positive concentration dependence is shown in the comparison of s_w-values as a function of loading concentrations or loading composition, or if s_w is higher than that estimated for solely repulsive hydrodynamic interactions [240], there is an attractive interaction or complex formation. For weak self-associations, only relatively small changes in s_w may be observed, comparable to the error in the isotherm analysis. Therefore, the determination of error bars for s_w is particularly helpful.

Using s_w isotherms, the question of whether or not there is an interaction can be answered more quantitatively in the context of K_D-values from fitting the isotherm, where the statistical analysis will reveal either a two-sided confidence interval for K_D (meaning there is an interaction), or only a lower limit for K_D (meaning there may not be an interaction at all, but based on the available data we can exclude any interaction stronger than the K_D limit).[16]

6.2.2 Hetero-Associations

As was the case for self-associations, in straightforward cases we may also recognize interactions in multi-component mixtures from the emergence of $c(s)$ peaks at higher s-values than the $c(s)$ peak positions of samples from each individual component. At the same time, in mixture samples at different concentrations, the peaks of the individual components may change in ratio, decline, or vanish altogether. The feature of new complex peaks in $c(s)$ may be especially evident, for example, in interactions between dissimilar-sized molecules, if the SV data record a signal specific to the component with low s-value: In this case, the mixture may exhibit clearly recognizable complex peaks at much higher s-values.

Overall, however, the assessment of the visible peak structure in $c(s)$ is again not the most sensitive approach, and will not always give unambiguous results. After $c(s)$ is integrated to provide s_w-values, the transport method provides a rigorous

---

[16]The opposite question — whether or not there is complex dissociation — can be answered in a similar way. The isotherm analysis then provides well-defined confidence limits for the K_D (as is the case, for example, for the dissociation of tubulin dimer [63]), or only an upper limit for K_D given the available data.

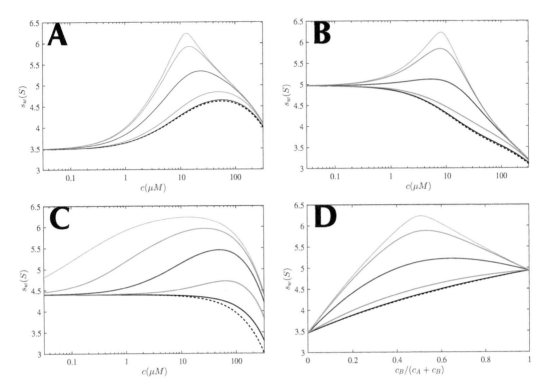

Figure 6.4 s_w isotherms for a bimolecular hetero-association reaction A + B ↔ AB following different concentration trajectories. The molecules are A with 3.5 S, 40 kDa and B with 5.0 S, 60 kDa forming a 6.5 S complex, and species signal increments are proportional to mass. Hydrodynamic nonideality was assumed uniform with $k_s = 0.01$ ml/mg. Curves shown are s_w isotherms in the absence of binding (dotted black), and with equilibrium dissociation constants K_D of 1 mM (blue), 100 μM (green), 10 μM (magenta), 1 μM (cyan), and 0.1 μM (orange). Trajectories are titration series with constant $c_{B,tot} = 8.3$ μM or 0.5 mg/ml (*Panel A*); titration series with constant $c_{A,tot}$ = 12.5 μM or 0.5 mg/ml (*Panel B*); equimolar dilution series (*Panel C*), and Job plot series with constant total molar concentration of 20 μM (*Panel D*).

framework independent of details of the boundary shapes and peak patterns or substructures. For interacting systems, typical isotherms for different concentration series are shown in Fig. 6.4. Given known s-values of the free species, noninteracting systems will follow the dotted black lines. Often the easiest to interpret are experiments with constant molar ratio (dilution series) where in the case of interactions the s_w-values will be increasing with higher concentrations, except for the effect of hydrodynamic interactions setting in at high concentrations. Titration series and Job series require quantitative modeling and well-defined, independently measured s-values of free components. As mentioned above in the discussion of self-associations, it is clear that we strictly can never determine the categorical 'absence' of binding or attractive interactions even if no increase in s_w as a function of concentration is apparent, but we can determine a lower limit for K_D.

The case where the signal reports only on a single component while the binding partner is silent warrants special consideration. In this case the s_w-isotherm from titration series of the detected component with increasing concentration of the silent component is very powerful, since s_w will be constant in the absence of an interaction. Conversely, interactions can be read from an increase of s_w in concentration series, which will be particularly large if the detectable binding partner is relatively small. This situation arises usually in fluorescence detection. Here the titration series has the additional advantage of reporting on signal changes of the complex.

Short of establishing a full isotherm, a powerful test for interactions between two components is the comparison of the experimental s_w measured for the mixture with the predicted $s_{w,\text{mix}}$ under the assumption that interactions are absent.

$$s_{w,\text{mix}} = (s_{w,1}f_1 + s_{w,2}f_2)/(f_1 + f_2) \,, \tag{6.2}$$

where $s_{w,1}$ and $s_{w,2}$ are the weighted-average s-value of the individual components measured separately (integrating $c(s)$ over the same s-range), with total signals f_1 and f_2, respectively (observed with the same optical signal as the mixture). If s_w of the mixture is larger than $s_{w,\text{mix}}$ then an attractive interaction is present. For data acquired side-by-side in the same run, the typical precision of s_w is on the order of 0.01 S, or 0.1% [262, 263], which presents a limit to this qualitative assessment.[17]

6.3 CLASSIFYING THE KINETICS − SLOW OR FAST?

Before proceeding with the detailed analysis of the interaction, it is useful to first study whether the reaction is slow or fast on the time-scale of sedimentation. Dependent on this, additional features of the data can be extracted and included in the quantitative analysis, and kinetic parameters may be initialized.

As we have seen in the theoretical sections, slow reactions are characterized by $c(s)$ distributions that have concentration-independent peak positions (Figs. 2.3B/D and 2.4B/D on p. 31/32), whereas fast reactions generally show concentration-dependent shifts in the peak position (Figs. 2.3A/C and 2.4A/C). Though this distinction is very powerful and clear in most cases, in practice more details must be considered.

When superimposing $c(s)$ distributions from experiments at different concentrations, we often face the problem that the resolution is lower at lower concentrations due to the difference in signal/noise ratio. The regularization may result in merging of what should be independent peaks, with the result that the combined broad peak has an intermediate peak position (see, for example, Fig. 5.12 of Part II [2]). This problem is easily recognized at the lowest concentrations from the low signal/noise ratio of the data, and at low to intermediate concentrations from the strong dependence of the peak structure on the regularization level. This issue can

---

[17]Another caveat is that interactions might be masked by conformational changes, for example, if small-molecule binding to a large protein stabilizes a conformational state with higher friction than the free protein. In this case, a method taking advantage of detection of the small molecule, if possible, would be preferable to detect binding.

be circumvented with Bayesian methods, but the result may still not be sufficiently informative to positively contribute to the distinction of the reaction kinetics.

For the same reason it can be problematic to compare the details of $c(s)$ distributions of dissimilar optical detection methods. Besides the different signal/noise levels, they may exhibit different sensitivity to trace impurities and buffer mismatch signals, which alone may result in slightly different peak structures at low concentrations.

As a consequence, for $c(s)$ distributions derived from data with low signal/noise, we should be aware of the possibly large uncertainty of the kinetic information. This problem arises similarly when using direct Lamm equation modeling. It is exacerbated by a low hydrodynamic resolution of species, e.g., involving small molecules. In this case, it is recommended to solely focus on the thermodynamics of the interaction *via* s_w. The broadening of $c(s)$ at low signal/noise is an important consideration also for fluorescence-detected SV experiments where titration series are often favored over dilution series. In titration series, when the fluorescent intensity is low, one may obtain a single broad peak for the each data set of the entire concentration series. In this case, even slow interactions might only show a peak that shifts position as concentration increases [255], and thus, the possibility of kinetics assessment will be limited. At high signal/noise ratios, by contrast, the ordinary distinction can be safely made.

For the following discussion, we will assume a reasonably high resolution for $c(s)$ distributions. We should be focused mainly on the major peaks that dominate the optical signal. This is because sometimes even stable species may exhibit some variability in the peak positions if they are present only in trace amounts [264]. This is true also for minor peaks that may compensate for imperfections of the approximation of the sedimentation patterns of rapid interacting systems with distributions of non-interacting Lamm equation solutions.

The distinction of slow or fast kinetics relies upon the recognition of boundary patterns. For this, we do need knowledge of the theoretically predicted boundary structure of rapidly interacting systems. For two-component rapid self-associations, as we have seen, there is only one reaction boundary, but at concentrations far above K_D, and for conditions close to the phase transition, a substructure of the reaction boundary is theoretically predicted (see Section 4.1.1). For self-associations, the dilute trailing edge of the diffusion-broadened boundary will always sediment at an s-value close to the monomer while the main boundary may exhibit much larger, concentration-dependent s-values (compare Figs. 2.5 on p. 38 and 3.12 on p. 96). In addition, due to the extremely broad sedimentation coefficient distributions of rapid monomer-n-mer associations with high $n > 2$, intermediate secondary peaks in $c(s)$ should be expected (Fig. 2.5). These may or may not occur at concentration-independent positions, and should not be taken as main indicators for a slow reaction kinetics. This problem can be exacerbated for systems with intermediate reaction kinetics, which may produce stronger secondary peaks between monomer and highest complex, and therefore tend to resemble slow systems.

Fast hetero-associations can also exhibit peaks that seem concentration-independent:

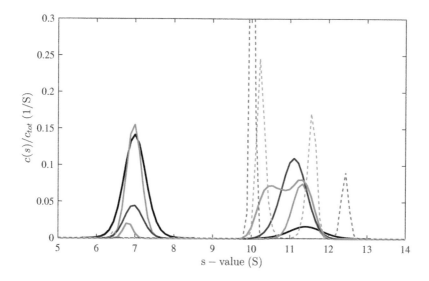

Figure 6.5 Example for a decreasing reaction boundary s-value in the titration of a rapidly reacting system A + B ↔ AB. Below the phase transition the major peaks at $\sim$ 11.5 S are not changing much in position with increasing concentration, which can make it harder to recognize this as a reaction boundary if data from a sufficiently large concentration range are not available. Shown are $c(s)$ traces derived from the simulated sedimentation boundaries for a component A with 100 kDa and 7 S, and a component B with 200 kDa and 10 S, together forming a complex with 13 S. Total loading concentrations of A are constant concentration at $K_D = 10$ μM, whereas B is at 1 μM (black), 3 μM (blue), 10 μM (magenta), 20 μM (green), 30 μM (cyan, dotted), and 100 μM (black, dotted). Conditions below or at the phase transition are shown with solid lines, and above the phase transition as dotted lines.

First and foremost, there is usually the undisturbed boundary at a constant s-value equal to one of the free species. The undisturbed s-value might switch from that of one free species to the other free species if the concentration series crosses the phase transition line.[18] Further, it is not widely appreciated (and it may seem counter-intuitive) that the reaction boundary itself can display s-values that show only relatively small changes with concentration — and even a small drop in s_{fast} — if a titration series configuration is used, and the phase transition point is not crossed (i.e., the variable concentration remains at lower concentrations; see the example of Fig. 6.5). Such a drop in s_{fast} should also not be confused with a concentration-dependence from repulsive hydrodynamic interactions. Hydrodynamic nonideality should be accompanied by boundary steepening and only occur at sufficiently high total concentrations; whereas the drop in s_{fast} from the reacting system should not lead to steeper boundaries and may occur at any concentration range.

For interactions with slow dissociation on the time-scale of sedimentation, if all peaks are resolved in $c(s)$ then these represent the individual free species and

---

[18] In theory, the undisturbed boundary may be entirely absent in a pathological concentration series exactly along the phase transition line.

their complexes. This makes the analysis more straightforward (see below). However, even slowly interacting systems with well-resolved peaks can show a negative concentration-dependence of the peak positions caused by nonideality. This should not detract, however, from the goal to use as wide a concentration range for the assessment of the reaction kinetics as possible.

A different situation occurs for conformational change reactions A→A*. These will not depend on concentration. But, similar to self-associations and hetero-associations, slow reactions may theoretically produce separate peaks for the different conformations whereas fast conformational equilibria result in a single peak. On the other hand, unless the conformational change is very large, most likely the two states will not produce resolvable peaks. Consequently, for slow interactions usually a single, broader peak is observed. As discussed in Section 1.3.1, a possibility for the experimental distinction of the reaction kinetics is then the study of the boundary width.

The above considerations allow us to recognize fast or slow reactions in most cases without ambiguity. While this is not essential for determining the equilibrium binding constants via $s_{w,\lambda}(\{c_{k,\text{tot}}\})$ isotherm analysis, it can be very useful to know whether the sedimentation boundaries and their s-, D-, f/f_0-, and M-values are those of sedimenting molecular species, or those of effective particles of the interacting system.[19] Of course, it is possible that the experimentally observed kinetics falls into the narrow range that is intermediate between slow and fast. How the corresponding $c(s)$ traces appear for hetero-associations in this case is illustrated in Figs. 2.14 and 2.15 on p. 58/59 [106]. If in doubt, simulations of SV data with Lamm equation solutions can be used to probe the behavior of the system in question, and answer, for example, to what extent noisy data can reveal the reaction kinetics.

6.4 DETERMINING THE ASSOCIATION SCHEME AND STOICHIOMETRY OF THE COMPLEXES

6.4.1 Apparent Molar Mass Estimates from $c(s)$ or $c(M)$ Peaks

For slow reactions, the apparent molar mass associated with $c(s)$ or $c(M)$ peaks provides the most straightforward approach to identify the complexes. The same potential and limitations as the molar mass analysis from SV of non-interacting systems applies (see Part II, Section 5.5 [2]). It rests on hydrodynamic scale relationships, fine-tuned by the analysis of boundary spread in combination with the peak s-values. The possibility to account for polydispersity and microheterogeneity in a distribution analysis is critical.

If the different species do not produce their own sedimentation boundary, diffusional deconvolution can usually only report a signal-average f/f_0-value, which limits the accuracy of the derived molar mass values. Of course, if the complex

---

[19]Incidentally, differences in the kinetic behavior can sometimes be a useful indicator for comparing different protein constructs such as native *vs.* modified (e.g., fluorescently labeled) [255].

produces a separate sedimentation boundary, then a segmented $c(s)$ or $c(M)$ with a freely adjustable f/f_0 for the complex boundary will be more accurate. For small oligomers the information from $c(M)$ is often sufficient to favor a model, since possible M-values of complexes are relatively far apart.

The use of the apparent molar mass has not yet been established for the analysis of fast reactions. However, considering Fig. 3.26 on p. 122, the apparent molar mass of the effective particle is expected to fall between the weight-average molar mass and that of the limiting complex. This may provide evidence for certain stoichiometries, and should be useful as a lower bound for the complex size. Obviously, as the highest complex state is attaining saturation at high concentrations, the determination of M-values from $c(M)$ peaks is trivially the same problem again as that of determining the molar mass of non-interacting species.

As mentioned above, the presence of significant hydrodynamic nonideality will strongly increase the measured M-values and make the boundary spread a poor estimator for the size of the complex.

6.4.2 Interpreting the s-values of Complexes

The identity of the complex may often be better assessed directly from the peak s-values, which are typically independent of the best-fit f/f_0-value.

For slow reactions, i.e., assuming that the observed s-values reflect those of the sedimenting species, traditionally the power-law $s \sim M^{2/3}$ (or $s_n \approx s_1 n^{2/3}$) has been used for estimating oligomer size [57], with the monomer s-value as a basis and — in the absence of evidence to the contrary — assuming that large conformational changes do not occur, and that the complex is not a product of end-to-end association of an already very elongated molecule. For not too high oligomer size this can often yield unambiguous results, considering that the molar masses of the complexes must be integral multiples of the free components.[20]

It is often very powerful to use the observed s-values, without any shape assumptions to calculate the minimum complex size and apply it to estimate a minimal stoichiometry. This can be achieved by calculating the mass of the equivalent sphere ($f/f_0 = 1.0$) that, for a given experimental condition (temperature, buffer density and viscosity), sediments with the experimentally observed s-value.

SEDFIT reports the minimum mass of the equivalent sphere in the PEAK INFO message box that appears when clicking on a molar mass button near the $c(s)$ peaks. The molar mass buttons can be generated with the Display ▷Show Peak Mw in c(s) command or the keyboard shortcut CTRL-M.

How can we find conditions that expose the limiting complex? Let us look at Fig. 6.6, which shows the s_w and effective particle s_{fast} isotherms as a function of loading

---

[20]Considering self-associations, the use of this power-law is actually equivalent to $c(M)$ with a pre-determined, fixed f/f_0 at the value of the monomer.

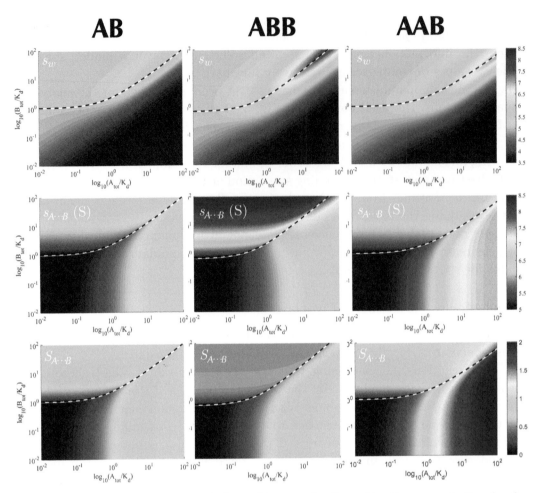

Figure 6.6 Isotherms of experimental observables as a function of loading concentration for the interaction of moderately-sized molecules A and B with fast kinetics, forming 1:1 complexes AB (*Left Column*), or 1:1 and 1:2 complexes AB and ABB (*Middle Column*), or 1:1 and 2:1 complexes AB and AAB (*Right Column*). The system parameters are s_A= 3.5 S, s_B= 5.0 S, s_{AB}= 6.5 S, s_{AAB}= 7.5 S, and s_{ABB} = 8.5 S, and $\varepsilon_A = \varepsilon_B$. Shown are the isotherm of the overall weight-average s_w (*Top Row*), the value s_{fast} of the reaction boundary (*Middle Row*), and the composition of the reaction boundary (*Bottom Row*). For the two-site models, non-cooperative symmetric sites were assumed. In all plots, the phase transition line is plotted as a bold black-and-white dashed line, and the line of equimolar concentrations is plotted as a thin gray dotted line. For each row, the color scheme of the contour plots remains the same for better comparison.

concentration in an analogous representation as Fig. 2.10B/C for 1:1, 1:2, and 2:1 reactions. (Isotherms like these can be generated using the **Effective Particle Explorer** tool in **SEDPHAT**, as shown in Fig. 6.2.) As can be discerned from comparing $s_{w,\lambda}(\{c_{k,\text{tot}}\})$ in the upper row of panels, the strategy of using equimolar series — trajectories along the diagonal — to push s_w close to the complex s-value may not be the optimal strategy, as it requires both the correct loading stoichiometry and fairly high concentrations of both components. Furthermore, even more exhaustive probing of the two-dimensional isotherm $s_{w,\lambda}(\{c_{k,\text{tot}}\})$ would not reveal very distinctive features between single- and two-site systems (different columns of Fig. 6.6).

Better opportunities arise when we exploit the reaction boundary isotherm $s_{\text{fast}}(\{c_{k,\text{tot}}\})$, which exhibits distinct asymmetries (e.g., comparing the plots in the middle row in Fig. 6.6). Titration series with moderate constant concentration of one of the binding partners at or below K_D, and increasing concentration of the other binding partner to values much greater than K_D, will achieve s_{fast}-values that can serve as a tight lower bound for the complex s-value. In contrast to s_w, the s_{fast} titrations can reveal the maximal stoichiometry from the different limiting complexes, with a titration of A with increasing B reporting on the maximum valence of A for B, and a titration of B with increasing A reporting on the maximum valence of B for A. However, in many cases it may be clear from *a priori* structural knowledge that one of the valences is unity.

Two additional features of the isotherms presented by the boundary pattern must be considered when designing the experimental concentration series. First, the phase transition line is asymmetric, therefore making it necessary always to use higher concentrations of the larger binding partner B to establish the limiting complex in the titration of constant A (vertical titration in Fig. 6.6) then the concentration of A necessary to establish the limiting complex in the titration of constant B (horizontal titration in Fig. 6.6). Second, this phenomenon is dependent on the relative size of the molecules: As shown in Fig. 2.12 on p. 54 and Fig. 4.8 on p. 143, the more dissimilar their s-value, the higher the asymmetry and the excess of B over K_D required to saturate the reaction boundary in the vertical titration. In the other extreme, for similar-sized molecules this problem disappears. In contrast, the horizontal titration of constant low B with increasing A is less dependent on relative component size; even very dissimilar sized molecules will yield reaction boundaries near to saturations at $c_{\text{A,tot}}$ greater than 10-fold K_D, although similar-sized interactions will still be saturated in this trajectory much earlier (Fig. 4.8).

In terms of the concentration needed to produce informative isotherm trajectories that achieve conditions close to saturation, it seems easier to deal with problems where the smaller component is monovalent for binding the bigger one, such that one can resort to the horizontal titration (constant B with increasing A). However, for a system with multiple small ligands binding a bigger macromolecule, the relative changes of the complex s-value can be very small, which can make interpretation more difficult, and hydrodynamic scaling laws may not be very useful.[21]

---

[21]For example, binding of a small ligand to a protein with an existing binding pocket accommodating the ligand may structurally not significantly increase the hydrodynamic friction at all, or even cause a decrease if binding is accompanied with, for example, some conformational rearrangements. These possibilities can be unraveled, for example, with the help of independent measurements of the diffusion coefficients by dynamic light scattering. An example can be found in [37].

The interpretation of interacting systems of large molecules — or at least where the small binding partner is the potentially multi-valent partner — is more straightforward thanks to the relatively higher hydrodynamic resolution. Unequivocal assignment of the limiting complex stoichiometry is often possible, even despite the range of theoretically sensible complex s-values. This is true, especially, when the direct experimental saturation of this complex is not assumed, and instead the entire binding isotherm is fitted with models of different stoichiometries (see below).

6.4.3 Determining the Stoichiometry from Multi-Component Distributions

A third, independent source of information from SV experiments on complex stoichiometry rests on the spectral and/or temporal signal contributions of the interacting partners [38, 39]. Principles were discussed earlier on different levels: Basic experimental aspects of data acquisition for MSSV were introduced in Section 4.4.4 of Part I [1]; fundamentals of multi-component distribution analyses MSSV and MCMC were treated in Chapter 7 of Part II [2]; and the potential and limitations of their application to the study of interacting systems were examined in detail in the Section 4.3 of the present volume. For illustration of MSSV, we have seen the examples of the 4:2 stoichiometry of the HLA-A2/U21 complex (Fig. 4.37 of Part I [1] and Fig. 4.7 of the present volume), a triple complex with 1:1:1 stoichiometry between a viral protein, its cell-surface receptor, and co-receptor (Fig. 7.5 of Part II [2]), the 1:1 complex between Tp35/bLF (Fig. 4.9 of the present volume), and the 1:1 interaction between two adaptor proteins (Fig. 4.6 of the present volume); many more examples can be found in the literature.

In comparison, the determination of complex stoichiometries from molar mass estimates (i.e., boundary spread), and the hydrodynamic considerations based solely on the complex s-value, both would fail to distinguish between a 1:2 and a 2:1 complex in an interacting system of similar-sized molecules, unless the entire isotherm is considered. With multi-component sedimentation coefficient distributions, however, these can usually be clearly identified.[22]

Nevertheless, results from MSSV and MCMC analyses can be interpreted together with mass-based and hydrodynamic considerations, to elevate the compositional information to reveal stoichiometry. Multi-component analysis should also be applied in the context of concentration series, to establish the kinetic type of the sedimentation process, and to what degree the $c_k(s)$ peaks reflect stable complex species or effective particles, as illustrated in the analysis of the dynamic Tp35/bLF complex on p.145. In the latter case, understanding the reaction boundary patterns will facilitate finding conditions of near-saturation.

As described in Section 4.3, even though MSSV can be applied best when the interacting components exhibit characteristic signal contributions, it is not limited

---

[22]Likewise, studies on the binding of small, but spectrally distinct ligands to larger macromolecules can be greatly facilitated in multi-signal mode, substituting for poor hydrodynamic resolution of macromolecules and their complexes with small ligands.

to this situation, since mass conservation constraints in the total concentration of each component (in MC-MSSV), as well as temporally distinguishable signals (in MCMC), can complement or even substitute for spectral discrimination. Finally, we note that MSSV is particularly helpful to unravel the nature of multiple co-existing complex peaks, and that it can be applied even to ascertain the average composition of broadly distributed complexes that may not hydrodynamically resolve.

6.4.4 The Job Plot

The Job plot (also known as method of continuous variation) is a method of plotting, at constant total molar concentration, the amount of complex *vs.* mole fraction of B in different mixtures of A and B. The Job trajectory will make an arc in the logarithmic plot of total loading concentrations (see the orange line in Fig. 6.1). Along this curve, the observed binding has a maximum at a position that can, under certain conditions, equal the stoichiometry of the reaction.

As shown in Fig. 6.7, this holds true if we plot an observable that is proportional to and dependent solely on the complex formed, and if the concentrations used are high — on the order of 100-fold — above K_D. Otherwise, the maximum can be hard to discern, and be shifted to molar ratios that do not correspond to the reaction stoichiometry. Unfortunately, SV experiments do not usually provide a signal proportional to the complex concentration. For example, using s_w as the observable will not produce results following the expectations of a Job plot, and instead exhibit traces strongly dependent on relative signal increments of the components, and relative species sizes.

In some cases, the choice of a concentration series describing an arc similar to the Job plot can still be an effective strategy to sample the two-dimensional binding isotherm, for example, to saturate the reaction boundaries with different component in molar excess, and/or making effective use of available stock concentrations. However, the interpretation of the results must be undertaken with a standard isotherm analysis of s_w, potentially incorporating other features of the boundary structure, or by global Lamm equation modeling.

6.4.5 Transition Point of the Excess Free or of the Undisturbed Boundary

Using a strategy similar to the Job plot, we can take advantage of the species resolution achieved in SV in order to alleviate some of its problems. However, the use of the entire boundary pattern as an indicator for the complex stoichiometry depends critically on the distinction between slow and fast kinetics on the time-scale of sedimentation.

In the case of slow kinetics, we expect to see boundaries of free species and complex species at their concentration invariant s-values, with amplitudes that correspond to the species populations. Therefore, assuming high-affinity binding that leads to stoichiometric complex formation, we can envision carrying out a series of experiments covering a wide range of molar ratios (similar to the Job

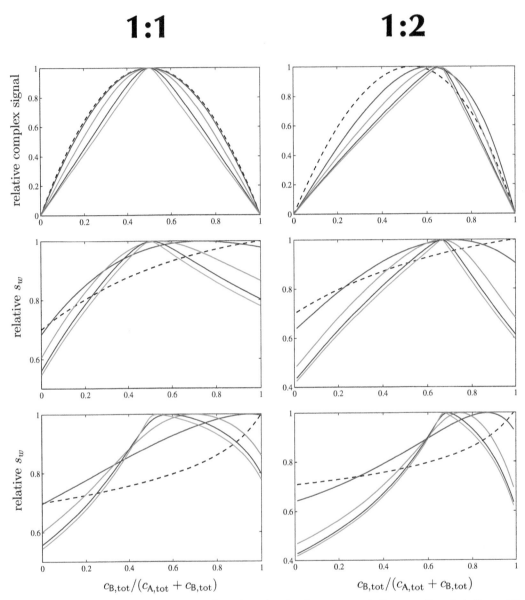

Figure 6.7 Job plots for s_w isotherms for a bimolecular hetero-association reaction $A + B \leftrightarrow AB$ (*Left Column*), and reaction $A + 2B \leftrightarrow AB+B \leftrightarrow ABB$ (*Right Column*), measuring only formed complexes (*Top Row*), overall s_w with weight-based signals (*Middle Row*), and s_w favoring the detection of A with $\varepsilon_A = 5\varepsilon_B$ (*Bottom Row*). Total molar concentrations $c_{A,tot} + c_{B,tot}$ were at 0.1-fold K_D (black), K_D (red), 10-fold K_D (green), 100-fold K_D (magenta), and 1000-fold K_D (cyan). For the calculation, A was assumed to be 60 kDa and to sediment with 3.5 S, B was assumed to be 100 kDa and to sediment with 5.0 S, and the 1:1 and 1:2 complexes were taken to sediment at 6.5 S and 8.5 S, respectively.

series). With concentrations high above K_D, either one or the other component will be entirely in the complex with no unbound form left. As the solution composition changes from molar ratio below the complex stoichiometry to above the complex stoichiometry, the identity of the excess free species switches. As a result, in such an experiment one can determine the complex molar ratio from the transition point of the excess free species.

Importantly, for fast interactions the same strategy cannot be applied to reaction boundaries, even though here, too, boundaries composed of free species of one or the other component appear, with an experimentally well-defined phase transition point. However, significant errors could ensue if the nature of the dynamic co-sedimentation process is not recognized. The undisturbed boundary does not simply reflect a molar excess relative to binding stoichiometry. While the phase transition point does indeed carry highly valuable information on the binding constant and the stoichiometry of the interaction, a quantitative interpretation needs the help of effective particle theory, Gilbert–Jenkins theory, or direct Lamm equation modeling. The difference between the molar ratio of the phase transition point and that of the reaction stoichiometry is strongly dependent on the particular conditions of the experiments, and errors could be very substantial.

For example, this is illustrated in Fig. 6.5 above, where despite substoichiometric concentrations B is dominant, and we see a peak at 7 S from A providing the undisturbed boundary. At a concentration of B of $\sim$2-fold K_D, the 7 S peak disappears and at 3-fold K_D a $\sim$10 S peak appears from B providing the undisturbed boundary for the now dominant A. If the molar concentration ratio at the transition point was considered the complex stoichiometry it would be in error by a factor of two.

6.4.6 Determining the Association Mode by Explicit Fitting with Different Lamm Equation Models

Of course, data from a Job plot — as well as any other trajectory in the parameter space of loading concentrations — can be subjected to explicit fitting of scan data with coupled systems of Lamm equations embedding different association models with certain complex stoichiometries (see Section 6.5.3 below).

Even though this approach is seemingly rigorous and can provide a safe route to distinguish alternative models, often not all features of the data can be easily fitted in a global Lamm equation model. A virtue of the sedimentation coefficient distribution analysis was that it could adapt to and unveil aspects of the data from unexpected or undesired imperfections, such as small boundaries from trace impurities, aggregates, and artificial broadening from polydispersity or micro-heterogeneity of the macromolecules of interest. The discrete nature of components on which the Lamm equation fits are predicated cannot account for these aspects, which often results in imperfect fits of the data. Unfortunately, in the presence of systematic misfits of the data, in turn, we have left the realm of rigorous analysis unless we reject these fits. Detailed Lamm equation modeling without rigorous criteria on the residuals is obviously the wrong choice.

On the other hand, clearly there are cases where Lamm equation models can achieve excellent fits — indistinguishable to residuals of $c(s)$ models — to the individual components and the mixtures alike. In this case, the performance of alternative models should be tested using F-statistics [265, 266], as a statistical guard against overly complex models. Also, the physical meaning of the best-fit parameter values must be judged, and how it is consistent with other biophysical data. For example, if a binding model requires an unphysically high s-value for a 1:1 complex, it may be rejected in favor of a 2:1 model.

However, for cases where models beyond the simplest two-step or two-component associations are likely, it may still not be convenient or even possible to formulate *a priori* a reasonably small number of models and test their ability to fit the data against each other. Therefore, it is sensible to examine first the number, shape, and concentration-dependence of $c(s)$ peaks, as well as the composition of multi-signal $c_k(s)$ peaks. These features can distill the most robust aspects of the SV experiments. Once these features are qualitatively and quantitatively understood, this can inform model-building and help to initialize parameter values for explicit Lamm equation fitting.

6.5 QUANTITATIVE ANALYSIS OF AFFINITY, LIFETIME, AND THE HYDRODYNAMIC PROPERTIES OF COMPLEXES

The following assumes that we have a suitably 'simple' association mode, such that an explicit model can be stated. Also, we assume that data from a wide range of loading concentrations are available. There are different approaches for extracting quantitative estimates of binding constants and other parameters. It can be useful to exploit them sequentially, using results from a less detailed approach to initialize parameters for the more detailed analysis.

6.5.1 Signal-Weighted Average Sedimentation Coefficients s_w

The principles of s_w for interacting systems were described in detail in Section 3.1. For interacting systems, s_w represents the average sedimentation property of the species in the sample under investigation. Because it is usually independent of the reaction kinetics,[23] as well as SV boundary patterns and shapes, s_w naturally can be used as a solid measure for characterizing the populations of species in the reaction mix. The essentials of the following subsections are covered in the detailed step-by-step protocol [243].

---

[23]The two caveats for rapid kinetics are (1) the case of quenching (Section 3.2.2); and (2) the small degree re-equilibration caused by radial dilution (Section 3.1.2).

6.5.1.1 Determining s_w from $c(s)$

An obvious prerequisite is for the $c(s)$ distribution to provide a good fit to the data. This is necessary to establish the link between distribution integrals and the transport method. To this end, all basic procedures for the $c(s)$ analysis in Chapter 8 of Part II should be followed [2]. In particular, the guidelines regarding the scan selection, meniscus position, and fitting limits are valid for non-interacting and interacting systems alike.

One feature specific to the application of $c(s)$ to interacting systems is that the sedimenting species may not be physical species but correspond instead to effective particles, i.e., systems of dynamically interconverting molecules. As described in Section 4.1, this will result in apparent frictional ratios that do not correspond to physical shapes. Therefore, unphysically low values of f/f_0 may be taken as a clue for the presence of an interaction between species that interconvert on the time-scale of sedimentation. This has no impact on the accuracy of s_w.

Integration of $c(s)$ should be carried out in the same way as for non-interacting species (Part II, Section 8.3.1 p.198 [2]), and should follow the same steps for the determination of confidence intervals. As the primary step of s_w determination, selecting an appropriate s-value range for integration is crucial. It should encompass all the $c(s)$ peaks of the interacting system, preferably such that the $c(s)$ values at the integration limits are zero. The same integration limits of $c(s)$ should be applied consistently across the different data sets.[24] For experiments with very low signal/noise ratio, the regularization-induced broadening of the peaks can make it difficult to find a common integration limit where $c(s)$ vanishes. In this case, regularization should be switched off for the low signal/noise data, or Bayesian modification should be used (e.g., using $c(s)$ at the higher concentrations as a prior), as described in Sections 5.7 and 8.2.5 of Part II [2].

An overlay plot of all $c(s)$ distributions can be very helpful. It may be composed with the companion software GUSSI [267], written and kindly provided to the SEDFIT/SEDPHAT community by Dr. Chad Brautigam. If correctly installed, a GUSSI instance can be spawned to plot the current $c(s)$ distribution from the Plot ▷GUSSI c(s) plot menu function of SEDFIT. $c(s)$ distributions can also be exported from each analysis using copy/paste functions, and accumulated in GUSSI. They can be graphed, normalized for inspection, and integrated as a group. The resulting s_w-values can be saved in a SEDPHAT data file of the isotherm type.

Usually it is reasonable to exclude impurities such as trace aggregation and degradation products, if they can be clearly distinguished from the macromolecules of interest. In this way, the focus is on the s-value range where the interacting species sediment. The integration should also exclude partial peaks toward s_{min} of

---

[24]Precise integration limits can be achieved using the SEDFIT function use c(s) integration ranges from file of the Size-Distribution Options (see Part II, p. 199 [2]).

the distribution.[25],[26] Finally, constant unrelated signal contributions common to all samples may need to be subtracted (see Eq. (6.1) on p. 189).

Integration of $c(s)$ will also allow the determination of the loading signal of the interacting material. For many experiments, this allows the calculation of loading concentrations on the basis of known signal increments. For example, this applies to most self-associations, unless fluorescence detection is used. It also applies to hetero-associations where one component is silent. A more complicated situation occurs if two or more components contribute to the signal. In any case, it is highly useful to compare the theoretically expected signal based on the pipetting schedule with the actually observed signal in order to catch unexpected errors, violations of signal conservation, or aggregation processes. If available, this actual concentration should be utilized in the isotherm file, rather than the nominal sample concentration.

6.5.1.2 Assembling the Isotherm

The s_w-values determined from integration of $c(s)$ at different concentrations must be compiled into a dedicated data file for further processing. An important consideration when dealing with hetero-associations is that s_w-values will generally be strongly dependent on the optical system used in the detection (see Fig. 3.1 on p. 70). Thus, s_w data from different optical systems or different data acquisition wavelengths cannot be combined into the same isotherm. Rather, one separate isotherm $s_{w,\lambda}(\{c_{k,\text{tot}}\})$ should be compiled for each signal, and isotherms should be globally modeled. This situation is different for self-associations, where the signal coefficients in the formula of s_w are identical and cancel out (see Eq. (3.8) on p. 69).

---

[25] An exception is the rare case when the free form of one binding partner has a very small sedimentation coefficient and contributes to $c(s_{\text{min}})$, in which case it should be included. However, to avoid artificial contributions from baseline correlations to $c(s_{\text{min}})$, in this case the baselines should be held fixed at zero (or an empirically determined realistic value) during fitting of $c(s)$. An alternative approach for the consideration of very small binding partners is their description as an extra discrete species, combined with suitable truncation of the continuous segment of $c(s)$ at low s-values to avoid overlap and correlations.

[26] Excluding s_{min} is straightforward when the peaks from the macromolecules of interest consistently clearly separate from s_{min}. Unfortunately, as Fig. 8.5 (p. 186) of Part II demonstrates, this may not always be the case [2]. As discussed in Section 8.2.1 of Part II, in such cases, the problem can be eliminated by using a Bayesian regularization [2]. Even though the result will re-shape the distribution, by design no errors in s_w outside the statistical limits given by the signal/noise ratio of the data can occur [127].

In SEDPHAT, the input file for isotherm analysis is a matrix of numbers in ASCII text format, delimited by tab or white space. Such a file can be created with the notepad.exe or wordpad.exe utility of the Windows operating system. The SEDPHAT name convention expects a file extension .isotherm.

Each line contains the data from one sample. The first K entries are the loading concentrations of the K components A, B, C, etc., in μM units. (For high-affinity interactions with sub-nanomolar K_D it is advantageous to enter concentrations in nM units, and to compensate by multiplying all reported binding constants in SEDPHAT by a factor 1000 [243].) The identity of components follows the definition in the selected interaction model; this is generally not the same as the size-ordered notation convention in the effective particle theory. For example, a self-associating system has only a single concentration entry; for a two-component system the first number represents μM concentration values for component A and the second is the μM concentration of component B. After the concentrations follows the s_w value in Svedberg units.

Assuming n samples and K components, these entries present an $n \times (K+1)$ matrix of numbers. It is helpful to sort the samples in ascending order of concentration. No concentration can be assigned 0.

Optionally, confidence intervals of s-values can be included in the isotherm file as an independent block of numbers, separated from the n lines with s_w-values by a new line containing the string CONFIDENCE INTERVALS. The next n lines each contain only two values — a lower bound and an upper bound for the s_w-value of the corresponding line in the data block above. The bounds must bracket the s_w-values, and entries must be in the same order.

It is advantageous not to carry out any buffer corrections to standard conditions (see $s_{20,w}$ in Part I, Chapter 1 [1]). Once loaded into SEDPHAT, buffer density and viscosity, along with a partial-specific volume, can be input and associated with the experimental data. SEDPHAT will use this information to calculate from the $s_{20,w}$-values of all species — which are the global model parameters — the s-values as they apply to the experimental conditions of the data. Experimental s-values will be plotted. This facilitates efficient data organization, and provides greater flexibility for global analysis.[27] Data acquired from different analytical centrifuges should be externally calibrated, in order to eliminate inconsistencies otherwise arising from temperature and other calibration errors (see Part I, Chapter 6 [1]).

---

[27]If this convention is not used, and instead $s_{20,w}$-values are entered directly into the isotherm file, then the buffer correction of SEDPHAT must be switched off by entering standard values in the parameter box associated with this experiment.

The `Data ▷Load New AUC Isotherm Data` function in `SEDPHAT` is used to load the isotherm. After selection of the `.isotherm` file, the `Experimental Parameters` box prompts the user to enter the type of isotherm (select the radio-button `sw isotherm` for s_w isotherms), as well as ancillary information of the experiment. The optical pathlength must be specified, along with signal coefficients in $M^{-1}cm^{-1}$ units for all relevant components, i.e., A for self-associating systems, A and B for two-component hetero-associations, etc. A checkbox allows the concentrations in the isotherm file to be subject to a correction factor as part of the model (see below).

Hydrodynamic model parameters are in units of standard conditions, which naturally allows global modeling of data with different conditions. Therefore, temperature, and ρ and η of the buffer solution at the experimental temperature, and apparent $\bar{v}_{xp}$-values under experimental conditions must be entered in the `Experimental Parameters`, and a global $\bar{v}_0$ parameter must be entered in the `Options ▷ Set VBAR20` menu.

These ancillary data will be kept in a separate ASCII file with extension xp. It also contains path information for the location of the isotherm data file. All future loading of this data can be accomplished by loading this xp-file, using the `Load Experiment` function. Multiple such experiments can be loaded side-by-side in `SEDPHAT`, in which case automatically a global analysis will be made.

When analyzing $s_{w,\lambda}(\{c_{k,\text{tot}}\})$ isotherms of hetero-associations in practice, it is worth considering the how the isotherm is graphically displayed. Even though the experimental data points can sample any regions in a multi-dimensional concentration space, they are customarily presented as a two-dimensional plot of s_w-values along a certain trajectory of concentration. A fitted line represents the best-fit model along the same trajectory. The trajectory is usually a straight line (see above); however, there is no requirement for experimental concentrations to be chosen in any systematic way at all, as long as they jointly sample the isotherm well. For example, it is desirable not to limit the degree of complex saturation that could be achieved in an experiment, and to compromise the information content of the experimental data, just to adhere to a systematic progression that facilitates a visually pleasing data display. The path connecting the data points in concentration space is irrelevant, and the shape of the model function in between data points has no impact on the analysis — only the model function values at the location of data points matter. Therefore, any arbitrary path connecting data points in concentration space may be chosen, along which model and data points are compared if a two-dimensional plot s_w-*vs.*-c is desired. Alternatively, if experimental data points fall along different rational paths, they can easily be separated into different isotherm files and jointly analyzed in a global analysis. This may even include isotherms with a single data point.

SEDPHAT will automatically attempt to recognize titration and dilution series from the concentrations in the isotherm files, and display them along an appropriate concentration variable. If multiple segments are recognized, they will automatically be assigned different branches, with model functions representing the appropriate slices of the multi-dimensional binding isotherm. Alternatively, SEDPHAT can be forced to find a single, curved trajectory to be displayed, through a toggle in the Display ▷Pretty Curves submenu.

It can be helpful to view the experimental data points in the concentration space as in Fig. 6.3 on p. 193, to examine how the concentrations relate to features of the isotherm that are not on the same trajectory. Such a two-dimensional color-contour map can be switched on using the 'microbutton' labelled s located in the button panel in the upper right corner of the isotherm plot; pressing the ESC button will return the display to the usual plots.

6.5.1.3 Modeling

Modeling of s_w isotherms proceeds on the basis of Eq. (3.7) after specification of a binding model that relates total loading concentrations with species concentrations *via* mass action law. Critical for the success of the s_w analysis is the precision and range of the data points. With careful experimentation, errors in s_w can be as low as ~0.01 S [243], which can be fully exploited on the solid foundation of the transport method. Accordingly, the match between data and best-fit model can be critically examined.

Fig. 6.8A highlights the importance of having a large enough concentration range that can define the model. In the example of Fig. 6.8A, more data points at much lower and much higher concentrations would eliminate ambiguity about the three models shown. However, to the extent that a large concentration range will also lead to greater variation in signal/noise ratios of the SV data, it is important to account for different size error bars of s_w across the isotherm. Unless compensated for by using different optical signals, the s_w-values obtained at lower concentration will usually have larger error. Therefore, errors should be specified in the isotherm file (see above), so that these appropriately gain lower statistical weight in the isotherm analysis.[28]

As part of the initialization of the model, buffer-corrected $s_{20,w}$-values must be specified for all species. At this point, they do not need to rise to the same level of accuracy as required for hydrodynamic shape modeling, because their values can generally be refined in the fit. Clearly, those that can be independently measured with good precision (i.e., usually s_A and s_B in hetero-associations) should be kept fixed at these values during the fit. A special case exists when the s-values of the free component are only known with the same precision as the s_w-values measured

---

[28]In cases where data are acquired with similar signal/noise ratio for all samples, such that all s_w values in the isotherm have nearly identical error bars, the specification of confidence intervals is not necessary if — as is usually the case — the error analysis of the fitting parameters is carried out using F-statistics [268].

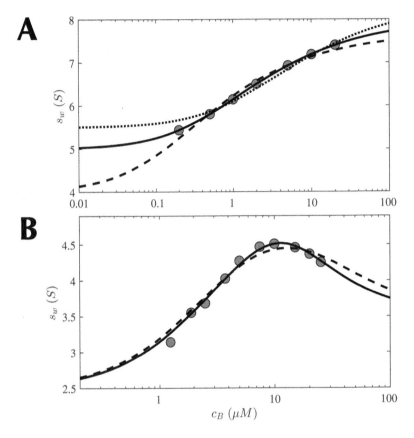

Figure 6.8 Examples for a close match between correct and incorrect s_w isotherm models. The examples highlight that for a precise characterization of self-association or hetero-association, it is important to acquire data over a large concentration range and/or to consider other hydrodynamic prior knowledge for the s-values of the different species. *Panel A*: Simulated isotherm of weight-average sedimentation coefficients s_w for a self-association of a 5 S monomer to form an 8 S dimer, with an equilibrium constant $K_D = 2$ μM, at concentrations of 0.2, 0.5, 1, 2, 5, 10, and 20 μM (circles). The solid line is the entire isotherm for this system. The dashed line is an isotherm for another model with a lower monomer s-value of 4 S and a dimer of 7.68 S, with an equilibrium constant $K_D = 0.5$ μM, and the dotted line represents the model with monomer s-value 5.5 S, dimer 8.35 S, and $K_D = 6.0$ μM. *Panel B*: Experimental data (circles) from a hetero-association titrating 5 μM of A with different concentrations of B (similar to [109, 269]). The best-fit model (solid line) using a correct A + 2B $\leftrightarrow$ AB+B$\leftrightarrow$ ABB with experimentally pre-determined s-values for A (2.47 S), B (3.5 S), a hydrodynamically modeled value for AB (4.5 S) leads to a best-fit value for ABB of 5.47 S and a K_D value for two symmetric non-cooperative sites of 2.4 μM. By contrast, an impostor fit with a single-site model (dashed line) leads to a best-fit s-value for AB of 8.87 S, with a K_D of 17.1 μM. The latter model is in conflict with hydrodynamic considerations.

for the mixtures. In this case, it is a better strategy to treat s_A and s_B as unknowns in the isotherm fit, but include the separately measured values as data points in the analysis [243].[29]

Good experimental estimates for free and complex s-values may also arise in favorable cases from $c(s)$ peaks of slowly reversible systems. Generally this is not possible for complexes of rapidly reversible systems due to the formation of reaction boundaries. Nonetheless, systems with rapid exchange may also provide estimates of s-values of one or both of the free species in the undisturbed boundary, as well as the s-values of saturating complexes in the reaction boundary. In rare cases, it is possible to modulate the kinetics and/or the assembly state with different solvent conditions allowing a better recognition of species s-values [99, 101], as in Fig. 2.2 on p. 30.

Finally, in the absence of other estimates for complex s-values, the latter may be estimated on the basis of hydrodynamic scaling laws, such as $s \sim M^{2/3}$ for compact particles, or using a `SEDFIT Calculator` function to predict s for a molecule with mass M at a certain frictional ratio. Of course, in cases where more detailed structural information is available, more advanced hydrodynamic modeling can be applied [270–272]. Alternatively, s-values from related molecules such as protein mutants may provide reasonable initial estimates.

Whichever method is used, in order to accommodate some degree of uncertainty about the precise value of species s-values, upper and lower bounds can be established and used as constraints in the analysis. For example, the dashed lines in Fig. 6.8 depict models that can be excluded on grounds of hydrodynamic considerations. For homo-dimerization, the dimer is unlikely — or impossible in the case of globular monomers — to sediment at close to twice the rate of the monomer, as is implied in the model of *Panel A* with the dashed line. Rather, in our experience values of s_2/s_1 of approximately 1.6 ± 0.2 should be expected for most proteins [255]. Similarly, the hetero-dimer of a 2.47 S and a 3.5 S species cannot sediment at 8.87 S (as in *Panel B*, dashed line).

After constraints are applied to species s-values, the possibility of achieving good but impostor fits will be strongly reduced, and parameter correlation between binding constants and s-values will be diminished.

---

[29]A simple work-around for the problem that `SEDPHAT` currently cannot accept zero concentrations, due to conflict with the log-scaled plotting, is the assignment of a vanishingly small concentration to the component that is not present. These data points may be added to the isotherm file alongside the mixtures, or be packed into separate isotherm files that are included with the mixtures in a global analysis.

The `Constraints` top-level menu in `SEDPHAT` leads to functions for establishing and modifying constraints for parameter values during refinement. Their input is piggybacking on the same parameter boxes as used for entering the parameter estimates, but temporarily cleared of refinement check-marks. In the present context, checking a box next to a parameter destines this value to be bounded, and a minimum value should be entered, first. The same parameter box will reappear a second time for input of the maximum value. Constraints will be effective only for parameters that are being refined in the fit.

After a `Fit`, the textual fit information in the `SEDPHAT` window will include either a note `CONSTRAINTS inactive` if all parameters refined within the bounds, or state `CONSTRAINTS ACTIVE` followed by a list of the parameter values that have reached, and were kept, at the limit indicated by an asterisk.

Concentration values warrant another discussion in the context of isotherm models. An important, yet easily overlooked difference between isotherm fitting and direct Lamm equation modeling of raw SV boundaries is that the isotherms conventionally treat the sample concentration as a known quantity, whereas they will usually be fitting parameters in Lamm equation modeling, based on known signal increments. This additional flexibility can present a distinct disadvantage of Lamm equation modeling, especially if only a single signal is acquired for multi-component systems. It is exacerbated if no additional concentration constraints from pipetting schemes can be embedded (see below). In this case, best-fit concentration assignments of similar-sized species will be strongly correlated, which propagates into larger error intervals for their products, i.e., the binding constants. For isotherm data sets, this flexibility is removed since concentration values are assigned *a priori*, and/or derived from $c(s)$ integration, and then typically kept fixed.

When studying hetero-associations, in either method, another concentration problem remains, which originates from unavoidable residual uncertainties in active concentrations (as discussed in Section 6.1.1 above) and potential errors in the signal increments (Part I, Section 4.3.2 [1]). These may lead to inconsistencies in relative concentrations, such that the assumed molar ratios, and therefore the predicted relative saturation of complexes, may be in error, resulting in distortions of the binding isotherm.[30] For example, considering the narrow ridge of the s_w isotherm along the equimolar line (see Fig. 3.1 on p. 70), it is easy to see how a moderate deviation from an equimolar molar ratio may amplify strongly into large deviations from expected s_w-values, producing a much more shallow binding isotherm. Likewise, titrations at concentrations very high above K_D that are

---

[30]The prevalence of concentration errors, be it from errors in the total component concentrations or incompetent fractions, is demonstrated by the experience with isothermal titration calorimetry, where non-integral 'n'-values are the rule rather than the exception [260, 273].

normally characterized by a maximum at the reaction molar ratio will be distorted to exhibit maxima at incorrect, non-integral concentration ratios.

Therefore, it can be appropriate to introduce a 'concentration correction factor' even for isotherms. Since errors in signal increment or stock concentrations apply uniformly to all samples of an experimental series of mixtures, the concentration correction factors are applied uniformly across all concentrations of that component in the isotherm. They can only be refined for one of the components in the non-linear regression, in order to avoid complete correlation with binding constants. This is the last step of refinement of the isotherm analysis and is normally not required and should be considered sparingly. Also, the value of the best-fit concentration factors must be critically considered for plausibility.

> Concentration correction factors will apply only to SV isotherms with the `allow concentration correction factors` flag checked in the **Experimental Parameters**. They can be introduced in the **Concentrations** menu, and/or by checking the box next to the component concentration in the **Global Parameters** box.

Concentration correction factors are different from incompetent fractions discussed in Section 6.1.1 above. The latter are fixed fractions of macromolecules that are present in solution — and contribute to s_w — but do not participate in binding.[31] Concentration correction factors and incompetent fractions should not both be refined, to avoid risk of parameter correlation.

Once all parameters that are to be refined in the non-linear regression are identified and, where possible, constrained, their iterative optimization is the next step. It corresponds to finding the lowest point in an error surface, that is, the smallest weighted root-mean-squared-deviation (rmsd) of a model as a function of adjustable parameter values. Isotherm models being non-linear, the error surface can potentially exhibit complex shapes with hills and valleys and multiple minima, and finding the physically reasonable global best-fit value is usually non-trivial. Establishing meaningful starting parameters that describe predicted values not too far from the experimentally observed values is key. Physical consideration should guide this process, not blind trial-and-error.

As a rule of thumb, we should classify parameters according to how uncertain their values are, and start with refining only the least known ones, gradually increasing the degrees of freedom in the fit until all the parameters are optimized. A typical optimization sequence will start at binding constants (utilizing potential symmetries in multi-site systems), then species s-values, cooperativity constants,

---

[31]In comparison, concentration errors and incompetent fractions cannot be distinguished in isothermal titration calorimetry [260, 273] or optical biosensing [274], which observe only reaction events. Such considerations can factor in to the success and refinement of global multi-method analyses.

and — if necessary at all — nonideality constants and concentration correction factors or incompetent fractions.

In SEDPHAT, a given set of model parameters can be tested in their performance to match the experimental data with the function RUN ▷Single Experiment Run or ▷Global RUN for single experiment or global analyses, respectively. Analogous Fit functions start non-linear optimization algorithms.

The optimization algorithm is chosen in the Options ▷Fitting Options submenu, and can be simplex, Marquardt–Levenberg, or simulated annealing [275]. Briefly, the simplex method has an advantage of randomly exiting shallow local minima, whereas the Marquardt–Levenberg algorithm has an advantage in homing into the optimal point based on local curvature of the error surface, and simulated annealing is advantageous for comprehensive but time-consuming searches of the parameter space. Often, a sequence of alternating simplex and Marquardt–Levenberg algorithms is the strategy of choice.

Once the best-fit parameters for a model have been found, and residuals are acceptable both with regard to the root-mean-square deviation as well as the systematicity, the veracity of the best-fit parameters should be critically examined, considering the possibility that, for some systems, other reasonable models may exist that might fit the data equally well. Further, the statistical significance of the best-fit parameter values of interest should be tested with a statistical error analysis. The established method of choice is the error projection method, where the error surface in the vicinity of the best fit is examined to find the extreme values of the parameter of interest that can be accommodated without the error surface exceeding a critical F-ratio (i.e., a relative increase given the degrees of freedom of the fit, related to the number of model parameters and data points) [265, 266, 276]. As mentioned above, it is possible that only one-sided limits exist. For example, for weak interactions, the result may be a statement that K_D must be higher than a certain value.

The most straightforward method to calculate a confidence interval in SEDPHAT proceeds *via* the function Statistics ▷Automatic confidence interval search w projection method. It reuses the model parameter boxes to identify the parameter for which the confidence interval is to be determined, requiring the user to uncheck the box next to the parameter value of interest (as if keeping it fixed). After specifying search parameters, such as step-size and search limits, this parameter will be fixed at sequentially higher values, while all other refining parameters are allowed to compensate, until a threshold value (e.g., two standard deviations) in F-statistics is exceeded, or the search range is exhausted. This is followed by a similar search at lower than best-fit parameter values. The critical values where the statistical thresholds are exceeded are then reported as confidence intervals.

6.5.2   Global Analysis with Isotherms of Boundary Features

The strength of the s_w isotherm analysis is that its foundation is the transport method, which is independent of any boundary features. This makes the analysis robust and often quite straightforward. However, the simplicity also represents its greatest weakness, due to the neglect of equally robust features of the sedimentation patterns [41, 106, 277]. Even the most cursory inspection of the sedimentation process will reveal a multi-modal boundary structure of multi-component interacting systems, which are ignored in the s_w-isotherm.

Therefore, it is very useful to improve the analysis through the incorporation of isotherms reflecting the amplitudes and s-values of the observed boundaries. The goal is a comprehensive set of isotherms that will be globally fitted. This will aid in reducing correlations between fitting parameters, alleviate the problem of predicting the isotherm endpoints, improve the precision of hydrodynamic parameters, and enhance the distinction between correct and incorrect boundary models. Which isotherm model must be used depends significantly on the kinetics of the reaction, which can usually be recognized from an overlay of the $c(s)$ distributions (see above).

6.5.2.1   Global Analysis

Even before considering boundary features, a first step to enhance the data interpretation is the global analysis of multiple s_w isotherms. In the simplest case, these may be replicates, but the concept of global analysis becomes much more powerful in the combination of complementary data sets, for example, from probing the concentration space in different trajectories. This may include dilution series at different molar ratios, and/or orthogonal titration series.[32]

When studying hetero-associations between proteins, more often than not the components differ in content of aromatic amino acids, which can be exploited by taking absorbance scans at different wavelengths in the UV (usually 230 nm, 250 nm, and/or 280 nm) and simultaneously acquiring interference optical data (see Part I, Section 4.4 [1]). Besides the potential for multi-signal $c_k(s)$ analysis to gain information on the composition of individual $c(s)$ peaks, the multi-signal data set also allows the transport method to be applied independently at each signal. Since s_w is strongly dependent on components' signal contributions (see Fig. 3.1 on p. 70), their combination will carry new information. In this way, multiple s_w isotherms can be determined — one at each signal — which can be usefully combined in a global analysis.

This is slightly different in self-associations, where, in the absence of oligomerization-induced signal changes, s_w data acquired at different wavelengths can be combined into a single isotherm. Here, different wavelengths can still be extremely useful, though, to access a larger concentration range.

---

[32]When different runs are made for the different concentration series, they should either be carried out on the same instrument, or on instruments that have been externally calibrated (Part I, Chapter 6 [1]) to avoid systematic errors causing mismatch of the data sets.

Simply repeating the data loading step in SEDPHAT will lead to a global analysis. Experimental data sets of different types, e.g., different isotherm types, can be seamlessly loaded side-by-side. Experiments will be distinguished by a number, which is displayed in the square in the upper right corner of the data plot.

All parameters in the Experimental Parameters boxes are 'local parameters.' A separate set of local parameters is introduced for each data set. By contrast, 'global parameters' related to the physical binding model are unique and apply to fits to each data set. They are entered in the Global Parameters boxes of the different models. An intermediate class of parameters are local but shared between different experiments, also referred to as 'linked' local parameters. They can be established with the redirect checkboxes, in combination with the number of the experiment from which this parameter will be copied.

Since interacting systems require modeling with molar concentration units, a crucial local parameter for each data set is the molar signal increment per cm (which can be shared among data sets using the same detection) and the optical pathlength in the experiment. These are specific to the particular experiment.

An important consideration in global modeling is the statistical weight given to each data set and each data point. The error surface to be minimized can be regarded as the global reduced χ^2,

$$
\chi^2_{r,\mathrm{glob}} = \frac{1}{\sum_{e=1}^{E} N_e} \sum_{e=1}^{E} \sum_{i=1}^{N_e} \frac{\left(y_i^{(e)} - f_i^{(e)}(\{p_{\mathrm{glob}}\}, \{p_{\mathrm{loc}}^{(e)}\}) \right)^2}{\left(\sigma_i^{(e)} \right)^2} , \tag{6.3}
$$

where E is the number of experiments, N_e the number of data points in experiment e, $y_i^{(e)}$ the ith data point of experiment e with standard deviation of data acquisition $\sigma_i^{(e)}$, and $f_i^{(e)}$ the fitting model of experiment e that depends on the set of local parameters $\{p_{\mathrm{loc}}^{(e)}\}$ and global parameters $\{p_{\mathrm{glob}}^{(e)}\}$ that are being refined. For a perfect fit of many data points with normally distributed noise of known magnitude, we arrive at $\chi_{r,\mathrm{glob}} = 1$. However, the problem arises that the uncertainties of data acquisition $\sigma_i^{(e)}$ are generally poorly known, and often include unknown systematic errors [268]. Therefore, it is useful to introduce *ad hoc* a weight-factor w_e for each experiment and substitute $\sigma_i^{*(e)} = w_e \times \sigma_i^{(e)}$ into Eq. (6.3). For a single-experiment fit this is irrelevant, as long as F-statistics if applied to calculate confidence intervals. In the global fit, this allows to compensate for very dissimilar numbers of data points, and for dissimilar influence of systematic imperfections in the data.

Noise parameters can be adjusted in the `Experimental Parameters` boxes to change the relative weight of data sets in global modeling. The `noise` field contains the average $\sigma_i^{*(e)}$ for all data points in the given experiment. Lowering this number by a factor x will give an experiment x^2-fold higher weight in the global fit. The checkbox `*sqrt(N1/Nx)` will apply an additional factor to compensate for differences in the number of data points relative to experiment #1.

If the $\sigma_i^{*(e)}$ used correspond to the true uncertainty of data points (as estimated, neglecting systematic errors, by the rmsd of a single data set) then no additional weight factors apply. Isotherms of s-values, such as s_w and s_{fast}, have typically similar precision, and the assumption of the same default nominal $\sigma_i^{*(e)}$ will minimize bias from weight factors. Some judgment will be necessary, however, when combining experiments with dissimilar types of measurement. If weight factors apply, as discussed in [268], sensitivity of the final parameter estimates to the weight factors should be probed, and propagated into the analysis of parameter uncertainty.

6.5.2.2 Species Population Isotherms and Average Sedimentation Coefficient of Complexes

As we have seen, for interactions where complexes on average persist for longer than the sedimentation from meniscus to bottom, sedimentation and binding become independent events. After ensuring that the system is in equilibrium with regard to mass action law prior to the start of sedimentation, the sedimentation coefficient distribution will ideally represent the hydrodynamically separated species, and reveal the population of species in different assembly states. The concentration-dependent population of species, in signal units, is referred to as 'population isotherm.' It can be modeled directly with mass action law, given component loading concentrations and extinction coefficients.

In order to achieve a good match between total concentrations and total signal, given known loading concentrations, signal increments may be refined in the analysis. *Vice versa*, given signal increments may be used to refine total component loading concentrations *via* concentration correction factors (see above).

Unfortunately, the resolution of $c(s)$ is dependent on the signal/noise ratio of the data acquisition and regularization. For example, species' peaks resolved at high concentrations may be merged into one broad peak when using standard regularization. In this case, Bayesian analysis (Part II, Section 5.7 [2]) can again be very useful: Because the peak positions are more discernable for the high signal/noise data, these s-values may be used as a prior for the lower concentration mixtures. This will help to pin-point contributions from the individual species at lower concentrations.

On the other hand, if the distinction between species' contributions is not significant, i.e., it does not lead to $c(s)$ peaks using standard regularization without prior probabilities of any s-values, the information content of these data sets may be limited to the s_w-value and it may be advantageous not to include data points for these mixtures in the population isotherm. This does not pose a problem, since in global analysis the number of data points for the different experiments do not need to match.

> The input file for population isotherm analysis in SEDPHAT is also in an ASCII format, similar to that of s_w isotherm. Here the first line must be PARTIAL CONCENTRATION ISOTHERM followed by the number of components and the number of species. For example, for a simple A + B ↔ AB system, the first line will therefore be "PARTIAL CONCENTRATION ISOTHERM 2 3" (meaning there are 2 components with 3 species). Accordingly, the 1st and 2nd columns indicate the total loading concentrations of the components, and the 3rd, 4th, and 5th columns indicate the measured population in signal units of the species, A, B, and AB, respectively. When importing this input file into SEDPHAT, it needs to be characterized as partial conc. type in the Experimental Parameters. A population isotherm will be plotted in logarithmic scale.

An intermediate case is sometimes found in multi-site systems, where the $c(s)$ of free species can be clearly discerned, but the different complexes cannot be hydrodynamically resolved [14]. In this case, integration of all complex species can be carried out, and divided by the total signal, to form an isotherm of type all complex signal fraction. Analogously, the signal-weighted average s-value of all complexes can be obtained by the same integration, and added as a separate isotherm specified in SEDPHAT as all complex sw.

6.5.2.3 Sedimentation Coefficients and Amplitudes of Reaction Boundaries

Perhaps the most obvious target for an improved data analysis is to exploit the clearly recognizable multi-modal sedimentation pattern of rapidly equilibrating hetero-associations.[33] The reaction boundaries can usually be clearly recognized, and their characteristic concentration-dependence can be used to define additional binding isotherms. These can be established through integration of $c(s)$, as discussed in Section 4.1.3.

---

[33]Unfortunately, at present this strategy cannot be used for rapidly reversible self-associating systems, even though they do exhibit broadened reaction boundaries, as predicted by Gilbert theory [277]. Attempts to generate more information from higher-order averages of the sedimentation coefficient, such as the z-average [278] or number-average, have not been fruitful. z-average sedimentation coefficients could be determined in principle, through differential sedimentation experiments [278], or from the slope $d(cs_w)/dc$ [279]. While the former is experimentally cumbersome, the latter is not really generating any new information beyond the information content of $s_w(c)$. Finally, the idea of simply using integrals over apparent sedimentation coefficient distribution functions $g^*(s)$ from dcdt [280] seems flawed due to its skew and the impact of diffusion on the averages and higher moments. Related, the proposal by Nichol and Winzor [123] to extrapolate from experimental data the asymptotic boundary shapes to compare with predictions of Gilbert theory has not been further examined or adopted in the literature. Therefore, in the case of rapidly reversible self-associations, the current method of choice for improving information content of the s_w analysis is to proceed directly to Lamm equation modeling [46, 47, 281].

The reaction boundary can be much closer to saturation, and therefore improve the determination of complex s-values. To the extent that the latter are correlated with parameter estimates for the binding constants, these, too can be improved. To optimally take advantage of this, titration series can be advantageous. On the other hand, a strategy of including just a few mixtures at concentrations with large excess of one component over the other will also exploit this feature.

The signal amplitudes of the undisturbed and reaction boundary, a_{slow} and a_{fast}, respectively, will likewise be easily identified and can be conveniently determined through integration of $c(s)$ as in Eq. (4.4) on p. 133.

In practice, when selecting the integration limits it is useful to keep in mind that under some conditions the reaction boundaries can be broad, as illustrated in Fig. 3.17 on p. 107, and may exhibit multi-modal features. Especially near phase transition points it may be difficult to distinguish the undisturbed from the reaction boundary (e.g., Fig. 4.2, p. 126). It may be prudent to exclude data points for such conditions from the isotherm.

The input file for s_{fast} isotherms in SEDPHAT is identical to that of s_w isotherms. To identify the data file as s_{fast} isotherm, the radio button for the isotherm type in the Experimental Parameters box must be set to EPT sw fast.

The isotherm of boundary amplitudes is similar to that of species population isotherms described above. The first line must read GILBERT PARTIAL CONCENTRATION ISOTHERM followed by the number of components and the number of boundaries. (Since rapidly reversible two-component systems will exhibit two boundaries, the two numbers will be the same.) For example, for a two-component hetero-associating system the header will be GILBERT PARTIAL CONCENTRATION ISOTHERM 2 2. As in the species population isotherm, the first n columns are the component loading concentrations, followed by a_{slow} and a_{fast}.

An example for the improved ability to discern the correct binding model with the global isotherm approach, as compared to the s_w isotherm alone (depicted in Fig. 6.8B), is shown in Fig. 6.9. In this case, the global analysis of s_w, s_{fast}, a_{slow}, and a_{fast} captures a comprehensive picture of the sedimentation boundaries, and provides essentially the same information as the Lamm equation model in Fig. 1.8. Another comprehensive example can be found in [106], where the two-site interaction of α-chymotrypsin binding soybean trypsin inhibitor was studied as a model system.

6.5.3 Lamm Equation Modeling

Lamm equation modeling is the direct fit of the recorded signal profiles across the solution column with solutions of the Lamm equation incorporating reaction terms, as described in Chapter 1. In contrast to the isotherm analyses described so far, it attempts to model boundary broadening, which besides diffusion may include some kinetic information. In our laboratory it is the last stage of the analysis, because it

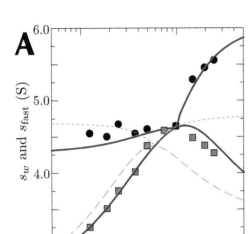 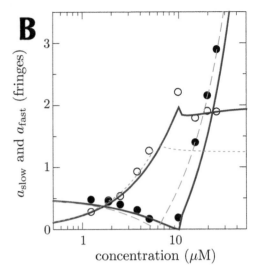

Figure 6.9   Global isotherm analysis of a titration series of constant 5 μM Ly49C (31 kDa) titrated with MHC class I molecules H-2Kb (45 kDa). Sedimentation at 50,000 rpm was recorded by interference optics. The $c(s)$ analysis shows clearly discernable undisturbed and reaction boundaries [269], including a phase transition at ∼10 μM from H-2Kb to Ly49C being the dominant component completely engulfed in the reaction boundary. *Panel A*: s_w (squares) and s_{fast} (circles); *Panel B*: a_{slow} (filled circles) and a_{fast} (open circles). Shown are the global best-fit 2:1 binding model (solid red lines), resulting in a binding constant of $K_D = 1.2$ μM and an s-value of 6.08 S for the 2:1 complex (fixing the s-value of the 1:1 complex at the hydrodynamically predicted value). The best-fit global model using an impostor 1:1 binding model is shown in cyan. Lamm equation models fitted directly to the boundaries from some of the mixtures are in Fig. 1.8 on p. 18, yielding very similar parameter estimates. Compare also with the analysis of only the s_w isotherm for the same system in Fig. 6.8B on p. 214. Reproduced from [109].

is more stringent in the requirement for sample purity and ideality, and an initial $c(s)$ analysis or individual components and mixtures may support or exclude this avenue. It is also more time-consuming than isotherm fitting, and, as mentioned above, can profit from parameter estimates that have already been optimized by isotherm models exploiting the overall boundary features. Practical aspects have been reviewed in [69, 70, 106].

In view of experimental design of concentration series, all considerations discussed in Section 6.1.1 apply. If possible, data from a large range of sample concentrations should be loaded into the global analysis (Fig. 1.8 on p. 18), similar to those concentrations used for isotherm analysis. It can also be profitable to include data acquired at different signals (e.g., as in Fig. 1.7). Data from conditions that lead to incomplete saturation of the reaction boundaries will provide more information on reaction kinetics than those under saturating conditions.

For the scan selection, it is crucial to load data representing the entire sedimentation process, following the same guidance as in Chapter 8 of Part II for studying non-interacting systems [2]. Partial boundary modeling (Part II, Section 8.1.4.1 [2]) may be used as a strategy to exclude sample imperfections, such as hydrodynamically well-separated aggregates, from the analysis. By contrast, for contaminations sedimenting slower than the system of interest, due to their large diffusion envelope, it is often advantageous to keep the full experimental data set but add a non-participating species to the model, i.e., to account for signal contributions of a macromolecule that does not interact with any macromolecules of interest (as in Figs. 1.4 and 1.7). Estimates for signal amplitude, s-value, and M-value of such a species can usually be retrieved from preliminary $c(s)$ analysis. Dependent on the optical detection system, additional factors contributing to the signal, such as baselines or gradients of signal amplification, may need to be accounted for the fitting; this is described in detail in Chapter 1 of Part II [1].

The combination of data sets from different sample concentrations is straightforward in SEDPHAT, simply requiring repeat loading operation of the raw scan files for each new data set. Each requires specification of ancillary experimental data in the Experimental Parameters Box, including signal increments, optical pathlength, nominal noise or data acquisition, as well as buffer conditions, all in the same way as described for isotherm experiments.

This also includes meniscus, bottom, fitting limits, and noise models (baseline/RI noise/TI noise), which are the same parameters found in the model parameter boxes of SEDFIT, with the additional option of partial boundary modeling (Part II, Section 8.1.4.1 [2]). Meniscus and bottom can be defined as shared local parameters between experiments, to be used for suitable pairs of data sets.

Loading concentration values are local parameters, by SEDPHAT convention in μM units, and are entered through the Concentrations menu, which can generate buttons superimposed to each data graph that display the local concentration of each component and, when left-clicked, allow changing these parameters. Importantly, constraints of shared local concentration parameters can be implemented by pressing the right mouse button and dragging a line from one button to the corresponding button of the concentration to be linked. (Circular constraints must be avoided.) To facilitate constraints for two-component samples prepared from stock mixtures, the second component concentration can be expressed as a molar ratio of B to A, which may be an unknown or linked parameter in dilution series.

Signal contributions from a single contaminating species that does not interact with the molecules of interest can be added to the Lamm equation model in SEDPHAT. This is achieved by checking the non-participating species flag in the Global Parameters box, which upon closing of the parameter box causes it to re-appear with additional section for entering the effective molar mass, s-value, and $\bar{v}$ of this species. The signal amplitudes are entered through the Concentrations ▷ Concentrations (non-participating species) menu function, which provides prompts for the signal amplitudes (in signal units) of the respective data set.

Establishing constraints can be extremely useful. One example is the meniscus/bottom position of the identical solution column scanned at different absorbance wavelengths, which (when using the same instrument) have to be identical. Another constraint can emerge from the design of the concentration series: The concentration of the constant component in titration series may be unknown, but must be the same across the series. Even better, the concentration ratio in a stock solution may be unknown, but will necessarily have to be identical across samples in a dilution series. In some cases where data are available from the same solution columns acquired with different signals, extinction or signal coefficients can be treated as unknowns if the concentration is a shared parameter across all data from the same cell; corresponding signal coefficients of the same component can be linked, in turn, across different cells (assuming data stem from the same instrument).

The non-linear regression of Lamm equation models is significantly more complex and time-consuming than the isotherms analysis. All available prior information should be used. For example, for hetero-associations the molar mass values of the free components should be set to the independently measured value. This apparent molar mass is usually not exactly the sequence molar mass due to the assumed partial-specific volume; this is inconsequential as long as it reflects the correct buoyant molar mass, and the same partial-specific volume is being used in all experiments (see Part I, Chapter 2 [1]). Usually a hierarchical relaxation of model parameters gives best results. Initially, s-values and binding constants should be fixed at values derived from the preceding isotherm analysis, and kinetic parameters be fixed to moderately slow or fast values (e.g., 0.001/sec) such as to facilitate later refinement. At this first stage, only concentration values should be fitted, to home the predicted signal amplitudes of the model in to the measured boundary amplitudes. Next, s-values and then equilibrium and kinetic binding constants should be added to the parameters that are being refined. For self-associating systems, the last parameter to optimize is the effective molar mass of the monomer.

Several considerations are useful to expedite Lamm equation fits in SEDPHAT: (1) An initial series of fits can be carried out with temporarily reduced precision of the Lamm equation solution. This can be achieved by pressing ALT-G to display the Numerical Control of Lamm Equation Solution box, and setting the default FE grid size initially to a low value such as 300. A final round of optimization can later be carried out with a more conservative higher value of 500 or 1000. (2) Where easily possible without sacrificing significant information (see Part II, Chapter 8 [2]), the back-diffusion region should be excluded from the fit limits, and the shut off back-diffusion flag of the Numerical Control of Lamm Equation Solution box should be set. (1) A computer with a large number of computing cores should be used, and the Compute With Multiple Threads flag in the Options menu should be set, as well as the flag multi-threaded kinetics in the Lamm equation options box.

The statistical weights of the data sets warrant additional consideration. When the data comprise sets with very dissimilar signal/noise ratio and number of data points, it is often observed that the data with high signal/noise ratio (e.g., those measuring higher concentrations) dominate the global fit. Systematic errors in the data acquisition or sample imperfections may lead to relatively minor deviations between model and data for the high concentration set, but the absolute magnitude of the rmsd caused by such imperfections can force a gross mismatch between model and data for the lower concentration set. In order to counteract such undue impact of minor imperfections, it can be useful to change the relative weight of the data sets, so that high concentration sets are de-emphasized and low concentration sets are enhanced in their importance to the global fit. This can be achieved by biasing the nominal noise $\sigma_i^{*(e)}$ attributed to the experiments as described on p. 220.

6.5.4 Free Pool Analysis

This model occupies a special place in the SV analysis, in that it is minimal in terms of the information extracted from the experiments — quite opposite to the Lamm equation modeling. The typical scenario where the free pool analysis has been used is the binding of a relatively smaller ligand to a very large particle. After incubation and quickly sedimenting the large particles with bound ligands, the 'supernatant' will reflect the unbound ligand. The goal is to model only the supernatant concentration as a function of total loading concentrations, without reference to the s-values of the large particles. It can be productively applied to distributions of heterogeneous particles, or multi-site binding, as long as all sites on all large particles are independent and equivalent.

Such an experiment can be carried out in a preparative centrifuge, using any quantitative method to determine supernatant concentrations after pelleting of complexes. It can, of course, also be carried in the analytical ultracentrifuge, and it is interesting to view it through the prism of theoretical framework we have established. When the life-time of the complexes is on the time-scale of sedimentation or longer, standard methods based solely on mass action law can be applied. In this case the assumption holds that the supernatant reflects the free ligand [282, 283].

However, for fast kinetics the problem arises that some of the free small binding partner will partition into the reaction boundary [41]. If not accounted for, this leads to an underestimate of the free concentration, at a magnitude that depends on the relative s-values of the interacting species. The magnitude of this effect can be quantified by effective particle theory. The errors are increasing with lower concentration of the small component. Fig. 6.10 shows the transition from conditions of very dissimilar-sized particles, where the mass action law provides an excellent approximation, to the case of two-fold s-value differences, where significant deviations are observed. The fact that free ligand can co-sediment in the fast-moving boundary can lead to an error of K_D and of the complex stoichiometry. On the other hand, the predictions for the concentration in the undisturbed boundary Eq. (2.22) on p. 52 can be used instead of the expressions of free ligand from mass action law to account for co-sedimentation and to extend the applicability of the free pool analysis to rapidly reversible systems of any size.

Even though it seems that only fitting the data from the free pool is very restrictive, since the sedimentation experiment should also have complementary information on the amplitude and s-value of the reaction boundary, such data may not always be available. The particular strength of the free pool approach is that it works when the fast boundary cannot be easily measured, perhaps caused by too fast sedimentation, when it is obscured by impurities, when the signal of these boundaries is outside the linear range of detection, or even when the large component exhibits some heterogeneity. In fact, for very dissimilar sized systems where the free pool approach is strong, the information content from the reaction (or complex) boundary becomes smaller than usual, since binding creates only small changes in large s-values ($s_{fast} \approx s_B$ at all times), and therefore the free pool approach still captures the essential information.

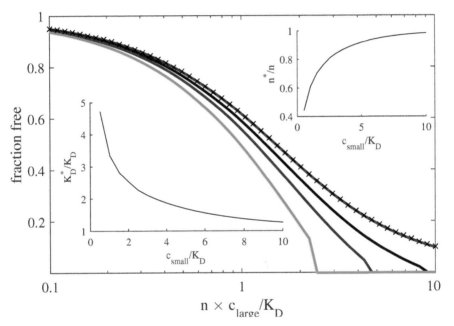

Figure 6.10 Effect of molecular sizes on the isotherm of the free pool. The ordinate plots the concentration of the 'non-sedimenting' material, expressed as a fraction of the total loading concentration of the smaller ligand, dependent on the concentration of the larger binding partner c_{large} with n sites per particle plotted in the abscissa. The concentration of the small ligand is constant at K_D. The ideal case governed solely by mass action law is illustrated by black crosses. The different lines are for pairs of binders with different s-values: a 0.5 S-species binding to a 100 S species forming a 100.1 S complex (red); a 2 S-species binding to a 10 S species forming an 11 S complex (blue); a 1.5 S-species binding to a 5 S species forming a 6 S complex (magenta); and a 1.5 S-species binding to a 3 S species forming a 4 S complex (green). If the co-sedimentation of free species in the reaction boundary is unaccounted for, this concentration will be underestimated. In a naïve fit assuming we know neither the binding constant nor the number of equivalent sites, we would arrive at incorrect answers for the binding constant and number of sites n. The insets show the apparent K_D (left lower inset) and the stoichiometry (upper right inset) for the system shown in blue (a 2 S-species binding to a 10 S species forming an 11 S complex) as a function of the total loading concentration of the small ligand.

6.6   GLOBAL MULTI-METHOD ANALYSIS

It is possible to further enhance the SV analysis of interacting systems through incorporation of complementary data from other techniques in a global multi-method analysis (GMMA) [268]. It has been shown that complementary observations can lead to synergy in the analysis and allow a more precise characterization of the interacting system. For example, in the study of α-chymotrypsin (CT) binding at two sites of soybean trypsin inhibitor (SBTI) we found that the incorporation of data from isothermal titration calorimetry (ITC), surface plasmon resonance (SPR), and fluorescence anisotropy as in Fig. 6.11, in addition to the SV titration, improve the parameter uncertainty not only of the binding constants, but also of the species s-values. Due to the added constraints from the combination of multi-method data, GMMA has the potential to expand the practical limits of studying multi-site and multi-component systems with cooperativity.

> SEDPHAT allows the seamless combination of data from many different techniques into a global multi-method analysis, simply by loading the different data sets into the SEDPHAT window, and specifying their type and experimental parameters.

Though a comprehensive practical discussion of GMMA is beyond the current scope, careful experimentation is essential. For example, the interaction should be studied for the same molecules, i.e., between unmodified molecules, or molecules carrying the same modifications. Planning tools for predicting useful data combinations and experimental parameters seamlessly extend from those described above in Section 6.1.4.3 to other techniques [260, 261, 284].

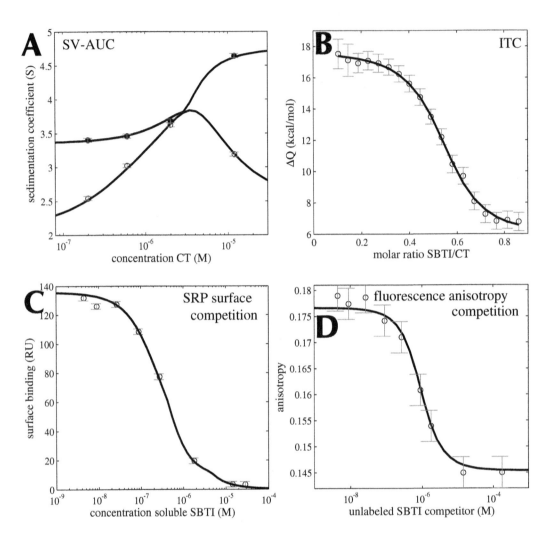

Figure 6.11 Global multimethod analysis of the two-site interaction of CT binding SBTI. *Panel A*: From an SV titration series s_w (open circles) and s_{fast} (closed circles) for 1.8 μM SBTI with different concentrations of CT; *Panel B*: Normalized heats of reaction measured via calorimetry from the titration of 20 μM CT with aliquots of 84 μM SBTI; *Panel C*: Steady-state SPR biosensor signals from the binding of a 0.3 μM CT to the surface-immobilized SBTI in the presence of different concentrations of soluble SBTI; *Panel D*: Fluorescence anisotropy of a mixture of 0.09 μM DyLight488-labeled SBTI and 1 μM CT with a range of concentrations of unlabeled SBTI. The best-fit GMMA model is shown as a solid line. Reproduced from [268].

Numerical Solution of Coupled Systems of Lamm Equations for Interacting Systems

$\mathbf{F}$ UNDAMENTALS of different methods for the numerical solutions of Lamm equations were sketched in Appendix A of Part II [2]. Here we extend this discussion and describe in more detail the finite element approach, originally introduced into the simulation of sedimentation velocity by Claverie [25, 285], with specific application to interacting systems. We follow the numerical scheme introduced in [28] and implemented in SEDPHAT.[1]

A.1 FINITE ELEMENT DISCRETIZATION

As we have seen in Eq. (1.2), the net chemical reaction flux q_i couples the Lamm equations for the migration and diffusion of species i

$$
\frac{\partial \chi_i}{\partial t} + \frac{1}{r}\frac{\partial}{\partial r}\left(r J_{i,\mathrm{tr}}\right) = q_i
$$
$$
\text{with} \quad J_{i,\mathrm{tr}} = \chi_i s_i \omega^2 r - D_i \frac{\partial \chi_i}{\partial r}\,,
$$
(A.1)

where we have used the abbreviation $J_{i,\mathrm{tr}}$ for the transport flux. Given a radial grid of N points r_n from meniscus $r_1 = m$ to bottom $r_N = b$ (see the discussion in Appendix A of Part II for possible choices of radial grids [2]), triangular elements

---

[1]The results were verified against independently programmed Lamm equation solutions in the software BPCFIT [26, 87].

are introduced as basis functions

$$
P_n(r) = \begin{cases} \dfrac{r - r_{n-1}}{r_n - r_{n-1}} & \text{for } r_{n-1} < r \le r_n \\[2ex] \dfrac{r_{n+1} - r}{r_{n+1} - r_n} & \text{for } r_n < r \le r_{n+1} \\[2ex] 0 & \text{else} \end{cases}
\tag{A.2}
$$

for $n = 2 \ldots N - 1$, with the special cases at either ends

$$
P_1(r) = \begin{cases} \dfrac{r_2 - r}{r_2 - r_1} & \text{for } r_1 < r \le r_2 \\[2ex] 0 & \text{else} \end{cases}
\tag{A.3}
$$

for $n = 1$ and

$$
P_N(r) = \begin{cases} \dfrac{r - r_{N-1}}{r_N - r_{N-1}} & \text{for } r_{N-1} < r \le r_N \\[2ex] 0 & \text{else} \end{cases}
\tag{A.4}
$$

for $n = N$. For an illustration of these 'hat' functions, see Fig. A2 of Part II [2]. They allow us to approximate the species' concentration profiles $\chi_i(r, t)$ by piecewise linear segments as

$$
\chi_i(r, t) \approx \sum_n \tilde{\chi}_n^{(i)}(t) P_n(r) ,
\tag{A.5}
$$

reducing the problem to the determination of suitable coefficients $\tilde{\chi}_n^{(i)}(t_j)$ at certain time-points j. Similar approximations can be made for the reaction fluxes

$$
q_i(r, t) \approx \sum_n \tilde{q}_n^{(i)}(t) P_n(r) .
\tag{A.6}
$$

The coefficients can be determined *via* the mass balance of flows in and out of the regions of specific elements P_n through radial integration of the Lamm equation Eq. (A.1)

$$
\int_m^b \frac{\partial \chi_i}{\partial t} P_n(r) r \, dr = - \int_m^b \frac{1}{r} \frac{\partial}{\partial r} \left(r J_{i,\mathrm{tr}} \right) P_n(r) r \, dr + \int_m^b q_i P_n(r) r \, dr .
\tag{A.7}
$$

The first term on the r.h.s. can be integrated by parts, exploiting that the flux vanishes at meniscus and bottom (see also Section A.3), and we arrive at

$$
\int_m^b \frac{\partial \chi_i}{\partial t} P_n(r) r \, dr = \int_m^b r J_{i,\mathrm{tr}} \frac{\partial P_n(r)}{\partial r} dr + \int_m^b q_i P_n(r) r \, dr .
\tag{A.8}
$$

With the radial approximation Eq. (A.5) and (A.6) (renaming the summation index to j) we arrive at a matrix equation

$$\sum_j \frac{\partial \tilde{\chi}_j^{(i)}(t)}{\partial t} \mathbf{B}_{jn} = -s\omega^2 \sum_j \tilde{\chi}_j^{(i)}(t) \mathbf{A}_{jn}^{[2]} + D \sum_j \tilde{\chi}_j^{(i)}(t) \mathbf{A}_{jn}^{[1]} + \sum_j \tilde{q}_j^{(i)}(t) \mathbf{B}_{jn} , \quad (A.9)$$

where

$$\mathbf{B}_{jn} = \int_m^b P_j(r) P_n(r) r \, dr ,$$

$$\mathbf{A}_{jn}^{[1]} = \int_m^b P_j(r) \frac{\partial P_n(r)}{\partial r} r \, dr , \quad \text{and} \quad (A.10)$$

$$\mathbf{A}_{jn}^{[2]} = \int_m^b \frac{\partial P_j(r)}{\partial r} \frac{\partial P_n(r)}{\partial r} r \, dr .$$

These matrix elements can be calculated analytically and their numeric values stored. At a given rotor speed, for species i we can contract the propagation matrices

$$\mathbf{A}_{jn}^{(i)} = s_i \omega^2 \mathbf{A}_{jn}^{[2]} - D_i \mathbf{A}_{jn}^{[1]} \quad (A.11)$$

and with the vector notation $\vec{\chi}^{(i)}$ for the array of $\tilde{\chi}_j^{(i)}$ for $j = 1, \ldots, N$, and likewise using $\vec{q}^{(i)}$ as a shortcut for all $\tilde{q}_j^{(i)}$, we arrive at the matrix equation

$$\mathbf{B} \frac{\partial \vec{\chi}^{(i)}}{\partial t} = -\mathbf{A}^{(i)} \vec{\chi}^{(i)} + \mathbf{B} \vec{q}^{(i)} . \quad (A.12)$$

A.2 TIME-STEPS OF COUPLED TRANSPORT AND REACTION

Since the reaction fluxes are dependent on the local concentration changes, the propagation from a time t_1 to t_2 is carried out by approximating a transport step without reaction, initially neglecting the second term of Eq. (A.12), followed by an estimate of the ensuing chemical fluxes [79].

For the transport step, a Crank–Nicholson scheme [286] provides numerical stability by basing the change on the average fluxes during the time-interval $\Delta t = t_2 - t_1$, as [28]

$$\vec{\chi}_{\text{trans},1}^{(i)}(t_2) = \left(2\mathbf{B} - \Delta t \mathbf{A}^{(i)} \right)^{-1} \left(2\mathbf{B} + \Delta t \mathbf{A}^{(i)} \right) \vec{\chi}^{(i)}(t_1) . \quad (A.13)$$

An estimate of the additional concentration changes due to the interaction is $\Delta \vec{\chi}_{\text{react}}^{(i)} = \Delta t \vec{q}^{(i)}$, but balancing the reaction fluxes to the average concentrations in the middle of the time-step, and accounting for the fact that these concentration changes are already contributing to changes in transport, we can write [28]

$$\vec{\chi}^{(i)}(t_2) = \vec{\chi}^{(i)}_{\text{trans},1}(t_2) + \frac{\Delta t^2}{2}\mathbf{B}^{-1}\mathbf{A}^{(i)}\vec{q}^{(i)}\left(\vec{\chi}^{(i)}(t_1)\right) . \tag{A.14}$$

For reactions that are quasi-instantaneously equilibrating, the reaction fluxes can be estimated from the difference between the local chemical equilibrium at t_1 and the re-equilibrated species after projected transport $\vec{\chi}^{(i)}_{\text{trans},1}(t_2)$.

Time-steps are limited such that the relative changes in concentrations from either reaction or transport remain small during Δt.

A.3   SIMPLIFIED SOLUTIONS FOR A SEMI-INFINITE SOLUTION COLUMN

One of the greatest challenges in the numerical solution of the Lamm equation are the steep concentration gradients that occur at the bottom of the solution column at high rotor speeds. These can dominate the performance of the algorithm and become the limiting factor for the precision as well as the length of the allowable time-steps.

At the same time, the end of the solution column will not influence the molecules far away from the region of back-diffusion. As described in Appendix A, Section A1 of Part II, for non-interacting species it is straightforward to predict the maximal back-diffusion region for given sedimentation and diffusion coefficients as the projected sedimentation equilibrium profile [2]. Similar is possible for interacting systems. If at the given experimental conditions back-diffusion does not extend into the radial region of experimental data that need to be fitted, then the boundary condition of the Lamm equation solution at the bottom may be changed without impacting the analysis. Thus, in many circumstances — especially when studying molecules > 10 kDa at rotor speeds of 50,000 rpm — permeable boundary conditions can be used. This will lead to concentration profiles for semi-infinite solution columns that are much more efficient to calculate due to the absence of back-diffusion.

Boundary conditions become important when considering the contributions from the transport flux in Eq. (A.7). The transition from Eq. (A.7) to Eq. (A.8) was based on integration by parts, utilizing that $J_{i,\text{tr}} = 0$ at the meniscus and the bottom. For permeable boundary conditions, this is not true, and an extra term $bJ_{i,\text{tr}}(b)P_N(b)$ appears in Eq. (A.8). After insertion of the flux defined Eq. (A.1), it becomes apparent that the semi-infinite boundary conditions require adjustments to the elements in the last row of matrices $\mathbf{A}^{[1]}$ and $\mathbf{A}^{[2]}$:

$$\begin{aligned}
\mathbf{A}^{[1*]}_{N-1,N} &= \mathbf{A}^{[1]}_{N-1,N} + b/(b - r_{N-1}) \\
\mathbf{A}^{[1*]}_{N,N} &= \mathbf{A}^{[1]}_{N,N} - b/(b - r_{N-1}) \\
\mathbf{A}^{[2*]}_{N,N} &= \mathbf{A}^{[1]}_{N,N} - b^2 ,
\end{aligned} \tag{A.15}$$

with all other matrix elements as in Eq. (A.10) [28].

Effective Particle Theory for Multi-Component Systems

$\mathbb{S}$ ALIENT features of sedimentation boundaries of rapidly reacting multi-component mixtures can be derived from the coupled Lamm equation solutions in the framework of effective particle theory [41]. Extending the scope from two-component systems (Section 3.2) to the general case, we now assume a mixture of K components that form E complexes leading to the coupled sedimentation of $N = K + E$ species. Complexes may include homomeric and heteromeric assemblies alike. We exclude flotation of any species, and complexes that sediment slower than any of their constituent free species. These conditions will be fulfilled in the vast majority of macromolecular mixtures.

B.1   BINDING EQUILIBRIA AND REACTION FLUXES

At first we need to establish nomenclature for bookkeeping. In the following we use indices $i = 1 \ldots N$ for counting species, which we enumerate such that indices with $i \leq K$ refer to the free species of the components $\kappa = 1 \ldots K$, and species indices $i > K$ are the complexes $e = 1 \ldots E$. Further, an index b will be used to distinguish the different sedimentation boundaries, ordered by increasing number of components and velocity.

We express the linked equilibria in an unambiguous way by considering, for each complex, only the reaction pathway between the complex and the free species of the constituent macromolecular components of that complex. This leads to a set of E reactions, one for each complex, of the form

$$n_{1,e}[\text{component } 1] + n_{2,e}[\text{component } 2] + \cdots + n_{K,e}[\text{component } K] \leftrightarrow [\text{complex } e] \,,$$
$$\text{(B.1)}$$

where the brackets indicate concentrations, and $n_{\kappa,e}$ is the number of copies of molecule κ participating in the complex e.[1]

---

[1]This is in slight deviation from the terminology of Section 1.1. The stoichiometries $n_{\kappa,e}$ are

Because attainment of chemical equilibrium is assumed to be quasi instantaneous on the time-scale of sedimentation, nowhere in the solution can complexes be separated from their free constituent components. Given the geometry of sedimentation with initially uniform solutions, it immediately follows that there will be at most K boundaries (Fig. B.1) separating K zones of the solution column. The first one will consist of a single-component solution, the second of a two-component mixture, and so on, up to a K-component mixture in the last (and fastest) boundary that has a composition identical to the loading mixture.

Species concentrations will generally be different in regions of the solution column, i.e., the various zones covered by different sedimentation boundaries (Fig. B.1), and therefore carry indices for both the species number and the boundary (or zone), $\chi_{i,b}$. All local species concentrations are related by mass action law at all times and in all zones throughout the solution column. For example, the concentration of the species $i = K + e$ in the zone b describing the complex formed in reaction e through assembly of $n_{\kappa,e}$ copies each of free forms of components κ is

$$\chi_{K+e,b} = \exp\left(-\frac{\Delta G_e}{RT}\right) \prod_{\kappa=1}^{K} (\chi_{\kappa,b})^{n_{\kappa,e}} . \tag{B.2}$$

where $\exp\left(-\frac{\Delta G_e}{RT}\right) = K_e$ is a cumulative association equilibrium constant for complex e. It is important to note that Eq. (B.2) reduces the independent concentrations to the free components in each boundary, $\chi_{\kappa,b}$, which, along with the boundary velocities w_b are the $K^2 + K$ unknowns that need to be determined.

Locally, as stated in Eq. (1.2), the coupled sedimentation/diffusion/reaction process follows a set of Lamm equations that are linked with total reaction fluxes q_i describing the local formation or depletion of species i due to the chemical reactions Eq. (B.1). In general, the reaction fluxes are composed of terms arising from each reaction

$$q_i = \sum_{e=1}^{E} q_{i,e} . \tag{B.3}$$

It is convenient to extend the copy number $n_{\kappa,e}$ from above to all species and to define $n_{i>K,e} = -\delta_{i-K,e}$ (with the Kronecker symbol producing values of -1 for $n_{e+K,e}$ and 0 otherwise), such that reactants and products of a reaction can be accounted for in the same way: With this notation the $q_{i,e}$ take the general form

$$q_{i,e} = -n_{i,e}\left(k_{\mathrm{on},e} \prod_{\kappa=1}^{K} \chi_\kappa{}^{n_{\kappa,e}} - k_{\mathrm{off},e}\chi_e\right) =: -n_{i,e}r_e . \tag{B.4}$$

describing changes in the molar species concentrations arising from binding and dissociation reactions. The dissociation rate constants $k_{\mathrm{off},e}$ are high, for fast local

---

identical with the S_i^κ in Chapter 1, renamed in the present context to avoid confusion with other symbols.

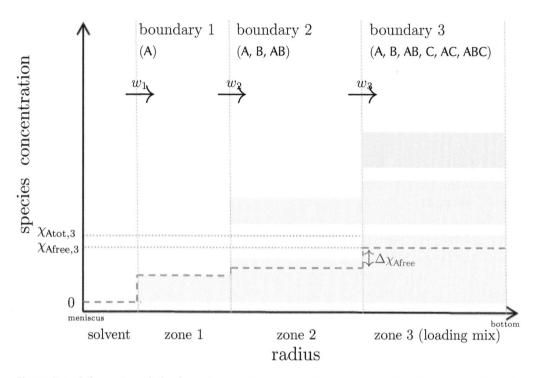

Figure B.1 Schematics of the boundary pattern for a three-component system depicted in the particular configuration where A provides the undisturbed boundary, B is dominant component in the middle reaction boundary, and C is dominant in the fastest reaction boundary. After the solution column is initially uniformly loaded at the start of sedimentation, the boundaries move with the velocities w_1, w_2, and w_3, respectively, forming different zones in the solution. The boundary velocities are determined by the composition in the respective zones, illustrated here with the species from a system where molecule A has independent sites for B and C. The ordinate depicts a concentration scale, with the components A, B, and C indicated as blue, green, and red patches, respectively, with the height of the patches reflecting the total component concentration in each zone. Molecules of each component will assume states of all possible species in each zone, dependent on other components in the same zone as well as the equilibrium constants. As a particular example of a species, highlighted as blue dashed line is the concentration of the free form of A. The difference in species concentration to the next higher zone is said to be co-sedimenting with the boundary separating the zones. For example, the concentration of free A co-sedimenting in boundary 3 is highlighted as $\Delta\chi_{\text{Afree}}$. As derived in the text, the $\Delta\chi_i$ correspond to the subset of molecules that must be attributed to the faster boundary to maintain identical time-average s-values of molecules from all components co-sedimenting in that boundary. It is governed by the average s-value of the component exclusive to this zone.

equilibration, but unknown. The value of $k_{\text{off},e}$ is implicitly defined by their ratio leading to the given equilibrium constant $k_{\text{on},e}/k_{\text{off},e} = K_e = \exp\left(-\frac{\Delta G_e}{RT}\right)$. The second identity in Eq. (B.4) expresses the fact of mass conservation: No matter what the reaction rates — abbreviated with the shortcut r_e — the changes in molar concentration of the constituents matches, after accounting for stoichiometry, the molar concentration change in the product. Thus, for all reactions contributing to the change in the concentration of a species, Eq. (B.3) becomes

$$q_i = -\sum_{e=1}^{E} n_{i,e} r_e \,. \tag{B.5}$$

In matrix/vector notation for all species, this is

$$\vec{q} = \mathbf{S}\vec{r} \,, \tag{B.6}$$

with the $(N \times E)$ matrix $\mathbf{S}$ composed of the stoichiometries $S_{i,e} = -n_{i,e}$ for each reaction, and the $(E \times 1)$ column vector $\vec{r}$ containing the unknown reaction rates r_e for each reaction.

An effective tool to eliminate the unknown reaction fluxes are linear combinations of reaction fluxes such that they cancel out. For example, for two-component systems this strategy was used on p. 76 to generate Eq. (3.15) from the coupled equations Eq. (3.12). In the general case, we are looking for sets of coefficients $A_{i,\kappa}$ such that

$$\sum_{i=1}^{N} q_i A_{i,\kappa} = 0 \,. \tag{B.7}$$

Therefore, we seek a matrix $\mathbf{A}$ such that $0 = \mathbf{A}^T \vec{q}$, or with Eq. (B.6) $0 = \mathbf{A}^T \mathbf{S}\vec{r}$, which leads to the requirement

$$\mathbf{S}^T \mathbf{A} = 0 \,. \tag{B.8}$$

In principle, such a matrix can be readily found using singular value decomposition of $\mathbf{S} = \mathbf{U}\mathbf{\Sigma}\mathbf{V}$, where $\mathbf{U}$ is a $(N \times N)$ matrix with columns $E + 1 \dots N$ spanning a basis of the null-space of $\mathbf{S}$, which is of dimension $N - E = K$ [287]. Thus, the choice $A_{i,\kappa} = U_{i,E+\kappa}$ satisfies Eq. (B.7), as will any linear combination of the $U_{i,E+\kappa}$.

However, there is a particular choice of coefficients that will turn out to be most useful. It is based on joining the identity matrix and the stoichiometries (hence the superscript '(st)'):

$$A_{i,\kappa}^{(st)} = \begin{pmatrix} 1 & \cdots & 0 \\ \vdots & 1 & \vdots \\ 0 & \cdots & 1 \\ n_{11} & \cdots & n_{K1} \\ \vdots & & \vdots \\ n_{1E} & \cdots & n_{KE} \end{pmatrix} \tag{B.9}$$

satisfies Eq. (B.8) since

$$
\mathbf{S}^T \mathbf{A}^{(st)} = \begin{pmatrix} -n_{11} & \cdots & -n_{K1} & 1 & & 0 \\ \vdots & & \vdots & & 1 & \\ -n_{1E} & \cdots & -n_{KE} & 0 & & 1 \end{pmatrix} \begin{pmatrix} 1 & & & 0 \\ & 1 & & \\ 0 & & 1 & \\ n_{11} & \cdots & n_{K1} \\ \vdots & & \vdots \\ n_{1E} & \cdots & n_{KE} \end{pmatrix}
$$

$$
= \begin{pmatrix} -n_{11} + n_{11} & \cdots & -n_{K1} + n_{K1} \\ \vdots & & \vdots \\ -n_{1E} + n_{1E} & \cdots & -n_{KE} + n_{KE} \end{pmatrix} = 0 \, .
$$

(B.10)

Therefore, we can make use of $A_{i,\kappa}^{(st)}$ to eliminate reaction fluxes *via* Eq. (B.7). It should be noted that the structure of $A_{i,\kappa}^{(st)}$ is such that it has non-zero entries in a column k only for species containing component κ. Thus, the total constituent concentration (in protomer units) of any component k can be conveniently calculated as

$$
\overline{\chi}_k = \sum_{i=1}^{N} \chi_i A_{i,k}^{(st)} \, ,
$$

(B.11)

and the constituent velocity as[2]

$$
\overline{\mathrm{v}_k} = \frac{\sum_{i=1}^{N} \chi_i v_i A_{i,k}^{(st)}}{\sum_{i=1}^{N} \chi_i A_{i,k}^{(st)}} \, .
$$

(B.12)

Another physical interpretation of $A_{i,\kappa}^{(st)}$ is that the products $\chi_i A_{i,\kappa}^{(st)}$ reflect the relative time a molecule from component κ spends in the state of species i (after normalization relative to total concentration of κ.

B.2 DIFFUSION-FREE COUPLED TRANSPORT

As introduced in Section 3.2.1, a time-honored simplification of the coupled sedimentation process is a diffusion-free picture. This can be achieved by setting $D_i = 0$ in the coupled Lamm equations of the reacting system in Eq. (1.2).[3] Further, we apply the model of a constant driving force in a rectangular solution column, eliminating complications from the radial geometry. This simplifies the coupled Lamm

---

[2]Compare Eqs. (1.9) and (1.11) on p. 6 in Chapter 1.

[3]We reiterate from Section 4.1, the rationale for this is that the diffusion-deconvoluted sedimentation coefficient distributions $c(s)$ — provided all boundaries can be resolved and a good fit can be achieved — allows the determination of s-values consistent with the transport method, which is equivalent, in turn, to the s-value of a diffusion-free boundary migrating at the second-moment position (Part II, Section 2.3 [2]).

equations Eq. (1.2) to

$$\frac{\partial \chi_i}{\partial t} + v_i \frac{\partial \chi_i}{\partial x} = q_i \,, \tag{B.13}$$

which is a general form of Eq. (3.12) previously introduced in Section 3.2.1 for a simple bimolecular reaction. Here, we have again used the symbols x and v to indicate the spatial coordinate and linear velocity, respectively, and $\chi_i(x,t)$ is the spatial concentration distribution evolving with time, and q_i is the chemical reaction flux.

A key step is the Ansatz for the species' concentration distribution to be composed of migrating step-functions $H(x - w_b t)$ [41]. This is naturally commensurate with the diffusion-free picture (see Part II, Chapter 4 [2]). More basic facts arise from the geometry of the experiment. Based on the instantaneous chemical equilibria we have concluded there can be only K zones separated by the sedimentation boundaries (Fig. B.1), the first containing a single component, the second containing two components, and so on, up to the last zone containing K components. Since the solution column is initially uniformly filled, the boundaries migrate with increasing velocities $w_1 < w_2 < \cdots < w_K$. In the absence of flotation the species' concentrations are increasing from zone to zone $\chi_{i,1} \leq \chi_{i,2} \leq \cdots \leq \chi_{i,K}$, and the number of components increases sequentially from the slowest to the fastest boundary (Fig. B.1). Therefore, in generalization of Eq. (3.13), we consider the concentration distributions

$$\chi_i(x,t) = \sum_{b=1}^{K} (\chi_{i,b} - \chi_{i,b-1}) H(x - w_b t) \,, \tag{B.14}$$

with K steps of sizes corresponding to the species concentration increment associated with the transition from zone $b-1$ to b (defining $\chi_{i,0} = 0$). The $\chi_{i,b}$ and w_b are unknowns to be determined; however, many will be zero due to the overall organization of chemical components as constituents of the boundaries, and all complex species are linked to the free constituents *via* Eq. (B.2).

To solve the Lamm equation we take the spatial and temporal derivative, which turns the Heaviside step-functions into Dirac's delta-functions [119]

$$\frac{\partial \chi_i(x,t)}{\partial t} = \sum_{b=1}^{K} (\chi_{i,b} - \chi_{i,b-1})(-w_b)\delta(x - w_b t)$$

$$\frac{\partial \chi_i(x,t)}{\partial x} = \sum_{b=1}^{K} (\chi_{i,b} - \chi_{i,b-1})\delta(x - w_b t) \,, \tag{B.15}$$

as illustrated for the simple bimolecular reaction in Eq. (3.14). Inserted in Eq. (B.13) we find

$$\sum_{b=1}^{K} (\chi_{i,b} - \chi_{i,b-1})(-w_b)\delta(x - w_b t) + v_i \sum_{b=1}^{K} (\chi_{i,b} - \chi_{i,b-1})\delta(x - w_b t) = q_i \tag{B.16}$$

for all $i = 1 \ldots N$, or short

$$\sum_{b=1}^{K} (\chi_{i,b} - \chi_{i,b-1})\,(v_i - w_b)\delta(x - w_b t) = q_i \quad \forall i = 1 \ldots N \;. \qquad (B.17)$$

At the analogous stage for the simple bimolecular system we have used an obvious choice to combine equations to eliminate q and get to Eq. (3.15). For the general case we use a formal approach taking advantage of the matrix $\mathbf{A}$ we found earlier in Eq. (B.9). Eq. (B.17) establishes N equations of migration coupled with the reaction fluxes q_i. The latter follow Eq. (B.7) above, and we insert Eq. (B.17) into Eq. (B.7) with the particular choice of coefficients $A_{i,\kappa}^{(st)}$ and observe

$$\sum_{i=1}^{N}\sum_{b=1}^{K} (\chi_{i,b} - \chi_{i,b-1})\,(v_i - w_b)\delta(x - w_b t)A_{i,\kappa}^{(st)} = 0 \quad \forall \, \kappa \;. \qquad (B.18)$$

Since this must be true for all x and t, it must hold for all boundaries separately

$$\sum_{i=1}^{N} (\chi_{i,b} - \chi_{i,b-1})\,(v_i - w_b)A_{i,\kappa}^{(st)} = 0 \quad \forall \, b, \kappa \;. \qquad (B.19)$$

Rearranging, we find the boundary velocities

$$w_b = \frac{\sum_{i=1}^{N} (\chi_{i,b} - \chi_{i,b-1})\, v_i A_{i,\kappa}^{(st)}}{\sum_{i=1}^{N} (\chi_{i,b} - \chi_{i,b-1})A_{i,\kappa}^{(st)}} \quad \forall \, b, \kappa \;, \qquad (B.20)$$

which is true for any choice of κ. As noted above, the products $\chi_i A_{i,\kappa}^{(st)}$ reflect the relative time a molecule from component κ spends in the state of species i. Therefore, it follows from Eq. (B.20) that any molecules contributing to the concentration steps in a boundary (e.g., including $\Delta\chi_{\text{Afree}}$ in Fig. B.1) — meaning all species 'co-sedimenting' with a boundary — exhibit the same time-average velocity w_b. This makes physical sense as a condition for a stable, concerted migration to form a boundary.[4]

It will be useful to express Eq. (B.19) with the help of the constituent concentrations of component k in a zone b, $\overline{\chi}_{k,b}$, and the corresponding constituent velocities $\overline{v_{k,b}}$ from Eqs. (B.11) and (B.12), respectively. After rearrangement, we find

$$(w_b - \overline{v_{k,b}})\,\overline{\chi}_{k,b} = (w_b - \overline{v_{k,b-1}})\,\overline{\chi}_{k,b-1} \;, \qquad (B.21)$$

---

[4]Different copies of one species co-existing in the same zone are indistinguishable. However, without consequence we may imagine a subset of them 'belonging' to the co-sedimenting fraction and others to a slower boundary that fills the solution column ahead. We allow for them to trade places, to restore microscopic indistinguishability. The physical interpretation of Eq. (B.20) becomes more obvious when lower boundaries vanish. A complementary viewpoint focuses on the molecules that cannot be 'dragged along' a boundary through transient complexation and thus fall behind. See also the discussion in Section 2.4.1.1.

which expresses a continuity in the flux density of molecules k crossing the moving boundary (if their constituent velocity in zone b is lower than that of the moving boundary), and emanating from it in zone $b - 1$ migrating there both at lower concentration and lower velocity.

B.3 BOUNDARY STRUCTURE

B.3.1 The Fastest Boundary

The task remains to calculate the component concentrations in each zone. From the geometry of sedimentation in an initially uniformly filled solution, it is obvious that the concentrations in the zone of the highest boundary are the initial loading concentrations

$$\chi_{i,K} = \chi_{i,\text{load}} \cdot \tag{B.22}$$

For principal reasons, one component must be contained exclusively in this zone, i.e., be 'dominant' for the highest boundary (Fig. B.1). Let us label this one as d_K. Since it is exclusively in zone K, there is no free species d_K in the lower boundaries, $\chi_{d_K,b<K} = 0$, and therefore all complexes to which it contributes vanish in the lower boundaries. Since our particular choice of $A_{i,\kappa}^{(st)}$ has non-zero elements only for species i that contain a component κ, it follows $\sum_i \chi_{i,K-1} A_{i,d_K}^{(st)}$ vanishes, and so does $\sum_i \chi_{i,K-1} v_i A_{i,d_K}^{(st)}$. Therefore, using the index $\kappa = d_K$ in Eq. (B.22) we find the velocity of the highest boundary to be

$$w_K = \frac{\displaystyle\sum_{i=1}^{N} \chi_{i,\text{load}} v_i A_{i,d_K}^{(st)}}{\displaystyle\sum_{i=1}^{N} \chi_{i,\text{load}} A_{i,d_K}^{(st)}} = \overline{v_{d_K}} \cdot \tag{B.23}$$

It is the time-average (or constituent average) velocity of the dominant component.

Which component is the dominant d_K? This can be answered through inspection of constituent quantities in Eq. (B.21) for $b = K$. Considering that there can only be positive concentrations, and that w_K must be faster than any constituent velocity in the boundary below, the l.h.s. must not be negative. As a consequence, w_K must be faster or equal to any constituent velocity. The answer is therefore straightforward and we can identify d_K with the component that exhibits the highest time-average velocity $\overline{v_{K,b}}$.

Having d_K, we know w_K through Eq. (B.23), and since we know the loading concentrations in Eq. (B.22), we can now calculate the co-sedimenting species from all other components, by rephrasing Eq. (B.19) to solve for the $\chi_{i,b-1}$ as

$$\sum_{i=1}^{N} \chi_{i,K-1}(w_K - v_i) A_{i,\kappa}^{(st)} = \sum_{i=1}^{N} \chi_{i,\text{load}}(w_K - v_i) A_{i,\kappa}^{(st)} \quad \forall \, \kappa \neq d_K , \tag{B.24}$$

where the r.h.s. is known. Since in the zone $K-1$ all complex species concentrations are linked to the free concentrations by mass action law, this is a system of $K-1$ equations for the $K-1$ unknown free concentrations $\chi_{\kappa \neq d_K, K-1}$ of non-dominant components in the lower boundary, from which all other species in the lower boundary are automatically determined through mass action law (Eq. B.2).[5] This completes the calculation of the highest boundary velocity, and the components and species concentrations not co-sedimenting and being left behind.

B.3.2 Next Lower Boundaries

The molecules left behind from the fastest boundary K have no knowledge of the boundary ahead, and they will behave like a new sedimenting system and readjusting equilibria to local component concentrations in the next lower zone $K-1$. Therefore, we can consider this the new highest boundary in the remaining system, and apply the same principles as above. This can be carried out recursively for all boundaries.

For efficiency, a new coefficient matrix for the smaller system can be found from $A_{i,\kappa}^{(st)}$ simply by deletion of the column d_K and all rows $K+e$ with non-zero entries for $n_{d_K,e}$.

A special case may arise where the remaining components are not completely linked anymore, for example, if components do not have any joint complexes. In this case, the system breaks apart into separately sedimenting sub-systems, which may consist of single non-interacting species, self-associating components, or sets of components with hetero-association. For nomenclature of the boundaries in keeping with the macroscopic velocities, we may derive d_{K-1} from the subsystem with the highest velocity, and trivially use $\chi_{i,K-2}$ for the other(s).

Finally, if there is only a single component left, this component constitutes the undisturbed boundary and it will sediment with its own constituent-average velocity. Typically this is simply the free species velocity, but may also be a concentration-dependent average velocity of a self-associating system.

B.3.3 Phase Transitions

From Eq. (B.21) we can see that for any non-dominant component x with

$$\overline{v_{x,b}} = w_b = \overline{v_{d_b,b}} , \tag{B.25}$$

i.e., any non-dominant component that has the same constituent-average velocity as the dominant component, the concentration in the next lower boundary $\overline{\chi}_{x,b-1}$

---

[5]An interesting consequence of coupled sedimentation is that even components that do not interact directly with the dominant component can appear to co-sediment: Consider a three-component system A+B+C↔AC+AB, where B does not bind C. Here, free A will co-sediment in the boundary dominated by C, therefore less A is left behind, causing lower amounts of AB to form, which will create a concentration step also in AB, as if it is co-sedimenting with AC. Obviously, the reason is that the entire system is coupled, even if indirectly. Any component that is not even indirectly linked to any other component should be considered separately upfront.

vanishes, meaning that this component is also entirely contained in the zone b co-sedimenting with the associated boundary, and fewer (or no) lower boundaries exist. In concentration space, the condition Eq. (B.25) defines a transition point: any lower concentration of x will decrease $\overline{v_{x,b}}$ below w_b and create fractions not co-sedimenting, and any higher concentrations of x will increase $\overline{v_{x,b}}$ and make it the new dominant component. Due to the change in propagation mode and the associated switch in the nature of the lower boundaries (often undisturbed boundary with free species) we refer to this as a 'phase transition.' All properties change continuously across this point in concentration space, though not the concentration first derivatives (such as slopes in isotherms).

There can be phase transitions for all reaction boundaries, including the highest boundary as well as the lower ones. For the phase transitions of the highest boundary, for two-component systems Eq. (B.25) defines a phase transition line. For three-component systems it defines a surface due to the additional dependence of species populations on the third component y with $y \neq x$ and $y \neq d_b$. The locations of the phase transitions in concentration space will be determined solely by the binding constants and species velocities.

The dimensionality of the concentration space that satisfies Eq. (B.25) for any component is reduced by 1. If two components satisfy Eq. (B.25) simultaneously, it will be reduced by 2, and so on. Finally, if all (non-dominant) components satisfy Eq. (B.25), the phase transitions will meet in a line that defines points in concentration space for which there is only a single boundary for the entire system. This line will be where the phase transition surfaces meet. An example can be found in Fig. 3.7 on p. 88.

Bibliography

[1] P. Schuck, H. Zhao, C.A. Brautigam, and R. Ghirlando, *Basic Principles of Analytical Ultracentrifugation.* Boca Raton, FL: CRC Press, 2015. ISBN 978-1-49-875115-5

[2] P. Schuck, *Sedimentation Velocity Analytical Ultracentrifugation: Discrete Species and Size-Distributions of Macromolecules and Particles.* Boca Raton, FL: CRC Press, 2016. ISBN 9781498768948

[3] T. Ando and J. Skolnick, "Crowding and hydrodynamic interactions likely dominate in vivo macromolecular motion." *Proc. Natl. Acad. Sci. USA*, vol. 107, no. 43, pp. 18 457–62, 2010. doi: 10.1073/pnas.1011354107

[4] A. Tardieu, F. Veretout, B. Krop, C. Slingsby, and F. Vérétout, "Protein interactions in the calf eye lens: interactions between beta-crystallins are repulsive whereas in gamma-crystallins they are attractive," *Eur. Biophys. J.*, vol. 21, no. 1, pp. 1–12, 1992.

[5] A. Stradner, G.M. Thurston, and P. Schurtenberger, "Tuning short-range attractions in protein solutions: from attractive glasses to equilibrium clusters," *J. Phys. Condens. Matter*, vol. 17, no. 31, pp. S2805–S2816, 2005. doi: 10.1088/0953-8984/17/31/005

[6] T.K. Dam and C.F. Brewer, "Lectins as pattern recognition molecules: the effects of epitope density in innate immunity." *Glycobiology*, vol. 20, no. 3, pp. 270–9, 2010. doi: 10.1093/glycob/cwp186

[7] A.J. Rowe, "Ultra-weak reversible protein-protein interactions." *Methods*, vol. 54, no. 1, pp. 157–66, 2011. doi: 10.1016/j.ymeth.2011.02.006

[8] T. Vorup-Jensen, "Integrin and Cell Adhesion Molecules," *Methods Mol. Biol.*, vol. 757, pp. 55–71, 2012. doi: 10.1007/978-1-61779-166-6. online at: http://www.springerlink.com/index/10.1007/978-1-61779-166-6

[9] J.E. Ladbury and S.T. Arold, "Noise in cellular signaling pathways: causes and effects." *Trends Biochem. Sci.*, vol. 37, no. 5, pp. 173–178, 2012. doi: 10.1016/j.tibs.2012.01.001

[10] M. Fried and D.M. Crothers, "Equilibria and kinetics of lac repressor-operator interactions by polyacrylamide gel electrophoresis." *Nucleic Acids Res.*, vol. 9, no. 23, pp. 6505–25, 1981. doi: 10.1093/nar/9.23.6505

[11] P.H. von Hippel and O.G. Berg, "On the specificity of DNA-protein interactions," *Proc. Natl. Acad. Sci.*, vol. 83, no. 6, pp. 1608–1612, 1986.

[12] J. Foote and C. Milstein, "Kinetic maturation of an immune response." *Nature*, vol. 352, no. 6335, pp. 530–2, 1991. doi: 10.1038/352530a0

[13] H. Wu and M. Fuxreiter, "The Structure and Dynamics of Higher-Order Assemblies: Amyloids, Signalosomes, and Granules," *Cell*, vol. 165, no. 5, pp. 1055–1066, 2016. doi: 10.1016/j.cell.2016.05.004

[14] H. Zhao, M.L. Mayer, and P. Schuck, "Analysis of protein interactions with pico-molar binding affinity by fluorescence-detected sedimentation velocity," *Anal. Chem.*, vol. 18, no. 6, pp. 3181–3187, 2014. doi: 10.1021/ac500093m

[15] F.E. LaBar and R.L. Baldwin, "The Sedimentation Coefficient of Sucrose," *J. Am. Chem. Soc.*, vol. 85, no. 20, pp. 3105–3108, oct 1963. doi: 10.1021/ja00903a012

[16] G.M. Pavlov, E.V. Korneeva, N.A. Smolina, and U.S. Schubert, "Hydrodynamic properties of cyclodextrin molecules in dilute solutions," *Eur. Biophys. J.*, vol. 39, no. 3, pp. 371–379, 2010. doi: 10.1007/s00249-008-0394-9

[17] S. Trachtenberg, P. Schuck, T.M. Phillips, S.B. Andrews, and R.D. Leapman, "A structural framework for a near-minimal form of life: Mass and compositional analysis of the helical mollicute Spiroplasma melliferum BC3," *PLoS One*, vol. 9, no. 2, p. e87921, 2014. doi: 10.1371/journal.pone.0087921

[18] T. Svedberg and K.O. Pedersen, *The Ultracentrifuge.* London: Oxford University Press, 1940. ISBN 0384588905

[19] H.K. Schachman, *Ultracentrifugation in Biochemistry.* New York: Academic Press, 1959. ISBN 1483270947

[20] H. Fujita, *Mathematical Theory of Sedimentation Analysis.* New York: Academic Press, 1962.

[21] H. Fujita, *Foundations of Ultracentrifugal Analysis.* New York: John Wiley & Sons, 1975.

[22] L.W. Nichol, J.L. Bethune, G. Kegeles, and E.L. Hess, "Interacting protein systems." in *Proteins*, H. Neurath, Ed. New York: Academic Press, 1964, ch. 9.

[23] J.R. Cann, *Interacting Macromolecules. The Theory and Practice of Their Electrophoresis, Ultracentrifugation, and Chromatography.* New York: Academic Press, 1970. ISBN 978-0-12-158550-1

[24] J.W. Williams, *Ultracentrifugation of Macromolecules.* New York: Academic Press, 1972.

[25] R. Cohen and J.-M. Claverie, "Sedimentation of generalized systems of interacting particles. II. Active enzyme centrifugation–theory and extensions of its validity range," *Biopolymers*, vol. 14, no. 8, pp. 1701–1716, 1975. doi: 10.1002/bip.1975.360140812

[26] C. Urbanke, B. Ziegler, and K. Stieglitz, "Complete evaluation of sedimentation velocity experiments in the analytical ultracentrifuge," *Fresenius Z. Anal. Chem.*, vol. 301, pp. 139–140, 1980. doi: 10.1007/BF00467783

[27] W.F. Stafford and P.J. Sherwood, "Analysis of heterologous interacting systems by sedimentation velocity: Curve fitting algorithms for estimation of sedimentation co-efficients, equilibrium and kinetic constants." *Biophys. Chem.*, vol. 108, no. 1-3, pp. 231–243, 2004. doi: 10.1016/j.bpc.2003.10.028

[28] J. Dam, C.A. Velikovsky, R.A. Mariuzza, C. Urbanke, and P. Schuck, "Sedimentation velocity analysis of heterogeneous protein-protein interactions: Lamm equation mod-eling and sedimentation coefficient distributions c(s)," *Biophys. J.*, vol. 89, no. 1, pp. 619–634, 2005. doi: 10.1529/biophysj.105.059568

[29] R. Cohen, B Giraud, and A Messiah, "Theory and practice of the analytical cen-trifugation of an active substrate-enzyme complex," *Biopolymers*, vol. 5, no. 2, pp. 203–225, 1967. doi: 10.1002/bip.1967.360050208

[30] D.L. Kemper and J. Everse, "Active enzyme centrifugation," *Methods Enzymol.*, vol. 27, pp. 67–82, 1973. doi: 10.1016/S0076-6879(73)27005-2

[31] R. Cohen and M. Mire, "Analytical-band centrifugation of an active enzyme substrate complex 1. Principle and practice of the centrifugation," *Eur. J. Biochem*, vol. 23, no. 2, pp. 267–275, 1971. doi: 10.111/j.1432-1033.1971.tb01618.x

[32] R. Cohen and M. Mire, "Analytical-band centrifugation of an active enzyme-substrate complex. 2. Determination of active units of various enzymes." *Eur. J. Biochem.*, vol. 23, no. 2, pp. 276–281, 1971. doi: 10.1111/j.1432-1033.1971.tb01619.x

[33] D.J. Llewellyn and G.D. Smith, "An evaluation of active enzyme centrifugation as a zonal and boundary technique by the analysis of simulated data," *Arch. Biochem. Biophys.*, vol. 190, no. 2, pp. 483–494, 1978. doi: 10.1016/0003-9861(78)90302-8

[34] L.K. Hesterberg and J.C. Lee, "Measurement of hydrodynamic properties of ac-tive enzyme by sedimentation," *Methods Enzymol.*, vol. 117, pp. 97–115, 1985. doi: 10.1016/S0076-6879(85)17010-2

[35] A. Modrak-Wójcik, K. Stepniak, V. Akoev, M. Zółkiewski, and A. Bzowska, "Molec-ular architecture of *E. coli* purine nucleoside phosphorylase studied by analytical ultracentrifugation and CD spectroscopy." *Protein Sci.*, vol. 15, no. 7, pp. 1794–1800, 2006. doi: 10.1110/ps.062183206

[36] C.Y. Chou, Y.H. Hsieh, and G.G. Chang, "Applications of analytical ultracentrifu-gation to protein size-and-shape distribution and structure-and-function analyses," *Methods*, vol. 54, no. 1, pp. 76–82, 2011. doi: 10.1016/j.ymeth.2010.11.002

[37] E. Yikilmaz, T.A. Rouault, and P. Schuck, "Self-association and ligand-induced con-formational changes of iron regulatory proteins 1 and 2," *Biochemistry*, vol. 44, no. 23, pp. 8470–8478, 2005.

[38] A. Balbo, K.H. Minor, C.A. Velikovsky, R.A. Mariuzza, C.B. Peterson, and P. Schuck, "Studying multi-protein complexes by multi-signal sedimentation velocity analytical ultracentrifugation," *Proc. Natl. Acad. Sci. USA*, vol. 102, no. 1, pp. 81–86, 2005. doi: 10.1073/pnas.0408399102

[39] H. Zhao, Y. Fu, C. Glasser, E.J. Andrade Alba, M.L. Mayer, G. Patterson, and P. Schuck, "Monochromatic multicomponent fluorescence sedimentation velocity for the study of high-affinity protein interactions," *Elife*, vol. 5, p. e17812, 2016. doi: 10.7554/eLife.17812

[40] T. Svedberg, "Molecular weight analysis in centrifugal fields," *Science*, vol. 79, no. 2050, pp. 327–332, 1934. doi: 10.1126/science.79.2050.327

[41] P. Schuck, "Sedimentation patterns of rapidly reversible protein interactions," *Biophys. J.*, vol. 98, no. 9, pp. 2005–2013, 2010. doi: 10.1016/j.bpj.2009.12.4336

[42] P. Schuck, "Diffusion of the reaction boundary of rapidly interacting macromolecules in sedimentation velocity." *Biophys. J.*, vol. 98, no. 11, pp. 2741–2751, 2010. doi: 10.1016/j.bpj.2010.03.004

[43] G.A. Gilbert and R.C.L. Jenkins, "Boundary problems in the sedimentation and electrophoresis of complex systems in rapid reversible equilibrium," *Nature*, vol. 177, no. 4514, pp. 853–854, 1956. doi: 10.1038/177853a0

[44] D.J. Cox, "Computer simulation of sedimentation in the ultracentrifuge. IV. Velocity sedimentation of self-associating solutes," *Arch. Biochem. Biophys.*, vol. 129, no. 1, pp. 106–123, 1969. doi: 10.1016/0003-9861(69)90157-X

[45] J.-M. Claverie, "Sedimentation of generalized systems of interacting particles. III. Concentration-dependent sedimentation and extension to other transport methods," *Biopolymers*, vol. 15, no. 5, pp. 843–857, 1976. doi: 10.1002/bip.1976.360150504

[46] P. Schuck, "Sedimentation analysis of noninteracting and self-associating solutes using numerical solutions to the Lamm equation," *Biophys. J.*, vol. 75, no. 3, pp. 1503–1512, 1998. doi: 10.1016/S0006-3495(98)74069-X

[47] P. Schuck, "On the analysis of protein self-association by sedimentation velocity analytical ultracentrifugation," *Anal. Biochem.*, vol. 320, no. 1, pp. 104–124, 2003. doi: 10.1016/S0003-2697(03)00289-6

[48] D.J. Cox, "Sedimentation of an initially skewed boundary," *Science*, vol. 152, no. 3720, pp. 359–361, 1966. doi: 10.1126/science.152.3720.359

[49] P. Schuck, "Size-distribution analysis of macromolecules by sedimentation velocity ultracentrifugation and Lamm equation modeling," *Biophys. J.*, vol. 78, no. 3, pp. 1606–1619, 2000. doi: 10.1016/S0006-3495(00)76713-0

[50] J.C. Gerhart and H.K. Schachman, "Allosteric interactions in aspartate transcarbamylase. II. Evidence for different conformational states of the protein in the presence and absence of specific ligands." *Biochemistry*, vol. 7, no. 2, pp. 538–552, 1968. doi: 10.1021/bi00842a600

[51] G.J. Howlett and H.K. Schachman, "Allosteric regulation of aspartate transcarbamoylase. Changes in the sedimentation coefficient promoted by the bisubstrate analogue N-(phosphonacetyl)-L-aspartate," *Biochemistry*, vol. 16, no. 23, pp. 5077–5083, 1977. doi: 10.1021/bi00642a021

[52] W.E. Werner, J.R. Cann, and H.K. Schachman, "Boundary spreading in sedimentation velocity experiments on partially liganded aspartate transcarbamoylase," *J. Mol. Biol.*, vol. 206, no. 1, pp. 231–237, 1989. doi: 10.1016/0022-2836(89)90536-6

[53] W.E. Werner and H.K. Schachman, "Analysis of the ligand-promoted global conformational change in aspartate transcarbamoylase. Evidence for a two-state transition from boundary spreading in sedimentation velocity experiments," *J. Mol. Biol.*, vol. 206, no. 1, pp. 221–230, 1989. doi: 10.1016/0022-2836(89)90535-4

[54] H. Schmeisser, I. Gorshkova, P.H. Brown, P. Kontsek, P. Schuck, and K. Zoon, "Two interferons alpha influence each other during their interaction with the extracellular domain of human type I interferon receptor subunit 2," *Biochemistry*, vol. 46, no. 50, pp. 14 638–14 649, 2007. doi: 10.1021/bi7012036

[55] M. Morris and G.B. Ralston, "A thermodynamic model for the self-association of human spectrin," *Biochemistry*, vol. 28, no. 1980, pp. 8561–8567, 1989.

[56] S. Yang, M.D.W. Griffin, K.J. Binger, P. Schuck, and G.J. Howlett, "An equilibrium model for linear and closed-loop amyloid fibril formation," *J. Mol. Biol.*, vol. 421, no. 2-3, pp. 364–377, 2012. doi: 10.1016/j.jmb.2012.02.026

[57] R.P. Frigon and S.N. Timasheff, "Magnesium-induced self-association of calf brain tubulin. I. Stoichiometry." *Biochemistry*, vol. 14, no. 21, pp. 4567–4573, 1975.

[58] L.P. Vickers and G.K. Ackers, "Analysis of calorimetric data for isodesmic systems," *Arch. Biochem. Biophys.*, vol. 174, no. 2, pp. 747–749, 1976. doi: 10.1016/0003-9861(76)90406-9

[59] J.M. González, M. Velez, M. Jiménez, C. Alfonso, P. Schuck, J. Mingorance, M. Vicente, A.P. Minton, and G. Rivas, "Cooperative behavior of *Escherichia coli* cell-division protein FtsZ assembly involves the preferential cyclization of long single-stranded fibrils," *Proc. Natl. Acad. Sci. USA*, vol. 102, no. 6, pp. 1895–1900, 2005. doi: 10.1073/pnas.0409517102

[60] D. Canzio, M. Liao, N. Naber, E. Pate, A. Larson, S. Wu, D.B. Marina, J.F. Garcia, H.D. Madhani, R. Cooke, P. Schuck, Y. Cheng, and G.J. Narlikar, "A conformational switch in HP1 releases auto-inhibition to drive heterochromatin assembly." *Nature*, vol. 496, no. 7445, pp. 377–381, 2013. doi: 10.1038/nature12032

[61] A.K. Attri, C. Fernandez, and A.P. Minton, "PH-dependent self-association of zinc-free insulin characterized by concentration-gradient static light scattering," *Biophys. Chem.*, vol. 148, no. 1-3, pp. 28–33, 2010. doi: 10.1016/j.bpc.2010.02.002

[62] R.C. Chatelier, "Indefinite isoenthalpic self-association of solute molecules," *Biophys. Chem.*, vol. 28, no. 2, pp. 121–128, 1987. doi: 10.1016/0301-4622(87)80081-9

[63] F. Montecinos-Franjola, P. Schuck, and D.L. Sackett, "Tubulin dimer reversible dissociation: Affinity, kinetics, and demonstration of a stable monomer," *J. Biol. Chem.*, vol. 291, no. 17, pp. 9281–9294, 2016. doi: 10.1074/jbc.M115.699728

[64] G. Krauss, A. Pingoud, D. Boehme, D. Riesner, F. Peters, and G. Maass, "Equivalent and non equivalent binding sites for tRNA on aminoacyl tRNA synthetases," *Eur. J. Biochem.*, vol. 529, pp. 517–529, 1975. doi: 10.1111/j.1432-1033.1975.tb02189.x

[65] M.P. Machner, C. Urbanke, M. Barzik, S. Otten, A.S. Sechi, J. Wehland, and D.W. Heinz, "ActA from *Listeria monocytogenes* can interact with up to four Ena/VASP homology 1 domains simultaneously," *J. Biol. Chem.*, vol. 276, no. 43, pp. 40 096– 40 103, 2001. doi: 10.1074/jbc.M104279200

[66] C. Urbanke, G. Witte, and U. Curth, "A sedimentation velocity method in the analytical ultracentrifuge for the study of protein-protein interactions," in *Protein-Ligand Interactions: Methods and Applications*, G.U. Nienhaus, Ed. Totowa: Humana Press, 2005, vol. 305, pp. 101–113.

[67] A. Plückthun and P. Pack, "New protein engineering approaches to multivalent and bispecific antibody fragments," *Immunotechnology*, vol. 3, no. 2, pp. 83–105, 1997. doi: 10.1016/S1380-2933(97)00067-5

[68] K. Garber, "Bispecific antibodies rise again," *Nat. Rev. Drug Discov.*, vol. 13, no. 11, pp. 799–801, 2014. doi: 10.1038/nrd4478

[69] J.J. Correia and W.F. Stafford, "Extracting equilibrium constants from kinetically limited reacting systems," *Methods Enzymol.*, vol. 455, pp. 419–446, 2009. doi: 10.1016/S0076-6879(08)04215-8

[70] C.A. Brautigam, "Using Lamm-equation modeling of sedimentation velocity data to determine the kinetic and thermodynamic properties of macromolecular interactions." *Methods*, vol. 54, no. 1, pp. 4–15, 2011. doi: 10.1016/j.ymeth.2010.12.029

[71] G.A. Gilbert, "Sedimentation and electrophoresis of interacting substances. I. Idealized boundary shape for a single substance aggregating reversibly," *Proc. R. Soc. London Ser. A*, vol. 250, no. 1262, pp. 377–388, 1959. doi: 10.1098/rspa.1959.0070

[72] J.L. Bethune and G. Kegeles, "Countercurrent distribution of chemically reacting systems II. Reactions of the type A + B ⇌ C," *J. Phys. Chem.*, vol. 65, no. 10, pp. 1755–1760, 1961. doi: 10.1021/j100827a018

[73] W.B. Goad and J.R. Cann, "Theory of sedimentation of interacting systems." *Ann. NY. Acad. Sci.*, vol. 164, no. 1, pp. 172–182, 1969. doi: 10.1111/j.1749-6632.1969.tb14039.x

[74] J.R. Cann and G. Kegeles, "Theory of sedimentation for kinetically controlled dimerization reactions," *Biochemistry*, vol. 13, no. 9, pp. 1868–1874, 1974. doi: 10.1021/bi00706a015

[75] D.J. Cox, "Computer simulation of sedimentation in the ultracentrifuge VI. Monomer-tetramer systems in rapid chemical equilibrium," *Arch. Biochem. Biophys.*, vol. 146, no. 1, pp. 181–195, 1971. doi: 10.1016/S0003-9861(71)80055-3

[76] D.J. Cox, "Calculation of simulated sedimentation velocity profiles for self-associating solutes," *Methods Enzymol.*, vol. 2, pp. 212–242, 1978. doi: 10.1016/S0076-6879(78)48012-7

[77] R. Trautman, "The impact of the desk-top computer on ultracentrifugation," *Ann. N.Y. Acad. Sci.*, vol. 164, pp. 52–65, 1969. doi: 10.1111/j.1749-6632.1969.tb14032.x

[78] R. Trautman, S.P. Spragg, and H.H. Halsall, "Absorption optics data processing with standards errors for sedimentation and diffusion coefficients from moving boundary ultracentrifugation." *Anal. Biochem.*, vol. 28, no. 1, pp. 396–415, 1969. doi: 10.1016/0003-2697(69)90195-X

[79] D.J. Cox and R.S. Dale, "Simulation of transport experiments for interacting systems," in *Protein-Protein Interactions*, C. Frieden and L.W. Nichol, Eds. New York: Wiley, 1981, pp. 173–211. ISBN 978-0471049791

[80] G.A. Gilbert and L.M. Gilbert, "Ultracentrifuge studies of interactions and equilibria: Impact of interactive computer modelling." *Biochem. Soc. Trans.*, vol. 8, no. 5, pp. 520–522, 1980. doi: 10.1042/bst0080520

[81] G.P. Todd and R.H. Haschemeyer, "General solution to the inverse problem of the differential equation of the ultracentrifuge," *Proc. Natl. Acad. Sci. USA*, vol. 78, no. 11, pp. 6739–6743, 1981.

[82] L.A. Holladay, "Simultaneous rapid estimation of sedimentation coefficient and molecular weight," *Biophys. Chem.*, vol. 11, no. 2, pp. 303–308, 1980. doi: 10.1016/0301-4622(80)80033-0

[83] L.C. Davis and M.S. Chen, "Computer simulations of mass transport for nonidentical interacting molecules," *Arch. Biochem. Biophys.*, vol. 194, no. 1, pp. 37–48, 1979. doi: 10.1016/0003-9861(79)90593-9

[84] G.A. Gilbert and L.M. Gilbert, "Detection in the ultracentrifuge of protein heterogeneity by computer modelling, illustrated by pyruvate dehydrogenase multienzyme complex," *J. Mol. Biol.*, vol. 144, no. 3, pp. 405–408, 1980. doi: 10.1016/0022-2836(80)90099-6

[85] J.R. Cann, "Effects of microheterogeneity on sedimentation patterns of interacting proteins and the sedimentation behavior of systems involving two ligands," *Methods Enzymol.*, vol. 130, no. 1964, pp. 19–35, 1986. doi: 10.1016/0076-6879(86)30005-3

[86] H.K. Schachman, "Analytical ultracentrifugation reborn," *Nature*, vol. 341, no. 21, pp. 259–260, 1989. doi: 10.1038/341259a0

[87] B. Kindler, "Akkuprog: Auswertung von Messungen chemischer Reaktionsgeschwindigkeit und Analyse von Biopolymeren in der Ultrazentrifuge," PhD Thesis. University Hannover, Germany, 1997.

[88] P. Schuck, C.E. MacPhee, and G.J. Howlett, "Determination of sedimentation coefficients for small peptides," *Biophys. J.*, vol. 74, no. 1, pp. 466–474, 1998. doi: 10.1016/S0006-3495(98)77804-X

[89] J.S. Philo, "An improved function for fitting sedimentation velocity data for low-molecular-weight solutes," *Biophys. J.*, vol. 72, no. 1, pp. 435–444, 1996. doi: 10.1016/S0006-3495(97)78684-3

[90] J. Dam and P. Schuck, "Calculating sedimentation coefficient distributions by direct modeling of sedimentation velocity concentration profiles," *Methods Enzymol.*, vol. 384, no. 301, pp. 185–212, 2004. doi: 10.1016/S0076-6879(04)84012-6

[91] A. Tiselius, "Electrophoresis and Adsorption Analysis as Aids in Investigations of Large Molecular Weight Substances and Their Breakdown Products." Nobel Lecture. 1948. online at: http://www.nobelprize.org/nobel_prizes/chemistry/laureates/1948/tiselius-lecture.pdf

[92] G.W. Schwert, "The molecular size and shape of the pancreatic proteases. Sedimentation studies on chymotrypsinogen and on alpha- and gamma-chymotrypsin," *J. Biol. Chem.*, vol. 179, pp. 655–664, 1949.

[93] V. Massey, W.F. Harrington, and B.S. Hartley, "Certain physical properties of chymotrypsin and chymotrypsinogen using the depolarization of fluorescence technique," *Discuss. Faraday Soc.*, vol. 20, no. 24, pp. 24–32, 1955. doi: 10.1039/DF9552000024

[94] J.L. Oncley, "Physical chemistry regarding the size and shape of protein molecules from ultracentrifugation, diffusion, viscosity, dielectric dispersion, and double refraction of flow," *Ann. N.Y. Acad. Sci.*, no. 121, pp. 48–57, 1941. doi: 10.1111/j.1749-6632.1941.tb35233.x

[95] J.L. Oncley, E. Ellenbogen, D. Gitlin, and F.R.N. Gurd, "Protein-protein interactions," *J. Phys. Chem.*, vol. 56, no. 1, pp. 85–92, 1952. doi: 10.1021/j150493a017

[96] R.F. Steiner, "Reversible association processes of globular proteins. I. Insulin," *Arch. Biochem. Biophys.*, pp. 333–354, 1952. doi: 10.1016/0003-9861(52)90344-5

[97] P.H. von Hippel and D.F. Waugh, "Casein: Monomers and polymers," *J. Am. Chem.*, vol. 4928, no. 1950, pp. 4311–4319, 1955. doi: 10.1021/ja01621a041

[98] E.O. Field and A.G. Ogston, "Boundary spreading in the migration of a solute in rapid dissociation equilibrium. Theory and its application to the case of human haemoglobin," *Biochem. J.*, vol. 6, pp. 661–665, 1955. doi: 10.1042/bj0600661

[99] M. Buisson, E. Valette, J.F. Hernandez, F. Baudin, C. Ebel, P. Morand, J.M. Seigneurin, G.J. Arlaud, and R.W. Ruigrok, "Functional determinants of the Epstein–Barr virus protease," *J. Mol. Biol.*, vol. 311, no. 1, pp. 217–228, 2001. doi: 10.1006/jmbi.2001.4854

[100] S. Boulant, C. Vanbelle, C. Ebel, F. Penin, and J.-P. Lavergne, "Hepatitis C virus core protein is a dimeric alpha-helical protein exhibiting membrane protein features," *J. Virol.*, vol. 79, no. 17, pp. 11 353–11 365, 2005. doi: 10.1128/JVI.79.17.11353

[101] N. Vunnam, J. Flint, A. Balbo, P. Schuck, and S. Pedigo, "Dimeric states of neural- and epithelial-cadherins are distinguished by the rate of disassembly." *Biochemistry*, vol. 50, no. 14, pp. 2951–61, 2011. doi: 10.1021/bi2001246

[102] S.K. Chaturvedi, H. Zhao, and P. Schuck, "Sedimentation of reversibly interacting macromolecules with changes in fluorescence quantum yield," *Biophys. J.*, vol. 112, no. 7, pp. 1374–1382, 2017. doi: 10.1016/j.bpj.2017.02.020

[103] G.A. Gilbert and R.C.L. Jenkins, "Sedimentation and electrophoresis of interacting substances. II. Asymptotic boundary shape for two substances interacting reversibly," *Proc. R. Soc. London Ser. A*, vol. 253, no. 1274, pp. 420–437, 1959. doi: 10.1098/rspa.1959.0204

[104] A.A. Sousa, S.A. Hassan, L.L. Knittel, A. Balbo, M.A. Aronova, P.H. Brown, P. Schuck, and R.D. Leapman, "Biointeractions of ultrasmall glutathione-coated gold nanoparticles: effect of small size variations." *Nanoscale*, vol. 8, no. 12, pp. 6577–88, 2016. doi: 10.1039/c5nr07642k

[105] P. Schuck, "Sedimentation velocity movies," 2011. online at: https://sedfitsedphat. nibib.nih.gov/tools/SedimentationDiffusionMovies/Forms/AllItems.aspx

[106] H. Zhao, A. Balbo, P.H. Brown, and P. Schuck, "The boundary structure in the analysis of reversibly interacting systems by sedimentation velocity." *Methods*, vol. 54, no. 1, pp. 16–30, 2011. doi: 10.1016/j.ymeth.2011.01.010

[107] G.A. Gilbert and R.C.L. Jenkins, "Analysis of reversible interacting systems by differential boundaries in electrophoretic and sedimentation experiments," *Nature*, vol. 199, p. 688, 1963. doi: doi:10.1038/199688a0

[108] L.M. Gilbert and G.A. Gilbert, "Molecular transport of reversibly reacting systems: Asymptotic boundary profiles in sedimentation, electrophoresis, and chromatography," *Methods Enzymol.*, vol. 48, pp. 195–211, 1978. doi: 10.1016/S0076-6879(78)48011-5

[109] J. Dam and P. Schuck, "Sedimentation velocity analysis of heterogeneous protein-protein interactions: Sedimentation coefficient distributions c(s) and asymptotic boundary profiles from Gilbert–Jenkins theory." *Biophys. J.*, vol. 89, no. 1, pp. 651–666, 2005. doi: 10.1529/biophysj.105.059584

[110] A. Tiselius, "The moving-boundary method of studying the electrophoresis of proteins," *Nov. Acta Regiae Soc. Sci. Ups. Ser. IV.*, vol. 7, no. 4, pp. 1–107, 1930.

[111] R.A. Alberty and H.H. Marvin, "The study of protein-ion interaction by the moving boundary method. Theory of the method." *J. Phys. Chem.*, vol. 54, no. 1, pp. 47–55, 1950. doi: 10.1021/j150475a004

[112] I.Z. Steinberg and H.K. Schachman, "Ultracentrifugation studies with absorption optics. V. Analysis of interacting systems involving macromolecules and small molecules," *Biochemistry*, vol. 5, no. 12, pp. 3728–3747, 1966. doi: 10.1021/bi00876a003

[113] R.F. Steiner, "Reversible association processes of globular proteins. V. The study of associating systems by the methods of macromolecular physics," *Arch. Biochem. Biophys.*, vol. 49, no. 2, pp. 400–416, 1954. doi: 10.1016/0003-9861(54)90209-X

[114] H. Zhao, P.H. Brown, and P. Schuck, "On the distribution of protein refractive index increments," *Biophys. J.*, vol. 100, no. 9, pp. 2309–2317, 2011. doi: 10.1016/j.bpj.2011.03.004

[115] M. Mason and W. Weaver, "The settling of small particles in a fluid," *Phys. Rev.*, vol. 23, no. 3, pp. 412–426, 1924. doi: 10.1103/PhysRev.23.412

[116] G.A. Gilbert, "In: General discussion," *Discuss. Faraday Soc.*, vol. 20, pp. 68–71, 1955. doi: 10.1039/DF9552000065

[117] R.F. Smith and D.R. Briggs, "The electrophoretic analysis of protein interaction. I. The interaction of bovine serum albumin and methyl orange," *J. Phys. Colloid Chem.*, vol. 54, no. 1, pp. 33–47, 1950. doi: 10.1021/j150475a003

[118] D. DeVault, "The theory of chromatography," *J. Am. Chem. Soc.*, vol. 65, no. 4, pp. 532–540, 1943. doi: 10.1021/ja01244a011

[119] R.F. Hoskins, *Delta Functions. Introduction to Generalized Functions*, 2nd ed. Coll House, U.K.: Horwood Publishing, 1999. ISBN 978-1-904275-39-8

[120] R.T. Hersh and H.K. Schachman, "Ultracentrifuge studies with a synthetic boundary cell. II. Differential sedimentation," *J. Am. Chem. Soc.*, vol. 77, no. 20, pp. 5228–5234, 1955. doi: 10.1021/ja01625a007

[121] L.M. Gilbert and G.A. Gilbert, "Sedimentation velocity measurement of protein association," *Methods Enzymol.*, vol. 27, pp. 273–296, 1973. doi: 10.1016/S0076-6879(73)27014-3

[122] D.J. Winzor, R. Tellam, and L.W. Nichol, "Determination of the asymptotic shapes of sedimentation velocity patterns for reversibly polymerizing solutes," *Arch. Biochem. Biphys.*, vol. 178, no. 2, pp. 327–322, 1977. doi: 10.1016/0003-9861(77)90200-4

[123] L.W. Nichol and D.J. Winzor, "Calculation of asymptotic boundary shapes from experimental mass migration patterns," *Methods Enzymol.*, vol. 130, no. 1968, pp. 6–18, 1986. doi: 10.1016/0076-6879(86)30004-1

[124] R.L. Baldwin, "Boundary spreading in sedimentation velocity experiments. VI. A better method for finding distributions of sedimentation coefficient when the effects of diffusion are large," *J. Phys. Chem.*, vol. 63, no. 10, pp. 1570–1573, 1959. doi: 10.1021/j150580a006

[125] K.E. van Holde and W.O. Weischet, "Boundary analysis of sedimentation-velocity experiments with monodisperse and paucidisperse solutes," *Biopolymers*, vol. 17, no. 6, pp. 1387–1403, 1978. doi: 10.1002/bip.1978.360170602

[126] L.W. Nichol and A.G. Ogston, "A generalized approach to the description of interacting boundaries in migrating systems," *Proc. R. Soc. London Ser. B*, vol. 163, no. 992, pp. 343–368, 1965. doi: 10.1098/rspb.1965.0073

[127] P.H. Brown, A. Balbo, and P. Schuck, "Using prior knowledge in the determination of macromolecular size-distributions by analytical ultracentrifugation." *Biomacromolecules*, vol. 8, no. 6, pp. 2011–2024, 2007. doi: 10.1021/bm070193j

[128] L.W. Nichol and D.J. Winzor, "The determination of equilibrium constants from transport data on rapidly reacting systems of the type $A + B \rightleftharpoons C$," *J. Phys. Chem.*, vol. 68, no. 9, pp. 2455–2463, 1964. doi: 10.1021/j100791a012

[129] J.R. Cann, "Effects of diffusion on the electrophoretic behavior of associating systems: The Gilbert-Jenkins theory revisited," *Arch. Biochem. Biophys.*, vol. 240, pp. 489–499, 1985. doi: 10.1016/0003-9861(85)90055-4

[130] P. Schuck and P. Rossmanith, "Determination of the sedimentation coefficient distribution by least-squares boundary modeling," *Biopolymers*, vol. 54, no. 5, pp. 328–341, 2000. doi: 10.1002/1097-0282(20001015)54:5¡328::AID-BIP40¿3.0.CO;2-P

[131] A. Bekdemir and F. Stellacci, "A centrifugation-based physicochemical characterization method for the interaction between proteins and nanoparticles," *Nat. Commun.*, vol. 7, p. 13121, 2016. doi: 10.1038/ncomms13121

[132] C.T. Chaton and A.B. Herr, "Elucidating complicated assembling systems in biology using size-and-shape analysis of sedimentation velocity data," *Methods Enzymol.*, vol. 562, pp. 187–204, 2015. doi: 10.1016/bs.mie.2015.04.004

[133] P.H. Brown and P. Schuck, "Macromolecular size-and-shape distributions by sedimentation velocity analytical ultracentrifugation," *Biophys. J.*, vol. 90, no. 12, pp. 4651–4661, 2006. doi: 10.1529/biophysj.106.081372

[134] W.B. Bridgman, "Some physical characteristics of glycogen," *J. Am. Chem. Soc.*, vol. 64, no. 10, pp. 2349–2356, 1942. doi: 10.1021/ja01262a037

[135] W.F. Stafford, "Boundary analysis in sedimentation transport experiments: A procedure for obtaining sedimentation coefficient distributions using the time derivative of the concentration profile," *Anal. Biochem.*, vol. 203, no. 2, pp. 295–301, 1992. doi: 10.1016/0003-2697(92)90316-Y

[136] P. Schuck, "Sedimentation coefficient distributions of large particles," *Analyst*, vol. 141, pp. 4400–4409, 2016. doi: 10.1039/C6AN00534A

[137] E. Brookes, W. Cao, and B. Demeler, "A two-dimensional spectrum analysis for sedimentation velocity experiments of mixtures with heterogeneity in molecular weight and shape." *Eur. Biophys. J.*, vol. 39, no. 3, pp. 405–14, 2010. doi: 10.1007/s00249-009-0413-5

[138] P. Schuck, "On computational approaches for size-and-shape distributions from sedimentation velocity analytical ultracentrifugation," *Eur. Biophys. J.*, vol. 39, no. 8, pp. 1261–1275, 2010. doi: 10.1007/s00249-009-0545-7

[139] B. Demeler and K.E. van Holde, "Sedimentation velocity analysis of highly heterogeneous systems," *Anal. Biochem.*, vol. 335, no. 2, pp. 279–288, 2004. doi: 10.1016/j.ab.2004.08.039

[140] P. Schuck, M.A. Perugini, N.R. Gonzales, G.J. Howlett, and D. Schubert, "Size-distribution analysis of proteins by analytical ultracentrifugation: strategies and application to model systems," *Biophys. J.*, vol. 82, no. 2, pp. 1096–1111, 2002. doi: 10.1016/S0006-3495(02)75469-6

[141] B. Demeler, H. Saber, and J.C. Hansen, "Identification and interpretation of complexity in sedimentation velocity boundaries," *Biophys. J.*, vol. 72, no. 1, pp. 397–407, 1997. doi: 10.1016/S0006-3495(97)78680-6

[142] S.B. Padrick, R.K. Deka, J.L. Chuang, R.M. Wynn, D.T. Chuang, M.V. Norgard, M.K. Rosen, and C.A. Brautigam, "Determination of protein complex stoichiometry through multisignal sedimentation velocity experiments," *Anal. Biochem.*, vol. 407, no. 1, pp. 89–103, 2010. doi: 10.1016/j.ab.2010.07.017

[143] S.B. Padrick and C.A. Brautigam, "Evaluating the stoichiometry of macromolecular complexes using multisignal sedimentation velocity," *Methods*, vol. 54, no. 1, pp. 39–55, 2011. doi: 10.1016/j.ymeth.2011.01.002

[144] N.A. May, Q. Wang, A. Balbo, S.L. Konrad, R. Buchli, W.H. Hildebrand, P. Schuck, and A.W. Hudson, "Human herpesvirus 7 U21 tetramerizes to associate with class I major histocompatibility complex molecules." *J. Virol.*, vol. 88, no. 6, pp. 3298–3308, 2014. doi: 10.1128/JVI.02639-13

[145] K.H. Minor, C.R. Schar, G.E. Blouse, J.D. Shore, D.A. Lawrence, P. Schuck, and C.B. Peterson, "A mechanism for assembly of complexes of vitronectin and plasminogen activator inhibitor-1 from sedimentation velocity analysis." *J. Biol. Chem.*, vol. 31, pp. 28 711–28 720, 2005.

[146] J.C.D. Houtman, H. Yamaguchi, M. Barda-Saad, A. Braiman, B. Bowden, E. Appella, P. Schuck, and L.E. Samelson, "Oligomerization of signaling complexes by the multipoint binding of GRB2 to both LAT and SOS1," *Nat. Struct. Mol. Biol.*, vol. 13, no. 9, pp. 798–805, 2006. doi: 10.1038/nsmb1133

[147] R.K. Deka, C.A. Brautigam, F.L. Tomson, S.B. Lumpkins, D.R. Tomchick, M. Machius, and M.V. Norgard, "Crystal structure of the Tp34 (TP0971) lipoprotein of *Treponema pallidum*: implications of its metal-bound state and affinity for human lactoferrin." *J. Biol. Chem.*, vol. 282, no. 8, pp. 5944–58, 2007. doi: 10.1074/jbc.M610215200

[148] N. Doan and P.G.W. Gettins, "α-Macroglobulins are present in some gram-negative bacteria: Characterization of the α_2-macroglobulin from *Escherichia coli*." *J. Biol. Chem.*, vol. 283, no. 42, pp. 28 747–56, 2008. doi: 10.1074/jbc.M803127200

[149] C.A. Brautigam, R.M. Wynn, J.L. Chuang, and D.T. Chuang, "Subunit and catalytic component stoichiometries of an in vitro reconstituted human pyruvate dehydrogenase complex," *J. Biol. Chem.*, vol. 284, no. 19, pp. 13 086–13 098, 2009. doi: 10.1074/jbc.M806563200

[150] J.K. Jensen, K. Dolmer, C. Schar, and P.G.W. Gettins, "Receptor-associated protein (RAP) has two high-affinity binding sites for the low-density lipoprotein receptor-related protein (LRP): Consequences for the chaperone functions of RAP." *Biochem. J.*, vol. 421, no. 2, pp. 273–82, 2009. doi: 10.1042/BJ20090175

[151] M. Barda-Saad, N. Shirasu, M.H. Pauker, N. Hassan, O. Perl, A. Balbo, H. Yamaguchi, J.C.D. Houtman, E. Appella, P. Schuck, and L.E. Samelson, "Cooperative interactions at the SLP-76 complex are critical for actin polymerization," *EMBO J.*, vol. 29, no. 14, pp. 2315–2328, 2010. doi: 10.1038/emboj.2010.133

[152] S.B. Padrick, L.K. Doolittle, C.A. Brautigam, D.S. King, and M.K. Rosen, "Arp2/3 complex is bound and activated by two WASP proteins." *Proc. Natl. Acad. Sci. USA*, vol. 108, no. 33, pp. E472–9, 2011. doi: 10.1073/pnas.1100236108

[153] X. Jiang, L.N. Kinch, C.A. Brautigam, X. Chen, F. Du, N.V. Grishin, and Z.J. Chen, "Ubiquitin-induced oligomerization of the RNA sensors RIG-I and MDA5 activates antiviral innate immune response," *Immunity*, vol. 36, no. 6, pp. 973–959, 2012. doi: 10.1016/j.immuni.2012.03.022

[154] I.C. Berke and Y. Modis, "MDA5 cooperatively forms dimers and ATP-sensitive filaments upon binding double-stranded RNA." *EMBO J.*, vol. 31, no. 7, pp. 1714–26, 2012. doi: 10.1038/emboj.2012.19

[155] M.P.M.H. Benoit, L. Imbert, A. Palencia, J. Pérard, C. Ebel, J. Boisbouvier, and M.J. Plevin, "The RNA-binding region of human TRBP interacts with microRNA precursors through two independent domains," *Nucleic Acids Res.*, vol. 41, no. 7, pp. 4241–4252, 2013. doi: 10.1093/nar/gkt086

[156] N.P. Coussens, R. Hayashi, P.H. Brown, L. Balagopalan, A. Balbo, I. Akpan, J.C.D. Houtman, V.A. Barr, P. Schuck, E. Appella, and L.E. Samelson, "Multipoint binding of the SLP-76 SH2 domain to ADAP is critical for oligomerization of SLP-76 signaling complexes in stimulated T cells," *Mol. Cell. Biol.*, vol. 33, no. 21, pp. 4140–4151, 2013. doi: 10.1128/MCB.00410-13

[157] C.A. Ydenberg, S.B. Padrick, M.O. Sweeney, M. Gandhi, O. Sokolova, and B.L. Goode, "GMF severs actin-Arp2/3 complex branch junctions by a cofilin-like mechanism," *Curr. Biol.*, vol. 23, no. 12, pp. 1037–1045, 2013. doi: 10.1016/j.cub.2013.04.058

[158] X. Zhang, J. Wu, F. Du, H. Xu, L. Sun, Z. Chen, C.A. Brautigam, X. Zhang, and Z.J. Chen, "The cytosolic DNA sensor cGAS forms an oligomeric complex with DNA and undergoes switch-like conformational changes in the activation loop," *Cell Rep.*, vol. 6, no. 3, pp. 421–430, 2014. doi: 10.1016/j.celrep.2014.01.003

[159] C.A. Brautigam, S.B. Padrick, and P. Schuck, "Multi-signal sedimentation velocity analysis with mass conservation for determining the stoichiometry of protein complexes," *PLoS One*, vol. 8, no. 5, p. e62694, 2013. doi: 10.1371/journal.pone.0062694

[160] B. Goldstein and B.H. Zimm, "Effect of concentration and intermolecular forces on the sedimentation of polystyrene spheres," *J. Chem. Phys.*, vol. 54, no. 10, p. 4408, 1971. doi: 10.1063/1.1674690

[161] G.K. Batchelor and C.-S. Wen, "Sedimentation in a dilute polydisperse system of interacting spheres. Part 2. Numerical results," *J. Fluid Mech.*, vol. 124, pp. 495–528, 1982. doi: 10.1017/S0022112082002602

[162] M. Watzlawek and G. Nägele, "Sedimentation of strongly and weakly charged colloidal particles: Prediction of fractional density dependence," *J. Colloid Interface Sci.*, vol. 214, no. 2, pp. 170–179, 1999. doi: 10.1006/jcis.1999.6181

[163] E. Lattuada, S. Buzzaccaro, and R. Piazza, "Colloidal swarms can settle faster than isolated particles: Enhanced sedimentation near phase separation," *Phys. Rev. Lett.*, vol. 116, no. 3, p. 038301, 2016. doi: 10.1103/PhysRevLett.116.038301

[164] S. Bucciarelli, J.S. Myung, B. Farago, S. Das, G.A. Vliegenthart, O. Holderer, R.G. Winkler, P. Schurtenberger, G. Gompper, and A. Stradner, "Dramatic influence of patchy attractions on short-time protein diffusion under crowded conditions," *Sci. Adv.*, vol. 2, p. e1601432, 2016. doi: 10.1126/sciadv.1601432

[165] R. Piazza, "Settled and unsettled issues in particle settling." *Rep. Prog. Phys.*, vol. 77, no. 5, p. 056602, 2014. doi: 10.1088/0034-4885/77/5/056602

[166] J.W. Swan and G. Wang, "Rapid calculation of hydrodynamic and transport properties in concentrated solutions of colloidal particles and macromolecules," *Phys. Fluids*, vol. 28, no. 1, p. 011902, 2016. doi: 10.1063/1.4939581

[167] Ai Díez, A. Ortega, and J. García de la Torre, "Brownian dynamics simulation of analytical ultracentrifugation experiments." *BMC Biophys.*, vol. 4, p. 6, jan 2011. doi: 10.1186/2046-1682-4-6

[168] S. Zorrilla, M. Jiménez, P. Lillo, G. Rivas, and A.P. Minton, "Sedimentation equilibrium in a solution containing an arbitrary number of solute species at arbitrary concentrations: Theory and application to concentrated solutions of ribonuclease," *Biophys. Chem.*, vol. 108, pp. 89–100, 2004. doi: dx.doi.org/10.1016/j.bpc.2003.10.012

[169] G.M. Pavlov, "The concentration dependence of sedimentation for polysaccharides," *Eur. Biophys. J.*, vol. 25, no. 5-6, pp. 385–397, 1997. doi: 10.1007/s002490050051

[170] G.M. Pavlov, I. Perevyazko, O.V. Okatova, and U.S. Schubert, "Conformation parameters of linear macromolecules from velocity sedimentation and other hydrodynamic methods," *Methods*, vol. 54, no. 1, pp. 124–135, 2011. doi: 10.1016/j.ymeth.2011.02.005

[171] S.E. Harding, G.G. Adams, F. Almutairi, Q. Alzahrani, T. Erten, M.S. Kök, and R.B. Gillis, "Ultracentrifuge methods for the analysis of polysaccharides, glycoconjugates, and lignins," *Methods Enzymol.*, vol. 562, pp. 391–439, 2015. doi: 10.1016/bs.mie.2015.06.043

[172] F.M. Almutairi, T. Erten, G.G. Adams, M. Hayes, P. McLoughlin, M.S. Kök, A.R. Mackie, A.J. Rowe, and S.E. Harding, "Hydrodynamic characterisation of chitosan and its interaction with two polyanions: DNA and xanthan," *Carbohydr. Polym.*, vol. 122, pp. 359–366, 2015. doi: 10.1016/j.carbpol.2014.09.090

[173] I. Nischang, I. Perevyazko, T.C. Majdanski, J. Vitz, G. Festag, and U.S. Schubert, "Hydrodynamic analysis resolves the pharmaceutically-relevant absolute molar mass and solution properties of synthetic poly(ethylene glycol)s created by varying initiation sites," *Anal. Chem.*, vol. 89, no. 2, pp. 1185–1193, 2016. doi: 10.1021/acs.analchem.6b03615

[174] F.M. Almutairi, J.G.H. Cifre, G.G. Adams, M.S. Kök, A.R. Mackie, J.G. de la Torre, and S.E. Harding, "Application of recent advances in hydrodynamic methods for characterising mucins in solution," *Eur. Biophys. J.*, vol. 45, no. 1, pp. 45–54, 2016. doi: 10.1007/s00249-015-1075-0

[175] G.M. Pavlov, A.J. Rowe, and S.E. Harding, "Conformation zoning of large molecules using the analytical ultracentrifuge zones," *Trends Anal. Chem.*, vol. 16, no. 7, pp. 401–405, 1997. doi: 10.1016/S0165-9936(97)00038-1

[176] S.A. Berkowitz, "Role of analytical ultracentrifugation in assessing the aggregation of protein biopharmaceuticals," *AAPS J.*, vol. 8, no. 3, pp. E590–605, 2006. doi: 10.1208/aapsj080368

[177] Y.-F. Mok, T.M. Ryan, S. Yang, D.M. Hatters, G.J. Howlett, and M.D.W. Griffin, "Sedimentation velocity analysis of amyloid oligomers and fibrils using fluorescence detection." *Methods*, vol. 54, no. 1, pp. 67–75, 2011. doi: 10.1016/j.ymeth.2010.10.004

[178] X. Wang, C. Zhang, Y.-C. Chiang, S. Toomey, M.P. Power, M.E. Granoff, R. Richardson, W. Xi, D.J. Lee, S. Chase, T.M. Laue, and C.L. Denis, "Use of the novel technique

of analytical ultracentrifugation with fluorescence detection system identifies a 77S monosomal translation complex." *Protein Sci.*, vol. 21, no. 9, pp. 1253–68, 2012. doi: 10.1002/pro.2110

[179] M.A. Olshina, L.M. Angley, Y.M. Ramdzan, J. Tang, M.F. Bailey, A.F. Hill, and D.M. Hatters, "Tracking mutant huntingtin aggregation kinetics in cells reveals three major populations that include an invariant oligomer pool." *J. Biol. Chem.*, vol. 285, no. 28, pp. 21 807–21 816, 2010. doi: 10.1074/jbc.M109.084434

[180] J.J. Hill and T.M. Laue, "Protein assembly in serum and the differences from assembly in buffer," *Methods Enzymol.*, vol. 562, pp. 501–527, 2015. doi: 10.1016/bs.mie.2015.06.012

[181] B. Kokona, C.A. May, N.R. Cunningham, L. Richmond, F.J. Garcia, J.C. Durante, K.M. Ulrich, C.M. Roberts, C.D. Link, W.F. Stafford, T.M. Laue, and R. Fairman, "Studying polyglutamine aggregation in *Caenorhabditis elegans* using an analytical ultracentrifuge equipped with fluorescence detection," *Protein Sci.*, vol. 25, pp. 1–13, 2015. doi: 10.1002/pro.2854

[182] J.B. Perrin, "Discontinuous Structure of Matter. Nobel Lecture." 1926. online at: http://www.nobelprize.org/nobel_prizes/physics/laureates/1926/perrin-lecture.html

[183] S. Ramaswamy, "Issues in the statistical mechanics of steady sedimentation," *Adv. Phys.*, vol. 50, no. 3, pp. 297–341, 2001. doi: 10.1080/00018730110050617

[184] S.E. Harding and P. Johnson, "The concentration-dependence of macromolecular parameters." *Biochem. J.*, vol. 231, no. 3, pp. 543–547, 1985. doi: 10.1042/bj2310543

[185] A. Solovyova, P. Schuck, L. Costenaro, and C. Ebel, "Non-ideality by sedimentation velocity of halophilic malate dehydrogenase in complex solvents," *Biophys. J.*, vol. 81, no. 4, pp. 1868–1880, 2001. doi: 10.1016/S0006-3495(01)75838-9

[186] W.G. McMillan and J.E. Mayer, "The statistical thermodynamics of multicomponent systems," *J. Chem. Phys.*, vol. 13, no. 7, p. 276, 1945. doi: 10.1063/1.1724036

[187] W.B. Russel, D.A. Saville, and W.R. Schowalter, *Colloidal Dispersions.* Cambridge, U.K.: Cambridge University Press, 1989. ISBN 0521426006

[188] C. Tanford, *Physical Chemistry of Macromolecules.* New York: Wiley, 1961. ISBN 0471844470

[189] K.S. Schmitz, *An Introduction to Dynamic Light Scattering by Macromolecules.* Boston, MA: Academic Press, 1990. ISBN 0126272603

[190] S. Newman and F. Eirich, "Particle shape and the concentration dependence of sedimentation and diffusion," *J. Colloid Sci.*, vol. 5, no. 6, pp. 541–549, 1950. doi: 10.1016/0095-8522(50)90046-8

[191] M.M. Kops-Werkhoven and H.M. Fijnaut, "Dynamic light scattering and sedimentation experiments on silica dispersions at finite concentrations," *J. Chem. Phys.*, vol. 74, no. 3, p. 1618, 1981. doi: 10.1063/1.441302

[192] G.K. Batchelor, "Sedimentation in a dilute polydisperse system of interacting spheres. 1. General theory," *J. Fluid. Mech.*, vol. 119, pp. 379–408, 1982. doi: 10.1017/S0022112082001402

[193] G.K. Batchelor, "Brownian diffusion of particles with hydrodynamic interaction," *J. Fluid. Mech.*, vol. 74, no. 1, pp. 1–29, 1976. doi: DOI10.1017/S0022112076001663

[194] G.K. Batchelor, "Sedimentation in a dilute dispersion of spheres," *J. Fluid. Mech.*, vol. 52, no. 2, pp. 245–268, 1972. doi: 10.1017/S0022112072001399

[195] W.B. Russel, "Brownian motion of small particles suspended in liquids," *Annu. Rev. Fluid Mech.*, vol. 13, no. 1, pp. 425–455, 1981. doi: 10.1146/annurev.fl.13.010181.002233

[196] D.R. Bauer, J.I. Brauman, and R. Pecora, "Molecular reorientation in liquids. Experimental test of hydrodynamic models," *J. Am. Chem. Soc.*, vol. 96, no. 22, pp. 6840–6843, 1974. doi: 10.1021/ja00829a004

[197] G.G. Stokes, "On the effect of the internal friction of fluids on the motion of pendulums," *Math. Phys. Pap.*, vol. 3, pp. 1–122, 1922. doi: 10.1017/CBO9780511702242.005. online at: http://adsabs.harvard.edu/abs/1851TCaPS...9....8S

[198] A.B. Glendinning and W.B. Russel, "A pairwise additive description of sedimentation and diffusion in concentrated suspensions of hard spheres," *J. Colloid Interface Sci.*, vol. 89, no. 1, pp. 124–143, 1982. doi: 10.1016/0021-9797(82)90127-8

[199] B. Cichocki and B.U. Felderhof, "Sedimentation and self-diffusion in suspensions of spherical particles," *Physica A*, vol. 154, pp. 213–232, 1989. doi: 10.1016/0378-4371(89)90010-1

[200] J.F. Brady and L.J. Durlofsky, "The sedimentation rate of disordered suspensions," *Phys. Fluids*, vol. 31, no. 4, pp. 717–727, 1988. doi: 10.1063/1.866808

[201] B. Cichocki, M.L. Ekiel-Jezewska, P. Szymczak, and E. Wajnryb, "Three-particle contribution to sedimentation and collective diffusion in hard-sphere suspensions," *J. Chem. Phys.*, vol. 117, no. 3, p. 1231, 2002. doi: 10.1063/1.1484380

[202] H. Hayakawa and K. Ichiki, "Statistical theory of sedimentation of disordered suspensions," *Phys. Rev. E*, vol. 51, no. 5, pp. R3815–R3818, 1995. doi: 10.1103/PhysRevE.51.R3815

[203] B. Cichocki and K. Sadlej, "Steady-state particle distribution of a dilute sedimenting suspension," *Europhys. Lett.*, vol. 72, no. 6, pp. 936–942, 2005. doi: 10.1209/epl/i2005-10341-6

[204] D.L. Koch and E.S.G. Shaqfeh, "Screening in sedimenting suspensions," *J. Fluid Mech.*, vol. 224, p. 275, 2006. doi: 10.1017/S0022112091001763

[205] W. Xu, A. Nikolov, and D.T. Wasan, "The effect of many-body interactions on the sedimentation of monodisperse particle dispersions," *J. Colloid Interface Sci.*, vol. 197, no. 1, pp. 160–9, 1998. doi: 10.1006/jcis.1997.5249

[206] P.N. Pusey and R.J.A. Tough, "Particle interactions," in *Dynamic Light Scattering: Applications of Photon Correlation Spectroscopy*, R. Pecora, Ed. New York: Plenum Press, 1985, ch. 4, pp. 85–179.

[207] A. Kholodenko, and J. Douglas, "Generalized Stokes-Einstein equation for spherical particle suspensions," *Physical Review E*, vol. 51, no. 2, pp. 1081–1090, 1995. doi: 10.1103/PhysRevE.51.1081

[208] R.H. Davis and M.A. Hassen, "Spreading of the interface at the top of a slightly polydisperse sedimenting suspension," *J. Fluid. Mech.*, vol. 196, pp. 107–134, 1988. doi: 10.1017/S0022112088002630

[209] P.N. Segrè, E. Herbolzheimer, and P.M. Chaikin, "Long-range correlations in sedimentation," *Phys. Rev. Lett.*, pp. 2–5, 1997. doi: 10.1103/PhysRevLett.79.2574

[210] R.E. Caflisch and J.H.C. Luke, "Variance in the sedimentation speed of a suspension," *Phys. Fluids*, vol. 28, no. 3, p. 759, 1985. doi: 10.1063/1.865095

[211] P.J. Mucha and M.P. Brenner, "Diffusivities and front propagation in sedimentation," *Phys. Fluids*, vol. 15, no. 5, p. 1305, 2003. doi: 10.1063/1.1564824

[212] A.J. Rowe, "The concentration dependence of transport processes: A general description applicable to the sedimentation, translational diffusion, and viscosity coefficients of macromolecular solutes," *Biopolymers*, vol. 16, pp. 2595–2611, 1977. doi: 10.1002/bip.1977.360161202

[213] Z. Dogic, A.P. Philipse, S. Fraden, and J.K.G. Dhont, "Concentration-dependent sedimentation of colloidal rods," *J. Chem. Phys.*, vol. 113, no. 18, pp. 8368–8380, 2000. doi: 10.1063/1.1308107

[214] I.L. Claeys and J.F. Brady, "Suspensions of prolate spheroids in Stokes flow. Part 1. Dynamics of a finite number of particles in an unbounded fluid," *J. Fluid Mech.*, vol. 251, pp. 411–442, 1993. doi: 10.1017/S0022112093003465

[215] S.E. Harding, "On the hydrodynamics of macromolecular conformation," *Biophys. Chem.*, vol. 55, pp. 69–93, 1995. doi: 10.1016/0301-4622(94)00143-8

[216] J.M. Creeth and C.G. Knight, "On the estimation of the shape of macromolecules from sedimentation and viscosity measurements," *Biochim. Biophys. Acta*, vol. 102, no. 2, pp. 549–558, 1965. doi: 10.1016/0926-6585(65)90145-7

[217] T.M. Laue, B.D. Shah, T.M. Ridgeway, and S.L. Pelletier, "Computer-aided interpretation of analytical sedimentation data for proteins," in *Analytical Ultracentrifugation in Biochemistry and Polymer Science*, S.E. Harding, A.J. Rowe, and J.C. Horton, Eds. Cambridge: The Royal Society of Chemistry, 1992, pp. 90–125.

[218] P.Y. Cheng and H.K. Schachman, "Studies on the validity of the Einstein viscosity law and Stokes law of sedimentation," *J. Polym. Sci.*, vol. 16, no. 81, pp. 19–30, 1955. doi: 10.1002/pol.1955.120168102

[219] M.M. Kops-Werkhoven, C. Pathmamanoharan, A. Vrij, and H.M. Fijnaut, "Concentration dependence of the selfdiffusion coefficient of hard, spherical particles measured with photon correlation spectroscopy," *J. Chem. Phys.*, vol. 77, no. 12, pp. 5913–5922, 1982. doi: 10.1063/1.443864

[220] H. Fujita, "Effects of a concentration dependence of the sedimentation coefficient in velocity ultracentrifugation," *J. Chem. Phys.*, vol. 24, no. 5, p. 1084, 1956. doi: 10.1063/1.1742683

[221] M. Dishon, G.H. Weiss, and D.A. Yphantis, "Numerical solutions of the Lamm equation. III. Velocity centrifugation," *Biopolymers*, vol. 5, no. 8, pp. 697–713, 1967. doi: 10.1002/bip.1967.360050804

[222] J.M. Creeth, "Approximate steady state condition in ultracentrifuge," *Proc. R. Soc. London Ser. A*, vol. 282, no. 139, p. 403, 1964. doi: 10.1098/rspa.1964.0242

[223] D.J. Scott, S.E. Harding, and D.J. Winzor, "Evaluation of diffusion coefficients by means of an approximate steady-state condition in sedimentation velocity distributions," *Anal. Biochem.*, vol. 490, pp. 20–25, 2015. doi: 10.1016/j.ab.2015.08.017

[224] R.L. Baldwin, "Boundary spreading in sedimentation-velocity experiments. 5. Measurement of the diffusion coefficient of bovine albumin by Fujita's equation," *Biochem. J.*, vol. 65, no. 3, pp. 503–512, 1957. doi: 10.1042/bj0650503

[225] G.M. Pavlov, D. Amoros, C. Ott, I.I. Zaitseva, J. García de la Torre, and U.S. Schubert, "Hydrodynamic analysis of well-defined flexible linear macromolecules of low molar mass," *Macromolecules*, vol. 42, no. 19, pp. 7447–7455, 2009. doi: 10.1021/ma901027u

[226] G.M. Pavlov, I. Perevyazko, and U.S. Schubert, "Velocity sedimentation and intrinsic viscosity analysis of polystyrene standards with a wide range of molar masses," *Macromol. Chem. Phys.*, vol. 211, no. 12, pp. 1298–1310, 2010. doi: 10.1002/macp.200900602

[227] G.M. Pavlov, K. Knop, O.V. Okatova, and U.S. Schubert, "Star-brush-shaped macromolecules: Peculiar properties in dilute solution," *Macromolecules*, vol. 46, no. 21, pp. 8671–8679, 2013. doi: 10.1021/ma400160f

[228] G.K. Batchelor and R.W.J. Van Rensburg, "Structure formation in bidisperse sedimentation," *J. Fluid. Mech.*, vol. 166, pp. 379–407, 1986. doi: 10.1017/S0022112086000204

[229] J.P. Johnston and A.G. Ogston, "A boundary anomaly found in the ultracentrifugal sedimentation of mixtures," *Trans. Faraday Soc.*, vol. 42, pp. 789–799, 1946. doi: 10.1039/TF9464200789

[230] J.J. Correia, M.L. Johnson, G.H. Welss, and D.A. Yphantis, "Numerical study of the Johnston–Ogston effect in two-component systems," *Biophys. Chem.*, vol. 5, no. 1-2, pp. 255–264, 1976. doi: 10.1016/0301-4622(76)80038-5

[231] T.N. Smith, "The sedimentation of particles having a dispersion of sizes." *Trans. Inst. Chem. Eng.*, vol. 44, pp. T153–157, 1966.

[232] R.H. Davis and A. Acrivos, "Sedimentation of noncolloidal particles at low Reynolds numbers," *Annu. Rev. Fluid Mech.*, vol. 17, no. 1, pp. 91–118, 1985. doi: 10.1146/annurev.fl.17.010185.000515

[233] W.F. Harrington and H.K. Schachman, "Analysis of a concentration anomaly in the ultracentrifugation of mixtures," *J. Am. Chem. Soc.*, vol. 1804, no. 10, 1953.

[234] S. Vesaratchanon, A. Nikolov, and D.T. Wasan, "Sedimentation in nano-colloidal dispersions: Effects of collective interactions and particle charge," *Adv. Colloid Interface Sci.*, vol. 134-135, pp. 268–278, 2007. doi: 10.1016/j.cis.2007.04.026

[235] A. Moncho-Jordá, A.A. Louis, and J.T. Padding, "Effects of interparticle attractions on colloidal sedimentation," *Phys. Rev. Lett.*, vol. 104, no. 6, p. 068301, 2010. doi: 10.1103/PhysRevLett.104.068301

[236] K.O. Pedersen, "On charge and specific ion effects on sedimentation in the ultracentrifuge," *J. Phys. Chem.*, vol. 62, no. 10, pp. 1282–1290, 1958. doi: 10.1021/j150568a028

[237] D. Stigter, "Primary charge effect on the sedimentation of long, colloidal rods," *J. Phys. Chem.*, vol. 86, no. 18, pp. 3553–3558, 1982. doi: 10.1021/j100215a013

[238] S. Allison, H. Wang, T.M. Laue, T.J. Wilson, and J.O. Wooll, "Visualizing ion relaxation in the transport of short DNA fragments." *Biophys. J.*, vol. 76, no. 5, pp. 2488–501, 1999. doi: 10.1016/S0006-3495(99)77404-7

[239] D.N. Petsev and N.D. Denkov, "Diffusion of charged colloidal particles at low volume fraction: Theoretical model and light scattering experiments," *J. Colloid. Interface Sci.*, vol. 149, no. 2, pp. 329–344, 1992. doi: 10.1016/0021-9797(92)90424-K

[240] T.R. Patel, S.E. Harding, A. Ebringerova, M. Deszczynski, Z. Hromadkova, A. Togola, B.S. Paulsen, G.A. Morris, and A.J. Rowe, "Weak self-association in a carbohydrate system," *Biophys. J.*, vol. 93, no. 3, pp. 741–749, 2007. doi: 10.1529/biophysj.106.100891

[241] H. Zhao, C.A. Brautigam, R. Ghirlando, and P. Schuck, "Current methods in sedimentation velocity and sedimentation equilibrium analytical ultracentrifugation," *Curr. Protoc. Protein Sci.*, vol. 7, p. 20.12.1, 2013. doi: 10.1002/0471140864.ps2012s71

[242] N. Naue and U. Curth, "Investigation of protein-protein interactions of single-stranded DNA-binding proteins by analytical ultracentrifugation." *Methods Mol. Biol.*, vol. 922, no. 2, pp. 133–49, 2012. doi: 10.1007/978-1-62703-032-8_8

[243] S.K. Chaturvedi, J. Ma, H. Zhao, and P. Schuck, "Using fluorescence detected sedimentation velocity for studying high affinity protein interactions," *Nat. Protoc.*, p. in press, 2017. doi: 10.1038/nprot.2017.064

[244] M. le Maire, B. Arnou, C. Olesen, D. Georgin, C. Ebel, and J.V. Møller, "Gel chromatography and analytical ultracentrifugation to determine the extent of detergent binding and aggregation, and Stokes radius of membrane proteins using sarcoplasmic reticulum Ca2+-ATPase as an example," *Nat. Protoc.*, vol. 3, no. 11, pp. 1782–1795, 2008. doi: 10.1038/nprot.2008.177

[245] A. Le Roy, H. Nury, B. Wiseman, J. Sarwan, J.-M. Jault, and C. Ebel, "Sedimentation velocity analytical ultracentrifugation in hydrogenated and deuterated solvents for the characterization of membrane proteins." *Methods Mol. Biol.*, vol. 1033, pp. 219–251, 2013. doi: 10.1007/978-1-62703-487-6_15

[246] A.G. Salvay, G. Communie, and C. Ebel, "Sedimentation velocity analytical ultracentrifugation for intrinsically disordered proteins." *Methods Mol. Biol.*, vol. 896, pp. 91–105, 2012. doi: 10.1007/978-1-4614-3704-8_6

[247] S. Polling, D.M. Hatters, and Y.-F. Mok, "Size analysis of polyglutamine protein aggregates using fluorescence detection in an analytical ultracentrifuge," *Methods Mol. Biol.*, vol. 1017, pp. 59–71, 2013. doi: 10.1007/978-1-62703-438-8_4

[248] W.B. Stine, "Analysis of monoclonal antibodies by sedimentation velocity analytical ultracentrifugation." *Methods Mol. Biol.*, vol. 988, pp. 227–240, 2013. doi: 10.1007/978-1-62703-327-5_15

[249] P. Job, "Recherches sur la formation de complexes minéraux en solution, et sur leur stabilité," *Ann. Chim. Fr.*, vol. 9, pp. 113–203, 1928.

[250] C.Y. Huang, "Determination of binding stoichiometry by the continuous variation method: the Job plot," *Methods Enzymol.*, vol. 87, pp. 509–525, 1982. doi: 10.1016/S0076-6879(82)87029-8

[251] I.K. MacGregor, A.L. Anderson, and T.M. Laue, "Fluorescence detection for the XLI analytical ultracentrifuge," *Biophys. Chem.*, vol. 108, no. 1-3, pp. 165–185, 2004. doi: 10.1016/j.bpc.2003.10.018

[252] H. Zhao, E. Casillas, H. Shroff, G.H. Patterson, and P. Schuck, "Tools for the quantitative analysis of sedimentation boundaries detected by fluorescence optical analytical ultracentrifugation," *PLoS One*, vol. 8, no. 10, p. e77245, 2013. doi: 10.1371/journal.pone.0077245

[253] H. Zhao, J. Ma, M. Ingaramo, E. Andrade, J. MacDonald, G. Ramsay, G. Piszczek, G.H. Patterson, and P. Schuck, "Accounting for photophysical processes and specific signal intensity changes in fluorescence-detected sedimentation velocity," *Anal. Chem.*, vol. 86, no. 18, pp. 9286–9292, 2014. doi: 10.1021/ac502478a

[254] J.S. Kingsbury and T.M. Laue, "Fluorescence-detected sedimentation in dilute and highly concentrated solutions," *Methods Enzymol.*, vol. 492, pp. 283–304, 2011. doi: 10.1016/B978-0-12-381268-1.00021-5

[255] H. Zhao, S. Lomash, C. Glasser, M.L. Mayer, and P. Schuck, "Analysis of high affinity self-association by fluorescence optical sedimentation velocity analytical ultracentrifugation of labeled proteins: Opportunities and limitations," *PLoS One*, vol. 8, no. 12, p. e83439, 2013. doi: 10.1371/journal.pone.0083439

[256] M.R. Marzahn, S. Marada, J. Lee, A. Nourse, S. Kenrick, H. Zhao, G. Ben-Nissan, R.-M. Kolaitis, J.L. Peters, S. Pounds, W.J. Errington, G.G. Privé, J.P. Taylor, M. Sharon, P. Schuck, S.K. Ogden, and T. Mittag, "Higher-order oligomerization promotes localization of SPOP to liquid nuclear speckles." *EMBO J.*, vol. 35, no. 12, pp. 1–22, 2016. doi: 10.15252/embj.201593169

[257] R.R. Kroe and T.M. Laue, "NUTS and BOLTS: Applications of fluorescence-detected sedimentation," *Anal. Biochem.*, vol. 390, no. 1, pp. 1–13, 2009. doi: 10.1016/j.ab.2008.11.033

[258] B. Demeule, S.J. Shire, and J. Liu, "A therapeutic antibody and its antigen form different complexes in serum than in phosphate-buffered saline: A study by analytical ultracentrifugation," *Anal. Biochem.*, vol. 388, no. 2, pp. 279–287, 2009. doi: 10.1016/j.ab.2009.03.012

[259] A. Le Roy, K. Wang, B. Schaack, P. Schuck, C. Breyton, and C. Ebel, "AUC and small-angle scattering for membrane proteins," *Methods Enzymol.*, vol. 562, pp. 257–286, 2015. doi: 10.1016/bs.mie.2015.06.010

[260] C.A. Brautigam, H. Zhao, C. Vargas, S. Keller, and P. Schuck, "Integration and global analysis of isothermal titration calorimetry data for studying macromolecular interactions," *Nat. Protoc.*, vol. 11, no. 5, pp. 882–894, 2016. doi: 10.1038/nprot.2016.044

[261] H. Zhao, G. Piszczek, and P. Schuck, "SEDPHAT A platform for global ITC analysis and global multi-method analysis of molecular interactions," *Methods*, vol. 76, pp. 137–148, 2015. doi: 10.1016/j.ymeth.2014.11.012

[262] N. Errington and A.J. Rowe, "Probing conformation and conformational change in proteins is optimally undertaken in relative mode," *Eur. Biophys. J.*, vol. 32, no. 5, pp. 511–517, 2003. doi: 10.1007/s00249-003-0315-x

[263] R. Ghirlando, A. Balbo, G. Piszczek, P.H. Brown, M.S. Lewis, C.A. Brautigam, P. Schuck, and H. Zhao, "Improving the thermal, radial, and temporal accuracy of the analytical ultracentrifuge through external references," *Anal. Biochem.*, vol. 440, no. 1, pp. 81–95, 2013. doi: 10.1016/j.ab.2013.05.011

[264] P.H. Brown, A. Balbo, and P. Schuck, "A Bayesian approach for quantifying trace amounts of antibody aggregates by sedimentation velocity analytical ultracentrifugation," *AAPS J.*, vol. 10, no. 3, pp. 481–493, 2008. doi: 10.1208/s12248-008-9058-z

[265] M.L. Johnson, "Why, when, and how biochemists should use least squares," *Anal. Biochem.*, vol. 225, pp. 215–225, 1992. doi: 10.1016/0003-2697(92)90356-C

[266] M.L. Johnson and M. Straume, "Comments on the analysis of sedimentation equilibrium experiments," in *Mod. Anal. Ultracentrifugation*, T.M. Schuster and T.M. Laue, Eds. Boston: Birkhäuser, 1994, pp. 37–65.

[267] C.A. Brautigam, "Calculations and publication-quality illustrations for analytical ultracentrifugation data," *Methods Enzymol.*, vol. 562, pp. 109–133, 2015. doi: 10.1016/bs.mie.2015.05.001

[268] H. Zhao and P. Schuck, "Global multi-method analysis of affinities and cooperativity in complex systems of macromolecular interactions." *Anal. Chem.*, vol. 84, no. 21, pp. 9513–9519, 2012. doi: 10.1021/ac302357w

[269] J. Dam, R Guan, K. Natarajan, N. Dimasi, L.K. Chlewicki, D.M. Kranz, P. Schuck, D.H. Margulies, and R.A. Mariuzza, "Variable MHC class I engagement by Ly49 NK cell receptors revealed by the crystal structure of Ly49C bound to H-2Kb," *Nat. Immunol.*, vol. 4, pp. 1213–1222, 2003. doi: 10.1038/ni1006

[270] A. Ortega, D. Amorós, and J. García de la Torre, "Prediction of hydrodynamic and other solution properties of rigid proteins from atomic- and residue-level models." *Biophys. J.*, vol. 101, no. 4, pp. 892–8, 2011. doi: 10.1016/j.bpj.2011.06.046

[271] S.R. Aragon and D.K. Hahn, "Precise boundary element computation of protein transport properties: Diffusion tensors, specific volume, and hydration," *Biophys. J.*, vol. 91, no. 5, pp. 1591–1603, 2006. doi: 10.1529/biophysj.105.078188

[272] E.-H. Kang, M. Mansfield, and J.F. Douglas, "Numerical path integration technique for the calculation of transport properties of proteins," *Phys. Rev. E*, vol. 69, pp. 1–11, 2004. doi: 10.1103/PhysRevE.69.031918

[273] T. Wiseman, S. Williston, J.F. Brandts, and L.N. Lin, "Rapid measurement of binding constants and heats of binding using a new titration calorimeter," *Anal. Biochem.*, vol. 179, no. 1, pp. 131–137, 1989. doi: 10.1016/0003-2697(89)90213-3

[274] P. Schuck, "Use of surface plasmon resonance to probe the equilibrium and dynamic aspects of interactions between biological macromolecules," *Ann. Rev. Biophys. Biomol. Struct.*, vol. 26, pp. 541–566, 1997.

[275] W.H. Press, S.A. Teukolsky, W.T. Vetterling, and B.P. Flannery, *Numerical Recipes in C*, 2nd ed. Cambridge: University Press, 1992.

[276] P.R. Bevington and D.K. Robinson, *Data Reduction and Error Analysis for the Physical Sciences*. New York: Mc-Graw-Hill, 1992.

[277] G.A. Gilbert, "Sedimentation and electrophoresis of interacting substances. III. Sedimentation of a reversibly aggregating substance with concentration dependent sedimentation coefficients," *Proc. R. Soc. London Ser. A*, vol. 276, no. 3, pp. 354–366, 1963. doi: 10.1098/rspa.1963.0211

[278] G. Kegeles, "Z-average sedimentation coefficient in reversibly self-aggregating systems," *Proc. Natl. Acad. Sci. USA*, vol. 69, no. 9, pp. 2577–2579, 1972.

[279] J.M. Beckerdite, C.A. Weirich, E.T. Adams, and G.H. Barlow, "Sedimentation coefficients of self-associating species. II. Tests with a simulated example and with beta-lactoglobulin A," *Biophys. Chem.*, vol. 17, no. 3, pp. 203–210, 1983. doi: 10.1016/0301-4622(83)87005-7

[280] J.J. Correia, "Analysis of weight average sedimentation velocity data," *Methods Enzymol.*, vol. 321, no. 1985, pp. 81–100, 2000. doi: 10.1016/S0076-6879(00)21188-9

[281] C.A. Sontag, W.F. Stafford, and J.J. Correia, "A comparison of weight average and direct boundary fitting of sedimentation velocity data for indefinite polymerizing systems," *Biophys. Chem.*, vol. 108, no. 1-3, pp. 215–230, 2004. doi: 10.1016/j.bpc.2003.10.029

[282] A. Revzin and P.H. von Hippel, "Direct measurement of association constants for the binding of *Escherichia coli* lac repressor to non-operator DNA," *Biochemistry*, vol. 16, no. 22, pp. 4769–4776, 1977. doi: 10.1021/bi00641a002

[283] G. Rivas, W.F. Stafford, and A.P. Minton, "Characterization of heterologous protein-protein interactions using analytical ultracentrifugation," *Methods*, vol. 19, no. 2, pp. 194–212, 1999. doi: 10.1006/meth.1999.0851

[284] H. Zhao and P. Schuck, "Combining biophysical methods for the analysis of protein complex stoichiometry and affinity in SEDPHAT," *Acta Crystallogr. Sect. D Biol. Crystallogr.*, vol. D71, pp. 3–14, 2015. doi: 10.1107/S1399004714010372

[285] J.-M. Claverie, H. Dreux, and R. Cohen, "Sedimentation of generalized systems of interacting particles. I. Solution of systems of complete Lamm equations," *Biopolymers*, vol. 14, no. 8, pp. 1685–1700, 1975. doi: 10.1002/bip.1975.360140811

[286] J. Crank and P. Nicolson, "A practical method for numerical evaluation of solutions of partial differential equations of the heat-conduction type," *Adv. Comput. Math.*, vol. 6, no. 2, pp. 207–226, 1947.

[287] G.H. Golub and C.F. van Loan, *Matrix Computations*. Baltimore, MD: The Johns Hopkins University Press, 1989. ISBN 0801854148

Index

Printed and bound by CPI Group (UK) Ltd, Croydon, CR0 4YY

01/11/2024

01782604-0008